Lecture Notes in Earth Sciences   62

Editors:
S. Bhattacharji, Brooklyn
G. M. Friedman, Brooklyn and Troy
H. J. Neugebauer, Bonn
A. Seilacher, Tuebingen and Yale

W0055438

Springer-Verlag Berlin Heidelberg GmbH

Henry V. Lyatsky

Continental-Crust Structures on the Continental Margin of Western North America

 Springer

Author

Dr. Henry V. Lyatsky
Lyatsky Geoscience Research & Consulting Ltd.
4827 Nipawin CR. NW
Calgary, Alberta, Canada T2K 2H8

Cataloging-in-Publication data applied for

Die Deutsche Bibliothek - CIP-Einheitsaufnahme

Lyatsky, Henry V.:
Continental crust structures of the continental margin of
western North America / Henry V. Lyatsky. - Berlin ;
Heidelberg ; New York ; Barcelona ; Budapest ; Hong Kong ;
London ; Milan ; Paris ; Santa Clara ; Singapur ; Tokyo :
Springer, 1996
 (Lecture notes in earth sciences ; 62)
 ISBN 978-3-540-60842-4
NE: GT

"For all Lecture Notes in Earth Sciences published till now please see final pages of
the book"

ISBN 978-3-540-60842-4   ISBN 978-3-540-49598-7 (eBook)
DOI 10.1007/978-3-540-49598-7

© Springer-Verlag Berlin Heidelberg, 1996
Originally published by Springer-Verlag Berlin Heidelberg New York in 1996

SPIN: 10528995   32/3142-543210 - Printed on acid-free paper

PREFACE

The aim of this volume is two-fold. At the more pragmatic level, it is to help answer the many questions about the structure of the Pacific continental margin of North America, which have arisen over the years as a result of continuing field mapping and geophysical surveys. The second objective is methodological - to illustrate the irreplaceable role of geological information among the various data sets used in earth-science studies.

The need to address these issues became apparent to the author during the several years he spent taking part in geological and geophysical studies on the west coast of Canada. All too often, results of geologic field mapping disagreed with tectonic predictions from too-straightforward local applications of global plate reconstructions, which due to their generality do not always take a full account of specific character of particular regions.

To be sure, the global approach has during the last quarter-century greatly expanded the vision of geoscientists, previously restricted to continental regions. However, a negative by-product of this expansion has been a decline of attention paid to local information, as tectonic studies have increasingly relied on simply fitting the development of a particular region into this or that prefabricated tectonic template.

Direct geological observations have limitations of their own. The observer in most cases deals with products of geologic processes, rather than with the processes themselves. Field mapping provides local information, and many years of effort are needed before a regional overview becomes possible. Geologic mapping is restricted to the ground surface, and even the deepest drillholes cannot sample more than the outermost shell of the Earth. The factual side of geologic mapping is usually limited to determination of rock types and their relationships in areas of exposure. Conclusions about the three-dimensional structure of a region and its evolution are still mostly inferential.

Broad incorporation into geological studies of geophysical data, assisted by ever-more-sophisticated modern computers, provides a huge volume of information unobtainable in other ways. Geophysical methods quickly afford regional coverage or images of the Earth's deep interior.

Geophysical methods have prompted the application in geological sciences of methodologies borrowed from exact sciences, such as mathematics and physics. Particularly important has been quantitative modeling, which allows a scientist to use the known parameters of a system to predict others. But in taking this approach too far, one encounters a dangerous pitfall.

A model is a simplified representation of a natural phenomenon. The quality of this or that representation is relative, and a representation is never perfect. To incorporate all characteristics of a geologic phenomenon, in a parametrized form, into a numerical or physical imitation is impossible. This requires one to rely on simplifying assumptions, and a model is no better than the assumptions at its base.

Unrealistic assumptions lead to unrealistic models. When a disagreement arises between model predictions and observations - such as those from geologic field mapping - a modeler may be tempted to downplay the differences or the significance of the offending observations. It becomes tempting to underestimate the role of an experienced geologist as a principal arbiter of the realism of a model.

But it is geological data and geological control that provide the ultimate means of testing abstract models. From this methodological position, the present study of the western North American continental margin is organized as follows:

1.   Geological information, available from field mapping and drilling, is gathered and summarized.
2.   Current geophysical models for this region are considered, with particular attention to their underlying assumptions.
3.   The available data, geological and geophysical, are synthesized into an internally consistent geologic-evolution concept.
4.   This concept is tested by comparison with direct geological observations from field mapping and drilling.

Because most current data sets and models cover northwestern Washington and western British Columbia, particular attention was paid to these areas. Fortunately, these areas contain many keys that help understand the structure of the entire western North American continental margin, which has baffled scientists for decades. The author does not claim to have resolved all these problems, but he does believe he has made a useful contribution to understanding continental-oceanic plate interrelations at this continental margin.

Rigidity of lithospheric plates is a critical assumption in current models of plate evolution. The lithophere of a plate is created at spreading centers manifested in the global system of mid-ocean ridges. It moves away from the place of its birth towards boundaries with other plates, with which it can interact in a variety of ways. Some interactions are of strike-slip type, with two plates simply sliding past each other. However, to compensate for the creation of new lithosphere at spreading centers, older lithosphere at some plate boundaries descends into the mantle as it is overriden by other plates. At such plate boundaries lie subduction zones.

If both regimes occur along a single plate boundary, the transition between them must be abrupt. Unless it can be tied to a change in orientation of the boundary, it must be associated with a junction of not two, but three different plates.

Such a template was used to interpret the structure and tectonic evolution of the western North American continental margin in the late 1960s and thereafter (Atwater, 1970; McManus et al., 1972; Barr and Chase, 1974; Riddihough and Hyndman, 1976). To satisfy the principles of rigid-plate tectonics, both regimes have to exist along this continental margin.

Also needed in rigid-plate reconstructions is a plate triple junction somewhere between the areas of proven ongoing subduction (in Oregon and southern Washington) and transform plate motion (along the southeastern Alaska margin; Atwater, 1970; McManus et al., 1972). Such a triple junction has been placed off Queen Charlotte Sound offshore British Columbia (Keen and Hyndman, 1979; Riddihough et al., 1983), where a spreading center has been postulated between the Pacific and Explorer oceanic plates (Hyndman et al. 1979; Riddihough, 1984). Off northern Vancouver Island, a transform boundary between the Explorer and Juan de Fuca oceanic plates has been postulated, but both these plates are assumed to be subducting beneath Vancouver Island (Hyndman et al., 1979; Riddihough and Hyndman, 1989).

With the assumed universality of the rigid-plate model, "broad similarity" has been suggested between the geology of western Oregon and that of western British Columbia, and the Cascadia zone of active subduction has been extended as far north as the mouth of Queen Charlotte Sound (Riddihough, 1979, 1984). An accretionary sedimentary prism (Yorath, 1980) - or even an accretionary complex containing several exotic "terranes" (Davis and Hyndman, 1989) - has been postulated off Vancouver Island.

Geological observations onshore and offshore (Shouldice, 1971; Tiffin et al., 1972) have come to be considered too "surficial" to be of major consequence for large-scale tectonic modeling (Yorath et al., 1985a,b; Yorath, 1987). Variants of the principal geophysical model for this area during the last decade (Clowes et al., 1987; Hyndman et al., 1990; Spence et al. 1991; Yuan et al., 1992; Dehler and Clowes, 1992) have become increasingly distant from geological observations. As new model variants emerged, they were checked for internal consistency, compatibility with neighboring local models and fidelity to the overall assumed tectonic picture.

However, detailed geological work continued, and many of its results proved incompatible with the conventional wisdom (Gehrels, 1990; Babcock et al., 1992, 1994; Allan et al., 1993; Lyatsky, 1993a). Importantly, questions arose about the applicability in this region of the conventional, simple rigid-plate assumption, as it was shown to be unable to account for all the geological and geophysical peculiarities in some areas (Carbotte et al., 1989; Allan et al., 1993; Davis and Currie, 1993). New solutions were made necessary by new findings and by rediscovery of forgotten old data (see Lyatsky et al., 1991; Lyatsky, 1993b).

Without aiming to resolve all the outstanding debates, tectonic implications of the geologic mapping and drilling results in this region are considered in the following chapters. These results are integrated with geochemical and geophysical data. Interpretations of these data, made by this author and by other workers, are verified by geological observations and by geologically plausible extrapolations from these observations. In searching for solutions consistent with all the information, the author has restricted himself to analyzing continental-crust structures along this continental margin. He believes, however, that future models for the offshore regions of the northeastern Pacific should consider the results obtained herein.

Acknowledgments

Through the support of the Geological Survey of Canada, Universities of British Columbia and Victoria, and NSERC, the author has been able to spend a total of several months in the field, examining first-hand the geology of the Canadian and .U.S. Cordillera and continental-margin areas.

The author is grateful to his colleagues whose encouragement and insightful questions helped him understand the geologic structure of the margin. Jim Monger, Bob Thompson and Glenn Woodsworth introduced him to the regional geology of the western Cordillera, and Jim Haggart, Cathie Hickson and Peter Mustard (all at the Geological Survey of Canada) acquainted him with specific problems in individual areas. Jim Murray (University of Alberta), Dick Chase (University of British Columbia), Dave Brew (U.S. Geological Survey) and Bob Crosson (University of Washington) offered useful discussions. Technical help was provided, at different times, by Art Haynes (Geological Survey of Canada) and Brian Fong (University of Calgary). Gerry Friedman (Brooklyn College) edited this volume. The responsibility for the scientific conclusions presented here, however, rests with the author alone.

CHAPTER 5 - SIGNIFICANCE OF THE TRANS-CORDILLERAN OLYMPIC-
WALLOWA ZONE IN GEOLOGIC EVOLUTION OF THE
WASHINGTON AND BRITISH COLUMBIA COASTAL REGIONS

CHAPTER 6 - CONTINENTAL MARGIN OFF SOUTHEASTERN ALASKA,
THE QUEEN CHARLOTTE ISLANDS, AND NORTHERN
VANCOUVER ISLAND

LIST OF FIGURES AND TABLES

CHAPTER 1 - OUTSTANDING ISSUES IN STUDIES OF CONTINENTAL MARGINS

Basic terminology related to margins of continents

Continental margin is one of the most common terms in modern geoscience literature. However, despite the existence of various classifications, no comprehensive definition of continental margin has yet been formalized, and this term is used loosely.

In a simple way, Dietz (1952, 1964) restricted this term to a set of submarine geomorphological features that rim continental landmasses: shelves, slopes and deep-water rises. This limitation had begun with Johnson (1919), who had originated the term. But whereas Johnson had considered continental margins to be products of accumulative processes, Dietz viewed them as results of tectonic activity in transition zones between continental landmasses (continents) and deep-ocean pelagic plains.

It is only a matter of perspective whether a continental margin or a continental terrace (shelf plus slope) belongs to a continent or to an ocean. Sedimentologically, they may be regarded as either margins of continental masses or parts of ocean basins (Friedman and Sanders, 1978). In tectonic terms, they commonly represent extensions of continental lithosphere from interiors of continents to their periphery.

In the conventional nomenclature, the shoreline is regarded as the inboard, upper boundary of the continental terrace. The outboard, lower boundary of the terrace is the foot of the continental slope or, where present, the trench. Confusingly, the term terrace is sometimes also used to describe any step-like bathymetric feature.

Continental shelves are shallow submarine plateaus in direct continuity with coastal lowlands onshore. They usually lie at less than 200 m water depth but may be deeper. Continental slopes, which lie outboard, are usually inclined 3°-10° but in some areas much more steeply. Most slopes are composed of two parts: steep upper part and gentle lower part, separated at mid-slope by a morphological break.

Often the term continental margin is used as synonymous with continental terrace. Confusingly, though, submerged and emergent marginal parts of continents are sometimes also regarded together. Tectonic movements and eustatic sea-level changes make the shoreline position ephemeral, and thus unreliable for demarcation of features that evolved over long periods of geologic time.

Complications may arise if generalized classifications are applied to local situations inflexibly. The Pacific continental margin of North America is usually classified as "active", as opposed to the "passive" Atlantic margin. As defined by Dickinson (1973), an active margin normally has a deep trench in front of a continental slope, and a magmatic arc onshore; these features are created due to subduction of an oceanic lithospheric plate beneath a continental plate. Where a magmatic arc is present, fore-arc and back-arc domains can be distinguished in relation to it, though no comprehensive demarcation yet exists for the inboard boundary of back-arc domains. Where a magmatic arc is absent, these domains lose their distinction.

In the absence of terminological consensus, the geologic term "continental margin" is used herein to include, in addition to the submerged areas outboard, onshore areas at least as far as the end of coastal plains. The uplands that begin there are usually created by cratonic or orogenic processes unrelated to present-day interactions of continental and oceanic plates. However, in active subduction settings where magmatic arcs have been erected, zones of continent-ocean transition may be extended inland far beyond coastal plains, to include the arcs and even the back-arc regions. Where practical, in this volume specific qualifiers are added for clarity, such as "submerged continental margin".

Definition of crustal type at continental margins

Continents are characterized by a specific type of lithosphere and crust, distinguished from oceanic lithosphere and crust by rock composition, layering, structure and thickness. Geological, geophysical and geochemical surveys the world over have revealed differences between continental and oceanic crust in rock composition (sialic vs. simatic), age (Early Archean to Recent vs. middle Mesozoic to Recent), and styles of structural deformation and reworking.

Variable types of magmatism occur in regions of continental crust. Felsic magmatism, which is completely absent in oceanic crust, is diagnostically continental. A broad range of metamorphic grades, from as low as zeolite to as high as granulite, is found only in continental crust; oceanic crust, by contrast, is only slightly altered. A wide diversity in styles of deformation is typical for continental regions, varying widely between cratonic and orogenic regimes. Oceanic crust, by all structural parameters, is simpler and more uniform. It is typically characterized by linear magnetic anomalies, which are absent in continental-crust regions.

Continental crust is usually 20 to 45 km thick, with density of crystalline rocks of 2,700 to 2,900 kg/m3 and seismic P-wave velocity of 6 to 7.5 km/s (e.g., Pakiser and Mooney, eds., 1989; Christiansen and Mooney, 1995). Oceanic crust is generally more uniform, 5 to 10 km thick, with densities of 2,900-3,000 kg/m3 and P-wave velocities exceeding 6.5 km/s.

Such generic descriptions are global averages from a multitude of observations. Large local deviations from these averages are common in continent-ocean transition zones, which sometimes contain crust with intermediate or uncertain velocity structure and rock properties (Couch and Riddihough, 1989). Bathymetric zoning of submarine parts of continental margins by itself may not permit unequivocal determination of crustal type. At some margins, continental crust, attenuated and thin, continues towards the ocean at considerable water depths (Grant, 1980, 1987; Rosendahl et al., 1992). Elsewhere, blocks of oceanic-type crust may lie under the continental slope and even shelf (Finn, 1990).

Crustal-type transitions may occupy broad zones where crust has geophysical properties intermediate between typically continental and typically oceanic. The conventional term "transtional crust" fails to distinguish between modified continental and modified oceanic crust. Blocks of both types make up large parts of the North American Pacific margin.

Historical outline of perspectives on Cordilleran geology
Pacific coastal areas of the North American continent are a part

of the Cordilleran orogenic system (Gabrielse and Yorath, eds., 1991; Burchfiel et al., 1992).

Since the first regional surveys, much of the Canadian Cordillera has been described as comprising five physiographic and geologic zones (Fig. 1). From east to west, they are: (1) Rocky Mountains (in the current literature, a fold-and-thrust belt) extending from Mexico to Alaska; (2) Omineca Belt, containing exhumed Precambrian rocks with links to the Archean to Early Proterozoic craton; (3) Intermontane Belt, once regarded as a median massif, containing a series of uplands with relatively subdued topography; (4) Coast Belt, the tallest and most rugged of all, with up to 80% granitoids (granodiorites and leucocratic diorites) as well as metamorphic rocks; and (5) the Insular Belt off the mainland coast, including Vancouver Island, Queen Charlotte Islands, and islands of southeastern Alaska. Thus subdivided, the Cordilleran orogenic system was described in the early geological synopses (e.g., Douglas, ed., 1970). No clear outboard boundary was demarcated at that time for the offshore part of the Insular Belt (see also King, 1969).

Change in perception began with a discovery that the magnetic field in the northeastern Pacific Ocean has a regular character, linear and stripe-like, distinctly different from that on the continent. Linear magnetic anomalies thousands of kilometers long remain parallel within broad domains, whereas stripe patterns change abruptly across well-defined domain boundaries (Raff and Mason, 1961). These magnetic lineations were soon explained as records of normal and reverse polarity of the changing geomagnetic field at the time of cooling of ocean-floor basalts erupted at and moving away from spreading centers (Vine and Matthews, 1963). Considered to be symmetrical relative to the spreading centers, these magnetic lineations came to be interpreted as isochrons which permit restoration of the history of sea-floor spreading and plate motions over hundreds of millions of years (Wilson, 1965; Vine and Wilson, 1965).

Early reconstructions of plate movements in the northeastern Pacific Ocean (Atwater, 1970) were pivotal to the revision of Cordilleran geology (e.g., Price and Douglas, eds., 1972). The idea that offshore plate interactions influenced the continental-margin geology has been accepted broadly and applied productively. At the extreme, however, plate movements have sometimes been considered the main factor in the genesis of continental orogens.

This ocean-based, Poseidonian perspective on continental geology came to dominance during the 1980s (see Burchfiel et al., eds., 1992). Now the development of the Cordillera is sometimes treated simply as a passive result of plate motions in the Pacific (Monger, 1993). This approach has the advantage of putting the Cordilleran geology into a global plate-tectonic context, but it risks ignoring self-development of continental lithosphere.

Off Vancouver Island, all rocks on the shelf and slope have been included by some workers into a Tertiary accretionary complex (Yorath, 1980; Hyndman et al., 1990; Dehler and Clowes, 1992). Such a complex of sedimentary and volcanic rocks presumably evolved as a result of accretion caused by subduction of oceanic plates (Duncan, 1982). On the Washington and Oregon continental margin, this complex was presumed to be very wide and even to extend into coastal areas onshore. Crust of oceanic origin was thought to underlie the Olympic Peninsula (MacLeod et al., 1977; Fig. 2; details of geology of the Olympic Peninsula are shown in Figs. 17 and 18 and in the corresponding chapter; the fault map of the submerged continental margin off Vancouver Island is presented in Fig. 50; the reader is encouraged to examine these and other diagrams before reading the following chapters, to acquaint himself/herself with the region).

Recent geologic summaries point to continental affinities of most western North American coastal regions (Burchfiel et al., eds., 1992). New geologic field evidence precludes an oceanic-lithosphere origin of the crust in basalt-rich areas in western Washington and Oregon (Babcock et al., 1992, 1994). Other workers considered the Insular Belt and, presumably, the continental crust to extend to the foot of the continental slope (von Huene, 1989; Gabrielse et al., 1991).

Shortcomings of current models of Cordilleran evolution
Disputes about the nature of the crust in western North America illustrate an undesirable side effect of the tectonic models deduced from a restricted geophysical data sets far offshore. This model-based approach relies mostly on global-scale plate reconstructions, at the expense of more-local geological investigations in areas of interest onshore. Magnetic anomalies in remote oceanic regions, rather than local geological observations obtained by mapping on land, all too often serve as a basis for reconstructions of the evolutionary history of marginal continental regions.

A casualty has been the old method of continental geological studies, which relied principally on factual observations from outcrops and drillholes. This traditional methodological cornerstone of geoscience has partly been displaced by the much simpler model-based approach. However, because modeling can take into consideration only a few parameters, it is no substitute for integrated studies based above all on factual observations.

The conventional model-based approach has at least two important shortcomings. First, the existing plate reconstructions are not everywhere conclusive. Errors in reconstructed motions are known to be substantial even for big plates (Engebretson et al., 1985; Stock and Molnar, 1988; DeMets et al., 1990). Off western North America, oceanic crust in many places is apparently too fragmented and deformed (Atwater, 1989) to be regarded as a rigid, coherent plate. Magnetic stripes, which serve as a basis for plate-motion reconstructions, are strongly curved or broken in deformed oceanic crust in the Gorda and Explorer parts of the Juan de Fuca plate (Atwater and Severinghaus, 1989; Couch and Riddihough, 1989; Davis and Currie, 1993).

A critical assumption in reconstructing plate motion is that plates are rigid. Principles of rigid-plate tectonics cannot be applied in these regions of deformed oceanic lithosphere (e.g., Carbotte et al., 1989; Allan et al., 1993).

The second shortcoming is the supposition that the continental lithosphere and crust of the North America plate only responded passively to plate interactions through time. As a result, tectonic studies have been reduced to accounting of presumably arbitrary events: terranes docking, rock deformation induced by stresses transmitted from far away.

Continental crust is rich in radioactive elements and thus has its own sources for self-development. Where studied in all its manifestations, continental tectonism, including that in marginal continental regions, cannot always be correlated with modeled plate motions. Rapid vertical movements that occurred in the Late Cretaceous and early Tertiary in the western Cordillera, including the Washington North Cascade Mountains and the British Columbia Coast Mountains (Figs. 1, 2; also Fig. 11) are not simply related to movements of the Juan de Fuca plate (Muller et al., 1992): geobarometry studies show that rocks of surface origin were first buried rapidly to depths as much as 30 km, then uplifted and exhumed (Brown et al., 1994). Tertiary felsic magmatism on the Olympic Peninsula is also puzzling if the crust in that area has an oceanic origin (Snavely, 1987). This puzzle is resolved if the Olympic Peninsula crust has continental affinities, as does the crust farther north, where felsic magmatism was widespread.

Horst-and-graben tectonics and development of fault-bounded depressions in the Late Cretaceous (e.g., the Nanaimo Basin; see the upcoming Fig. 6 and the corresponding sections for geologic details) was not correlative with the plate convergence usually modeled for that time (Pacht, 1984). The Queen Charlotte Basin was presumed in Poseidonian models to have been stretched greatly in the Tertiary (Yorath and Hyndman, 1983; Hyndman and Hamilton, 1993), but large extension in that area was later shown to be inconsistent with geological and geophysical observations (Thompson et al., 1991; Lyatsky, 1993a). The unusual pattern of

(a)   (b)

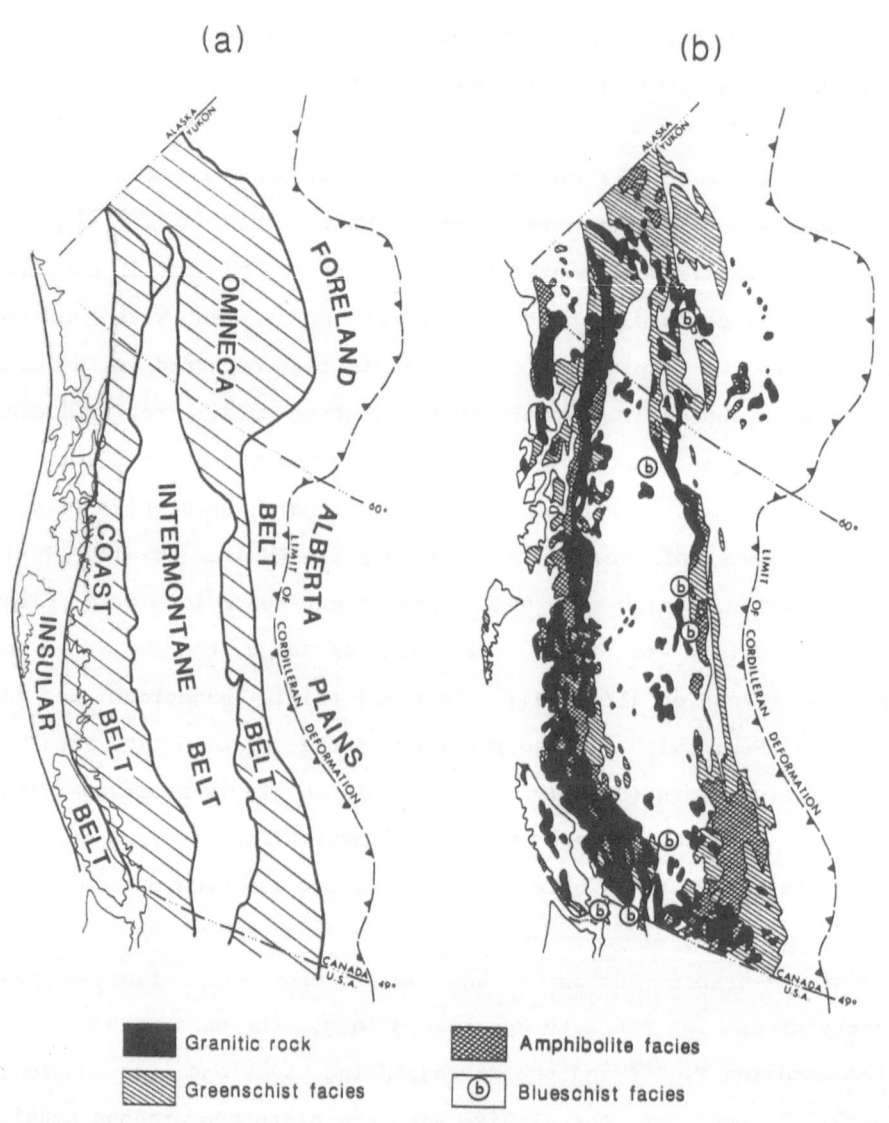

Figure 1. General zoning of the Canadian Cordillera: (a) morphogeologic belts; (b) simplified metamorphic map (courtesy J.W.H. Monger, 1992).

11

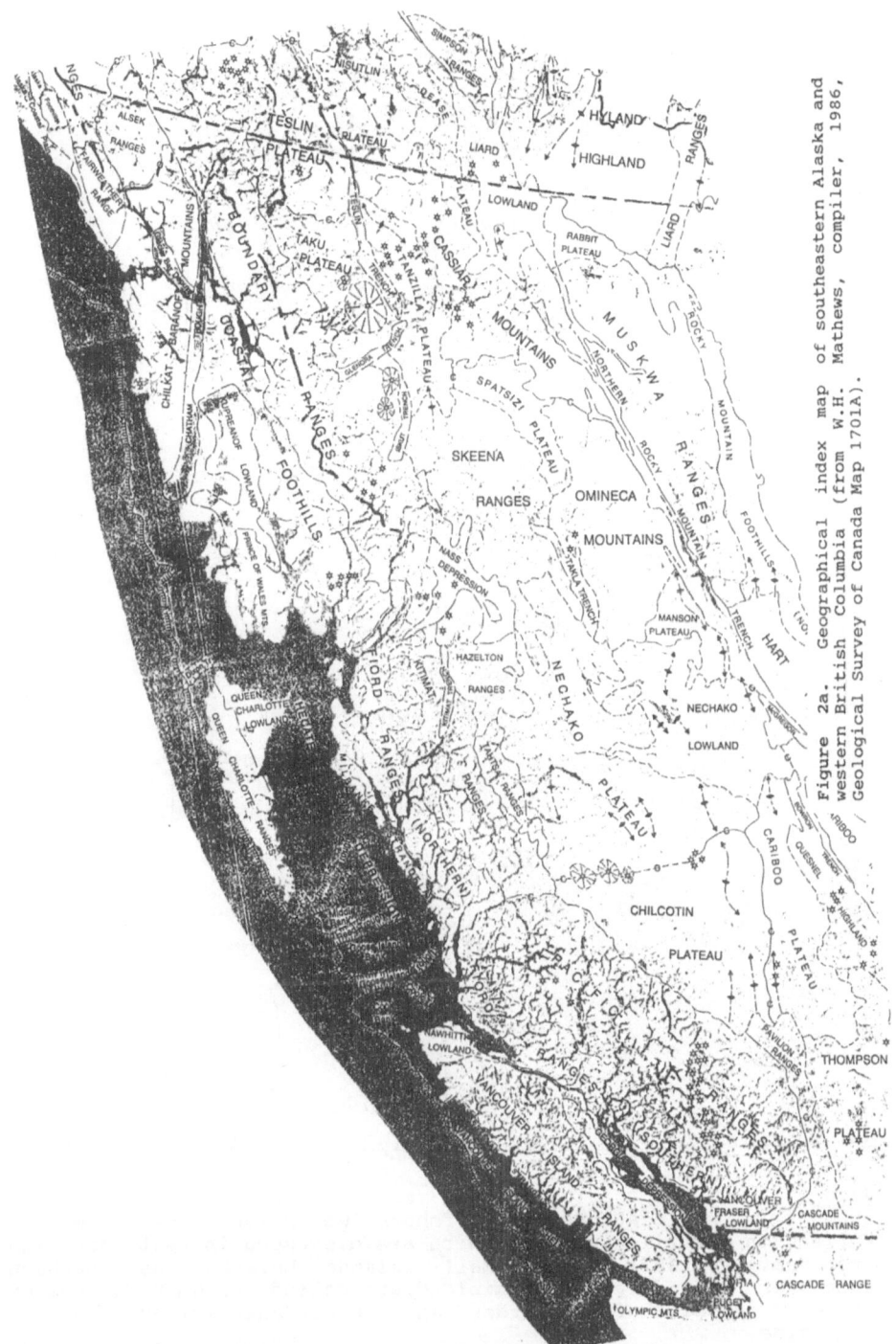

Figure 2a. Geographical index map of southeastern Alaska and western British Columbia (from W.H. Mathews, compiler, 1986, Geological Survey of Canada Map 1701A).

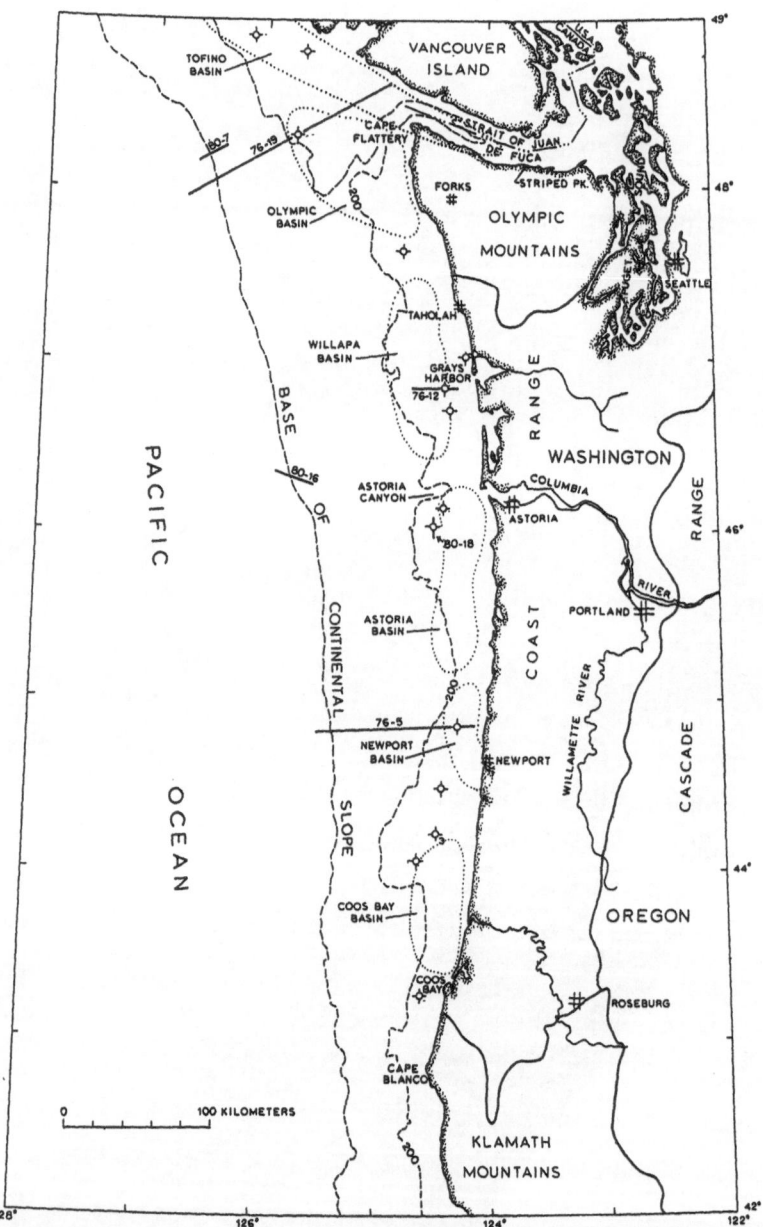

Figure 2b. Geographical index map of western Washington and Oregon, with locations of offshore wells and old seismic reflection profiles some of which are discussed in text (modified from Snavely, 1987). The small islands between the southern Vancouver Island and the mainland are called the Gulf Islands on the Canadian side of the border, and the San Juan Islands on the U.S. side.

13

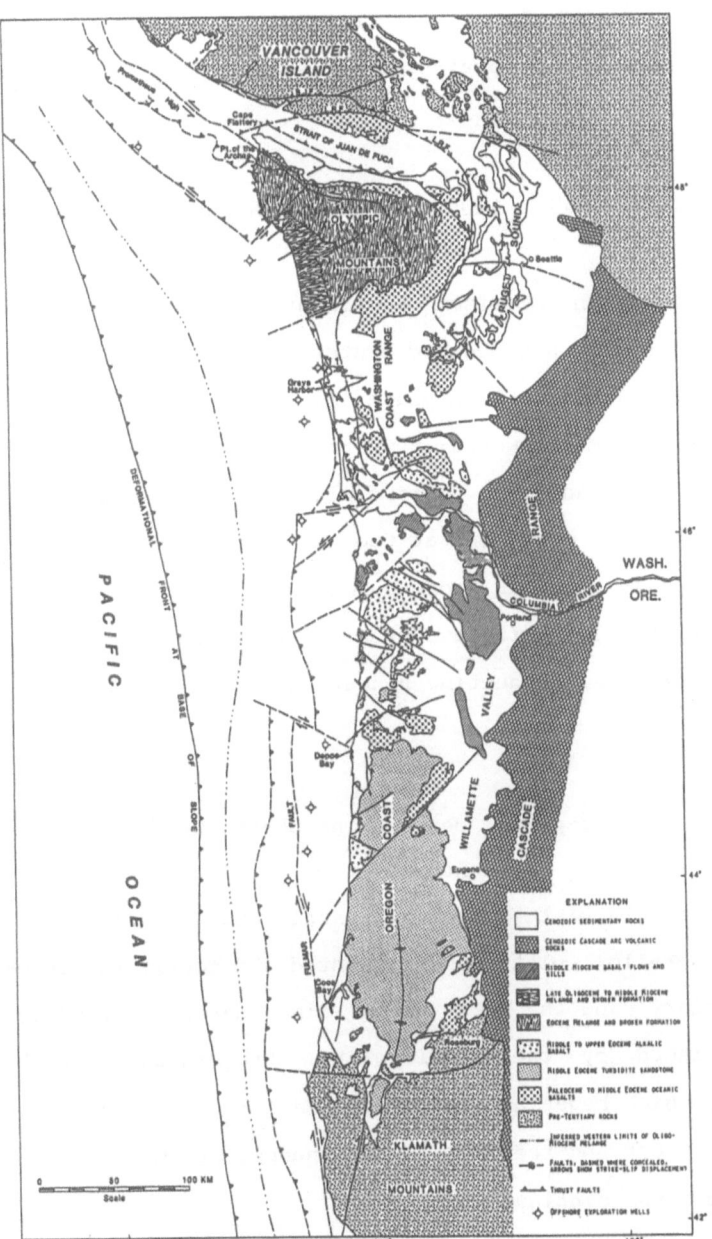

Figure 2c. Principal geologic features in the western Cordillera from the Klamath Mountains to Vancouver Island (modified from Snavely, 1987). Details of the geology of specific areas are presented in the subsequent chapters and figures.

seismicity onshore and offshore along the western North American continental margin still awaits a compelling, comprehensive explanation (e.g., Acharya, 1992).

Structure of western North America plate boundary in current models

At present, the North American continent interacts along its western margin with two main oceanic plates: the Pacific plate and the much smaller Juan de Fuca plate (Figs. 3, 4). The Pacific plate is the largest in the world, and its oceanic crust floors most of the Pacific Ocean. The Pacific plate is sliding dextrally past the North American continent at two NNW-trending segments of their shared boundary: the San Andreas fault in California and the Fairweather-Queen Charlotte fault system off southeastern Alaska. In the north, it is subducting beneath Alaska (Atwater, 1970, 1989; von Huene, 1989).

The Juan de Fuca oceanic plate, between northern California and British Columbia, is a remnant of the Farallon plate, which together with the Kula plate once dominated the Pacific Ocean region. The Farallon plate was fragmented and mostly subducted during the Tertiary, and the Kula plate disappeared completely. They were replaced by the Pacific plate. Fragmentation of the Farallon plate, which began at around 50 Ma, is continuing at present (also Stock and Lee, 1994). The number of small oceanic blocks or microplates continues to grow.

From detailed studies, sea-floor spreading is occurring between the Juan de Fuca and Pacific plates at the Juan de Fuca Ridge (Johnson and Holmes, 1989). Subduction of the Juan de Fuca

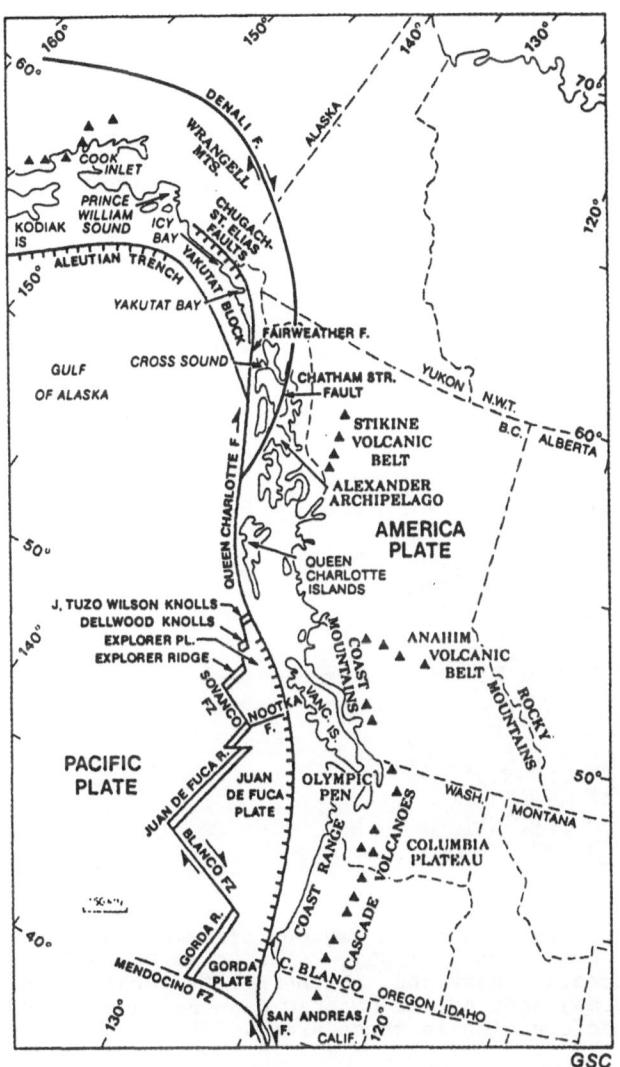

Figure 3a. Conventionally assumed plate boundaries off western North America and major late Cenozoic volcanic belts in the western Cordillera (after Riddihough and Hyndman, 1989, 1991). Distribution of Quaternary volcanoes in coastal areas is shown in more detail in Fig. 10.

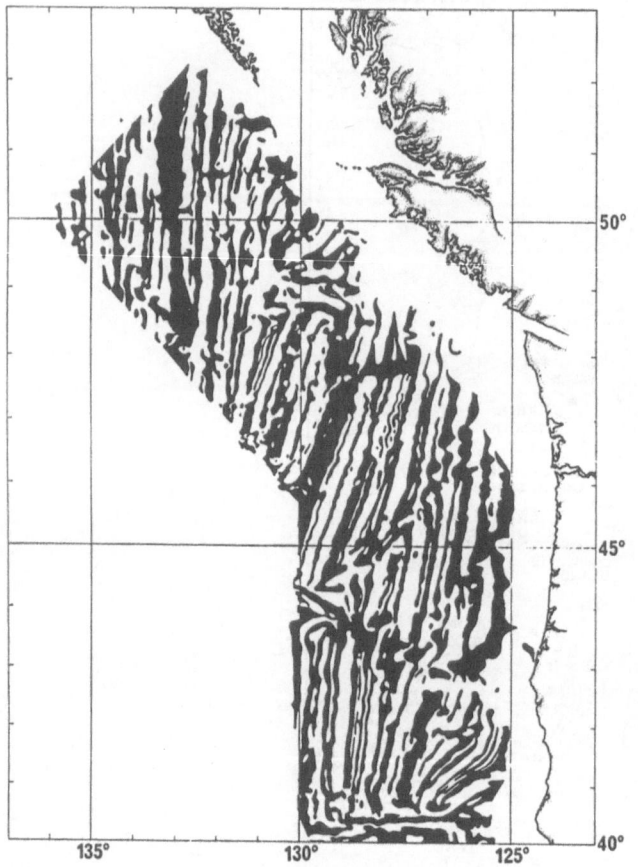

Figure 3b. Magnetic stripes in oceanic regions off western North America (after Raff and Mason, 1961). Shading marks positive anomalies. Chaotic anomalies mark the northern (Explorer; Fig. 3a) and southern (Gorda) ends of the oceanic Juan de Fuca plate, reflecting intraplate deformation in these areas.

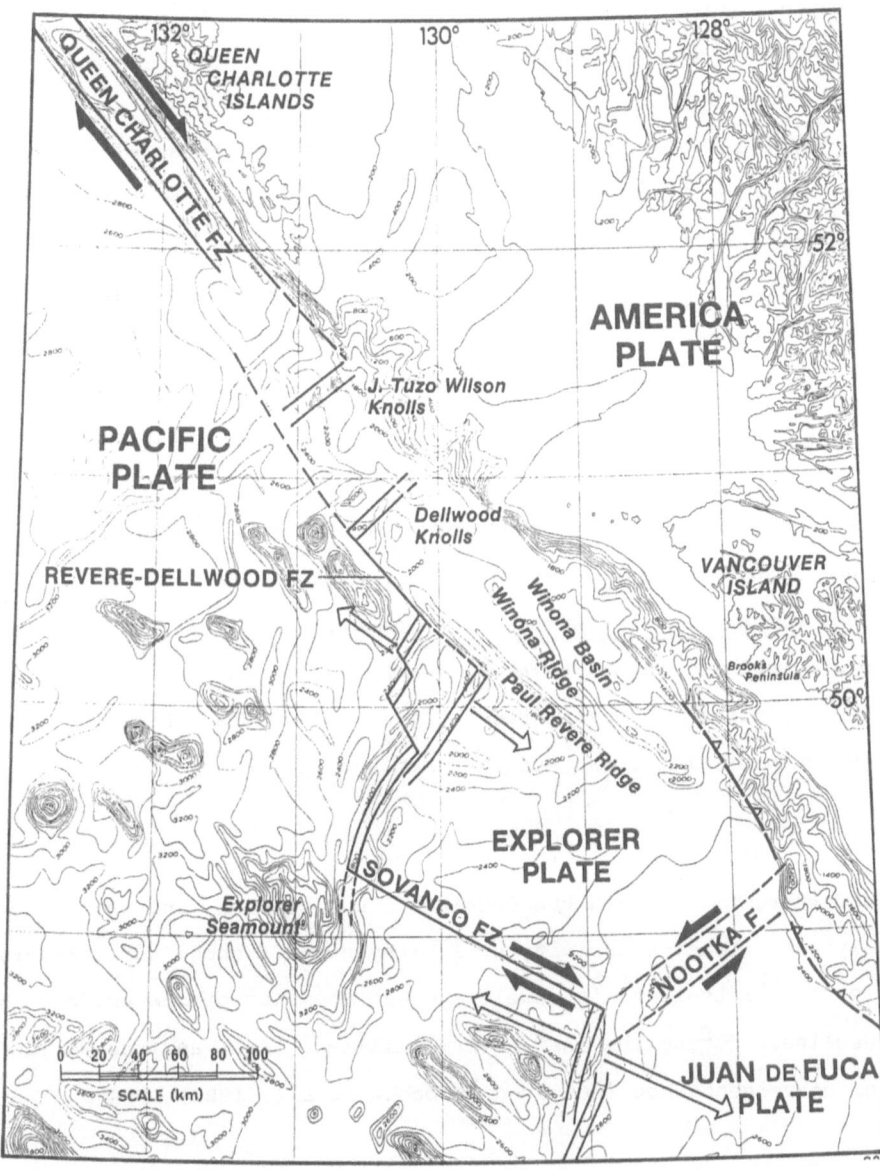

Figure 4. Conventionally assumed plate boundaries and major sea-floor features off western Canada (bathymetry in meters; modified from Riddihough and Hyndman, 1989).

oceanic plate under the North American continent is taking place at the Cascadia subduction zone.

Complications occur at the ends of the Juan de Fuca plate, where it is most fragmented - off northern California in the south and off British Columbia in the north (Riddihough, 1984). In the southern (Gorda) part of the Juan de Fuca plate, convergence with the North American continent is apparently accommodated entirely by internal deformation of the oceanic plate, and subduction in that area is thought to have stopped (Couch and Riddihough, 1989).

Less well understood is the geodynamics of the northern part of the Juan de Fuca plate off Vancouver Island (Fig. 4). It was proposed previously that an independent, small Explorer oceanic plate exists in that area, whose northern boundary with the Pacific plate was postulated to lie off Queen Charlotte Sound (Hyndman et al., 1979; Keen and Hyndman, 1979).

The middle part of the Juan de Fuca plate is still being underthrusted beneath the continent, but in an unusual manner, without thrust seismicity or a bathymetric trench (e.g., Acharya, 1992). Plate-tectonic models suggest the rate of convergence has declined during the last several million years, and the obliquity of convergence has increased (Babcock et al., 1992, 1994).

The least resolved are the interactions of northern Juan de Fuca plate (or its Explorer fragment) with the Pacific plate off northern Vancouver Island. The Pacific plate is believed, in all models, to be moving past North America dextrally in a NNW direction, perhaps with a very small component of convergence off the Queen Charlotte Islands (e.g., Minster and Jordan, 1978; DeMets et al., 1990). The situation off southeastern Alaska is simpler: no convergence has been inferred there, and the North America and Pacific plates are assumed to be separated by a right-lateral transform boundary. In reality, a complex fault system in a broad structural zone has been found along the plate boundary in that area (von Huene, 1989).

The logic of rigid-plate tectonics requires a ridge-trench-fault triple junction between the three plates, which has been modeled off Queen Charlotte Sound. Sea-floor spreading between the Pacific and Juan de Fuca plates (or the Explorer fragment) is supposedly taking place from two parallel ridges oriented at a right angle to the continental margin off Queen Charlotte Sound (Riddihough et al., 1980).

During the Cenozoic, periods of transtension were proposed to have caused rifting and large stretching of the continental crust in the Insular Belt, resulting in the creation of the Queen Charlotte Basin. By contrast, uplift of the Queen Charlotte Islands was ascribed to late Cenozoic transpression (Yorath and Hyndman, 1983; Hyndman and Hamilton, 1993). Deep seismic profiles across the Vancouver Island margin have been interpreted in terms of continentward-dipping thrust slices that presumably developed as a result of Cenozoic subduction (Yorath, 1980; Yorath et al., 1985a,b; Clowes et al., 1987; Hyndman et al., 1990).

Major pitfalls occur in uncritical application of theoretical models, which are based on generalized assumptions, to specific local geologic situations. Large uncertainties still bedevil the available plate reconstructions (Engebretson et al., 1985; Stock and Molnar, 1988; DeMets et al., 1990), so inferred plate interactions need to be tested by independent means.

Geologic field mapping shows that Late Cretaceous and early Tertiary thrust belts lie only between Vancouver Island and the mainland (Brandon et al., 1988; England and Calon, 1991). The island itself is characterized by a regular pattern of steep faults (Jeletzky, 1976; Muller, 1977a-c; Muller et al., 1981), and the geology of the Queen Charlotte Islands is similar (Thompson et al., 1991; Lewis et al., 1991a,b). Fault networks can be correlated between the western British Columbia mainland, the Queen Charlotte and Vancouver islands, and the interior shelf. This suggests reactivation of old steep faults was the main mode of tectonism in this region during the Tertiary. Structure of the entire Insular Belt was interpreted from gravity, magnetic and seismic data to be controlled by a network of steep faults bounding crustal blocks (Brew et al., 1991; Lyatsky, 1993a).

Subsequent chapters will show that two prominent, inter-regional structural zones meet off Vancouver Island. The Fairweather-Queen Charlotte fault system runs from the Alaskan interior along the continental margin off southeastern Alaska and the Queen Charlotte Islands. The Olympic-Wallowa zone continues from the Cordilleran interior in eastern Washington and Oregon into the Strait of Juan de Fuca. This structural configuration is not taken into account in the existing tectonic models. Still, these two fault systems, related in the Cordilleran interior to zones of weakness in the continental crust, also control large parts of the plate boundary with the adjacent oceanic plates.

CHAPTER 2 - EVALUATION OF THE DATA BASE

Direct geological observations - the main source of information

Only geological observation can provide direct information about rocks, their properties and field relationships. Observation yields the most reliable controls on any models, qualitative or numerical, used to predict unknown parameters from known ones. Because inferences about structure and composition of rocks from geophysical data are non-unique, geological observations are irreplaceable as a controlling tool.

This study of the western North American continental margin benefited from combining geological observations with geophysical data onshore and offshore. Constrained by geophysical data, geologic relationships observed on land by mapping were projected into submerged parts of the continental margin. A wealth of geological information, obtained by outcrop mapping in coastal areas and by offshore well drilling and sea-floor dredging, is available along the Pacific continental margin of northwestern North America. Onshore geology provided the primary constraints on geophysical interpretations and plate-tectonic models.

Geologic mapping has been carried out in many parts of Oregon and Washington, and ongoing programs offer a new look on the geology of the region. Though results are not yet summarized everywhere, reports of previous and recent surveys and of industrial drilling help elucidate the regional and local geologic structure. Offshore, geological information is provided by wells drilled for hydrocarbon exploration and research, including those of the previous Deep Sea Drilling Program (DSDP) and the ongoing Ocean Drilling Program (ODP).

Reconnaissance mapping was carried out on Vancouver and Queen Charlotte islands in the 1960s and 1970s. Samples of sedimentary and igneous rocks on the ocean floor were obtained by dredging on the shelf, slope and abyssal plain. Fourteen deep exploration wells were drilled in the 1960s on the interior and exterior shelf of western British Columbia. In bathyal areas, wells were drilled by the DSDP and the ODP. Important new drilling results were provided by ODP Leg 146.

Detailed geologic reports are available for parts of Vancouver Island. A program of detailed mapping of large parts of the Queen Charlotte Islands, in conjunction with geophysical surveys, has resulted in many reports by the Geological Survey of Canada published in the late 1980s and early 1990s.

In southeastern Alaska, ongoing detailed geologic mapping onshore has led to revision of a number of earlier ideas about the evolution of that area. However, offshore geologic information is scarce. Still, the new data have already produced a more comprehensive understanding of geology of southeastern Alaska and of the entire continental margin.

Physical parameters of rocks - constraints on interpretation of potential-field data

Rock magnetization

Magnetization is the rock property that relates a magnetic anomaly to its geologic source (Reynolds et al., 1990). The principal magnetic mineral in the study region is evidently magnetite (Coles and Currie, 1977; Arkani-Hamed and Strangway, 1988). It is mostly associated with igneous rocks, whose distribution largely controls the magnetic anomaly pattern.

Magnetic susceptibilities of volcanic rocks in the region have been reported by Currie and Muller (1976), Finn (1990), Dehler and Clowes (1992). Paleozoic volcanics in the Insular Belt have low susceptibility, rarely more than 100x10-6 emu (1,250x10-6 SI units).   Triassic and Jurassic volcanics, with values between only 40x10-6 and 2,000x10-6 emu (500x10-6 to 25,000x10-6 SI units), are variously magnetic.   Usually highly magnetic, >1,000x10-6 emu (>12,500x10-6 SI units), are Eocene basalts on Vancouver Island and in Washington, which are commonly marked by strong anomalies.

Most of the exposed granitoid plutons in the region are marked by prominent positive and negative magnetic anomalies. This makes it possible to locate, by analogy with such anomalies, plutons hidden under roof rocks or sea water (Arkani-Hamed and Strangway, 1988; Finn, 1990; Lyatsky, 1991a). Many high-grade metamorphic rocks, such as those of the Jurassic Westcoast complex on Vancouver Island, are associated with positive magnetic anomalies, but others, such as the Leech River complex on southern Vancouver Island, are consistently associated with negative anomalies.

Interpretation is complicated because magnetization of rocks may be induced or remanent, normal or reverse, and sometimes different magnetization vectors from the same source body interfere.   As a result, causative bodies produce a variety of anomaly forms which may be difficult to interpret in detail.   Alignment of magnetic anomalies and presence of steep linear gradient zones may indicate faults. In oceanic regions, magnetic-anomaly lineations indicate presence of blocks of oceanic crust.

Rock density

Information about rock densities was obtained from numerous publications (Stacey, 1975; Currie and Muller, 1976; MacLeod et al., 1977; Anderson and Greig, 1989; Finn, 1990; Sweeney and Seemann, 1991; Dehler, 1991). Another important source of density information is well logs. Most gravity anomalies between northern Olympic Peninsula and southeastern Alaska are produced by density contrasts between rocks of three main categories (Lyatsky, 1993a).

1. Low densities (1,800 to 2,500 kg/m3, increasing with depth) characterize Neogene sediments on the interior continental shelf (e.g., in the Queen Charlotte Basin). In the core of the Olympic Mountains onshore and in the Hoh and Tofino Basins on the exterior shelf, Tertiary sediments are denser (2,400-2,600 kg/m3).

2. Lower Jurassic clastic rocks are relatively light, ≤2,640 kg/m3, but Cretaceous clastics are about 2,700 kg/m3. Upper Triassic limestones have a higher density of ≥2,760 kg/m3. Volcanic rocks of Tertiary and Middle Jurassic age on the Queen Charlotte Islands have a density of 2,650 kg/m3, but Lower Jurassic volcanics on Vancouver Island are usually heavier (2,700-2,800 kg/m3).

3. High densities characterize Upper Triassic Karmutsen basalts: 2,880 kg/m3 on the Queen Charlotte Islands and around 2,950 kg/m3 on Vancouver Island. Eocene basalts in western Washington have densities between 2,200 and 2,950 kg/m3, and a density of 2,950 kg/m3 on southern Vancouver Island. Paleozoic rocks, depending on lithology and grade of metamorphism, have variable densities from 2,730 to 2,900 kg/m3.

Plutonic massifs in the Insular and Coast belts generally have densities between 2,600 kg/m3 for granite and 2,820 kg/m3 for diorite. Depending on their country rocks, many plutons in this region are not marked by strong gravity anomalies and are more readily interpereted from magnetic maps. Confusingly, some plutons cause negative gravity anomalies similar to those over Tertiary sediment-filled depressions. Prominent gravity lows along the entire length of the western North America continental margin are associated with Tertiary sedimentary basins (Couch and Riddihough, 1989). Yet, plutonic and metamorphic rocks of continental-crust crystalline basement may also contribute to some of those pronouced anomalies.

Processing of potential-field data
Fundamental notions

Magnetic and gravity data may reveal different aspects of rock composition and structure of the region. Magnetic anomaly at any given locality is the difference between the recorded magnetic-field intensity and the theoretical one predicted by the International Geomagnetic Reference Field. Magnetic maps reflect rock properties no deeper than the Curie isotherm, whereas gravity maps represent density contrasts at both shallow and deep levels in the lithosphere.

Gravity anomaly is the difference between the measured gravity field and a field computed for a given location from theory, assuming an idealized rotating, spheroidal Earth (Goodacre et al., 1987a). Density contrasts which cause gravity anomalies are located at various levels in the Earth: (1) below the base of the crust; (2) in the crystalline crust; and (3) in the volcano-sedimentary supracrustal cover. Measured gravity values are affected also by topography, as well as by recording-site elevation and latitude. Desirable for geological interptetation are those anomalies which reflect the Earth structure at crustal and supracrustal levels.

Magnetic and gravity coverage

Old magnetic maps in the region were based on profiles acquired by ship (MacLeod et al., 1977; Currie et al., 1983). High-quality aeromagnetic data are now available in British Columbia (Currie and Teskey, 1988) and Washington (Finn, 1990). For the purpose of this study, magnetic data were corrected for the International Geomagnetic Reference Field and resampled at a 812.8-m interval (i.e., two samples per mile).

Gravity station spacing varies across the region and is about 10 km on average. The best coverage is available offshore, whereas land areas are covered unevenly and generally in less detail (Currie et al., 1983; Finn et al., 1991; Sweeney and Seemann, 1991). These data were gridded at an optimal 2-km interval in British Columbia and a 5-km interval in Washington.

Reductions of gravity data

Dependence of gravity values on the distance from the center of the Earth, hence on elevation, is accounted for by the free-air reduction (Goodacre et al., 1987b). The Bouguer reduction (Goodacre et al., 1987c) onshore takes into account the attraction of the rock mass, assumed to be a horizontal slab, density 2,670 kg/m3, between the recording station and the sea level. Offshore, the Bouguer reduction involves "replacing" sea water (density 1,030 kg/m3) with an equivalent thickness of rock (density 2,670 kg/m3). Thus, anomalies in a Bouguer map mainly reflect crustal structure and variations in Moho depth.

Gravitational effects of variations in crustal thickness may be partly attenuated by the isostatic reduction. The Airy model of isostasy, assumed typically for the Earth's crust, requires that areas of positive topography, if in equilibrium, be underlain by crust of increased thickness; areas of negative topography must be underlain by abnormally thin crust. The isostatic reduction accounts for the gravitational attraction of these assumed crustal roots and antiroots, and an isostatic map represents crust-sourced anomalies better than does a Bouguer map (Simpson et al., 1986; Goodacre et al., 1987d; Simpson and Jachens, 1989).

However, isostatic maps may still contain anomalies other than those sourced by intracrustal or supracrustal density contrasts. These components of the gravity field may be related to crustal flexure, local variations in mantle density and heat balance, etc. Their wavelengths usually exceed those of geologic features of interest. Many such anomalies are correlative with topography, and they can be attenuated by the enhanced isostatic reduction (Sobczak and Halpenny, 1990).

This algorithm employs a least-squares procedure to linearly correlate conventional isostatic gravity values in a map area with topography onshore and imaginary rock-equivalent topography offshore. The latter is computed by "replacing" the mass of sea water (density 1,030 kg/m3) with an equivalent mass of rock (density 2,670 kg/m3) and adding the thickness of the simulated rock layer to the existing bathymetry. The topography-anomaly relationship is used to correct the isostatic data and thus produce an enhanced isostatic anomaly map. Anomalies in such a map are, in theory, caused largely by intracrustal and supracrustal sources, with other influences minimized.

Interpretation of maps resulting from gravity reductions requires caution. These procedures rely on specific assumptions, for example, that: the geoid is represented by the reference ellipsoid in the map area; isostatic compensation is one-dimensional, sensu Airy; mantle density is constant beneath the map area; a horizontal slab of uniform density 2,670 kg/m3 represents the rock mass between the sea level and the ground surface; lower-crustal roots and antiroots are plane masses at a depth of 30 km; crustal flexure produces a linear relationship between topography and isostatic anomaly values. Fortunately, errors arising from variations on these assumptions are usually small (Simpson et al., 1986; Goodacre et al., 1987a-d).

Along the continental margin, any interpretation errors due to edge effects are minimized by calibrating the interpretations with seismic refraction and gravity models. Where bathymetric relief is large, only coarse crustal structure is interpreted. Such interpretations are robust enough, and constrained well enough, to be relatively insensitive to edge effects.

Horizontal-gradient maps

Horizontal-gradient maps enhance short-wavelength features in gravity and magnetic data. They reflect lateral variations in the magnitude of a potential field; abrupt variations are emphasized. Horizontal-gradient maps help interpret shallow crustal structure.

Different methods of generating such maps have been used by different workers. The finite-difference method estimates the horizontal gradient at a grid node from differences with anomaly values at neighboring grid nodes (Cordell and Grauch, 1985).

Another common technique (Sharpton et al., 1987; Goodacre et al., 1987e) relies on fitting of a planar surface to a window of 5x5 potential-field values. The slope of this plane is a scalar quantity considered to represent the magnitude of the horizontal gradient of the potential field in the center of the window. Yet another method (Lyatsky et al., 1992a,b) involves fitting a third-order surface to a window of 5x5 potential-field values. A higher-order surface offers a more realistic representation of the field within the window, while the third order is still low enough for the best-fit surface not to be greatly affected by any spurious data points. The horizontal gradient computed at the center of the window is treated as a vector and displayed on a map as an arrow whose azimuth represents the direction of the gradient and whose length is proportional to the gradient's magnitude.

Contouring or color coding scalar gradient values can be used to produce maps, as was done with gravity data for Washington and southwestern British Columbia (Finn et al., 1991; Dehler and Clowes, 1992). An aeromagnetic horizontal-gradient vector map has been produced for western British Columbia from northern Vancouver Island to Dixon Entrance (Lyatsky et al., 1992a,b).

<u>Upward continuation of potential-field data</u>

To investigate large geologic features in the region, gravity and magnetic data were upward continued to nominal elevations ranging from 5 to 100 km. This procedure simulates the appearance of the potential-field data recorded at a specified elevation above the ground or sea level. Wavelength-filtered gravity maps of Washington state were produced by Finn et al. (1991), who presented two maps containing anomalies longer and shorter than the selected 100-km wavelength cut-off. Upward continuation requires a more complex filter but produces maps whose physical meaning is more intuitive.

The theory of upward continuation was discussed by Grant and West (1965), Blakely and Connard (1989), Teskey et al. (1989a). No sources or sinks of the potential field are assumed to exist between the real and nominal map levels. At sea, this assumption is justified because no rocks protrude above sea level, and onshore in the continental-margin region topographic elevations rarely exceed 1000 m. The Olympic Mountains are high (>2400 m) but localized. Such high elevations are only attained on a regional scale inland from the areas of interest, in the North Cascade and Coast Mountains. Regardless, the most geologically informative gravity maps in the region were found to be the ones upward continued to 20 km (Lyatsky, 1991a), many times higher than the tallest mountain peaks.

Upward continuation involves filtering the data on the basis of anomaly wavelength: short-wavelength anomalies are attenuated preferentially. If the potential field is measured on a horizontal plane and desired on a higher horizontal plane, upward continuation is given by (Blakely and Connard, 1989):

$$F[hU(x,y)] = F[h(x,y)]exp(-k\delta z) \qquad \delta z > 0,$$

where k is the anomaly wavenumber (the quantity $k/2\pi$ is the inverse of wavelength), δz is the distance of upward continuation, $F[h(x,y)]$ is the recorded potential field in the Fourier domain, and $F[hU(x,y)]$ is the upward-continued potential field in the Fourier domain.

Local, short-wavelength anomalies, which would not be observed at a high recording level, are suppressed. Broad anomalies of shallow origin are not excluded, and most features in upward-continued maps are caused by large sources at various depths in the subsurface. Upward continuation to 20 km gives a good picture of the large-scale structure of the upper crust, but is still detailed enough to permit correlation of anomalies with features in surface geology.

Assessment of seismic data

Overview of the data

Much of the available reflection and refraction data are old and have a low resolution. However, combined with the available modern seismic profiles in Oregon (Keach et al., 1989), Washington (Taber and Lewis, 1986) and British Columbia (Yorath et al., 1987; Clowes et al., 1987; Rohr and Dietrich, 1992; and others) and with the potential-field data, they help interpret the deep structure of different parts of the continental margin.

Modern controlled-source seismic profiles are available in places across the British Columbia continental margin. These reflection and refraction data have been acquired largely by the LITHOPROBE program. Other seismic data are available from the U.S. Geological Survey. As expected, shallow subsurface levels are imaged best, but resolution decreases rapidly with depth. Refraction data across the margin generally offer reasonably good constraints for modeling the structure of the upper crust, but the data for the lower crust and upper mantle are of lower quality.

In the reflection surveys, signal penetration is reduced due to scatter from structural and stratigraphic contacts between rock bodies with contrasting lithologies. Results are poor images of deep parts of sedimentary basins and uncertain definitions of the basement (Bruns and Carlson, 1987; Lyatsky, 1991b).

Coarse images of the lithosphere are provided by inversion of teleseismic arrivals from distant earthquakes. Numerous recording stations are in operation in western U.S. and Canada, and the first important summaries of results have already been published (e.g., Humphreys and Dueker, 1994a,b).

Ambiguities in seismic interpretation

Deep seismic data across the Vancouver Island margin are generally of good quality. But even so, care must be taken to resist the temptation to overinterpret. This caveat is important because in some influential papers these data have been cited as "the first direct evidence for the process of subduction underplating" (sic!) beneath the continental margin off southern Vancouver Island (Clowes et al., 1987, p. 31; see also Yorath et al., 1985a,b; Hyndman et al., 1990).

Such a view is overly optimistic. During a workshop of the International Association of Seismology and Physics of the Earth's Interior in 1987, alternative interpretations and velocity models derived from these data have been presented by investigators from national and foreign institutions (see papers in: Green, ed., 1990). This diversity of opinions arose despite the a priori restriction on interpretations that "the Juan de Fuca oceanic plate is being subducted beneath the collage of exotic terranes that constitute Vancouver Island and the western North American mainland" (Green, ed., 1990, p. 1).

All the same, analysis of the data led different groups of workers to very different conclusions. Many participants observed that interpretation suffered from variations in the recording parameters and quality of the data and from poor constraints on deep velocity structure. Most workers took care stress the general problem of non-uniqueness of geophysical interpretations.

Thybo (1990) noted that a subducted slab was only assumed to exist, not compellingly resolved. Subjectivity of seismic interpretations was also pointed out by Morgan and Warner (1990), who cautioned that their own refraction model across the margin is only "one of a series of solutions" (p. 40). Weber (1990) similarly acknowledged his interpretation as tentative because, "due to the non-uniqueness of modeling, other models may also explain the observed data" (p. 49).

Limited seismic coverage contributed to the uncertainties in refraction models (Morgan and Warner, 1990): only the top part of the southern Vancouver Island refraction profile is constrained well. This leaves uncertain the structure of lower-crustal levels, the position of the Moho, and the existence of a subducted oceanic slab.

Still very uncertain is the nature and even existence of a high-density, high-velocity body under Vancouver Island and adjoining shelf (cp. Stacey, 1973). This body was once presumed to be an old delaminated slice of subducted oceanic crust (Stacey, 1973; Spence et al., 1985) or newly underthrusted oceanic lithosphere (Riddihough, 1979). Confirming the non-uniqueness of geophysical models, such a slice in the models of Egger and Ansorge (1990), Drew and Clowes (1990) and others has different velocities and shapes. These models differ also in the position and dip of the subducted slab.

Thybo (1990) showed that the high-velocity mid-crustal sliver is not required by the refraction data at all. No high-velocity zone was shown in the model of Iwasaki and Shimumara (1990), and Fowler and Pandit (1990) also noted that such a sliver under Vancouver Island may not exist.

Seismic images were noted to be especially poor on the continental shelf and upper slope off Vancouver Island. Thybo (1990) found signal penetration to be reduced, probably due to folding and faulting of shallow sediments, and deep reflections to be masked by multiples. Iwasaki and Shimumara (1990) also found this part of the profile ill-constrained. Mereu (1990) modeled in the refraction profile three crustal zones: oceanic, transitional, and continental. The transitional zone, under the shelf and upper slope, was shown as a coherent crustal block about 20 km thick, its seismic velocity increasing downward from 6 to 7 km/s. Such properties typical for continental crust. The Moho in Mereu's model deepens towards the continent not smoothly but stepwise.

The dependence of seismic images on the parameters chosen for data processing and display is a common complication (Hawthorne, 1990). Indeed, detailed reprocessing of reflection data on Vancouver Island showed that many seismic events initially considered to be reflections from deep structures are in fact diffractions (Milkereit et al., 1990). Stacking and migration tests on these data showed that the deep event once regarded as the subducted oceanic Moho may be just off-line noise (Hawthorne, 1990; Levato et al., 1990).

Methodological principles of this study

To minimize uncertainties in the interpretation of geophysical data, several precautions were taken in this study. The first was to "stand on the continent", where the most reliable information, obtained from surface geologic mapping, is available. The observed geological and geophysical relationships were then projected to the periphery of the continent offshore, as far as marine data allowed. Results thus obtained, having an advantage of being consistent with all available information onshore and offshore, served as constraints for tectonic inferences.

Offshore, some direct geologic information is available from dredging the sea bed and drilling wells. Geophysical data include mostly sonar images of the ocean floor, seismic profiles and potential-field (gravity, magnetic, electromagnetic) survey results. Non-uniqueness in the interpretation of geophysical data was reduced in this study by using diverse data types jointly with geological facts, to generate an internally consistent interpretation (Lyatsky and Lyatsky, 1990).

In the past, a major hindrance to regional integration of data was jurisdictional and institutional barriers at the U.S.-Canada border. Interpretations in British Columbia often differed from those in the adjacent parts of Alaska and Washington, and vice versa. In this study, geology of coastal British Columbia is linked to that of the neighboring regions in U.S., in order to obtain a more comprehensive geologic model of the crustal structure of the western margin of the North American continent.

Pre-Tertiary stratigraphic record

Paleozoic

Early Cambrian plutonic rocks have been identified on the south islands of southeastern Alaska (Gehrels, 1990) and on the San Juan Islands near the British Columbia-Washington boundary (Brandon et al., 1988). The country rocks intruded by these plutons are suspected to be Precambrian.

Younger Paleozoic rocks of various lithologies and ages are also found in southeastern Alaska (Gehrels and Saleeby, 1987a,b; Gehrels et al., 1987; Brew et al., 1991). They decrease in abundance to the south, in western British Columbia. Only Middle(?) Devonian or younger Paleozoic rocks have been mapped along the mainland coast of Hecate Strait (Woodsworth and Orchard, 1985), on the Queen Charlotte Islands (Hesthammer et al., 1991) and on Vancouver Island (Muller, 1980a; Massey and Friday, 1989; Andrew et al., 1991). They include a variety of stratified units and plutons, variously metamorphosed and deformed. On Vancouver Island, Paleozoic rocks are at least 5.5 km thick (Massey and Friday, 1989).

A complex Paleozoic assemblage is found locally in the coastal provinces of western Washington (on the San Juan Islands; Brandon et al., 1988) is still poorly studied.

Mesozoic

Mesozoic rocks are more widespread. After a regional stratigraphic hiatus, this succession begins with massive basalts and associated tuffs of the Upper Triassic Karmutsen Formation (Table I). These basalts also occur in southern and southeastern Alaska (Jones et al., 1977), on the Queen Charlotte and Vancouver islands (Sutherland Brown, 1968; Muller, 1977a,b) and near the mainland shore of Hecate Strait (Woodsworth, 1988).

The Karmutsen Formation is extremely thick, in places up to 6 km (Andrew and Godwin, 1989d). It comprises mafic extrusive rocks of different types: submarine pillow lavas and subaerial flows, volcanic breccia and tuff. Their geochemistry is diverse (Barker et al., 1989) but in general similar to that of Cenozoic basalts of the Columbia River province in the Cordilleran interior in eastern and central Washington (Andrew and Godwin, 1989b). Since the Columbia River basalts are intracontinental (Reidel et al., 1994), so are many parts of the Karmutsen Formation. In the 1970s, the Karmutsen Formation was considered to be oceanic crystalline crust (Muller, 1977a), but that idea is inconsistent with the new evidence.

With a narrow but gradational contact, Karmutsen volcanics pass upsection into sedimentary strata of latest Triassic and Early Jurassic age, which consist of shallow marine limestone, shale and sandstone. On the Queen Charlotte Islands, these rocks are subdivided into two groups, Kunga and Maude, with a cumulative thickness exceeding 1000 m. The lower stratigraphic units in this package are mainly carbonate, whereas and the upper units are made up of shale and sandstone with minor tuff and volcanic flows (Cameron and Tipper, 1985; Thompson et al., 1991).

These carbonates and clastics of Late Triassic age extend, without

39

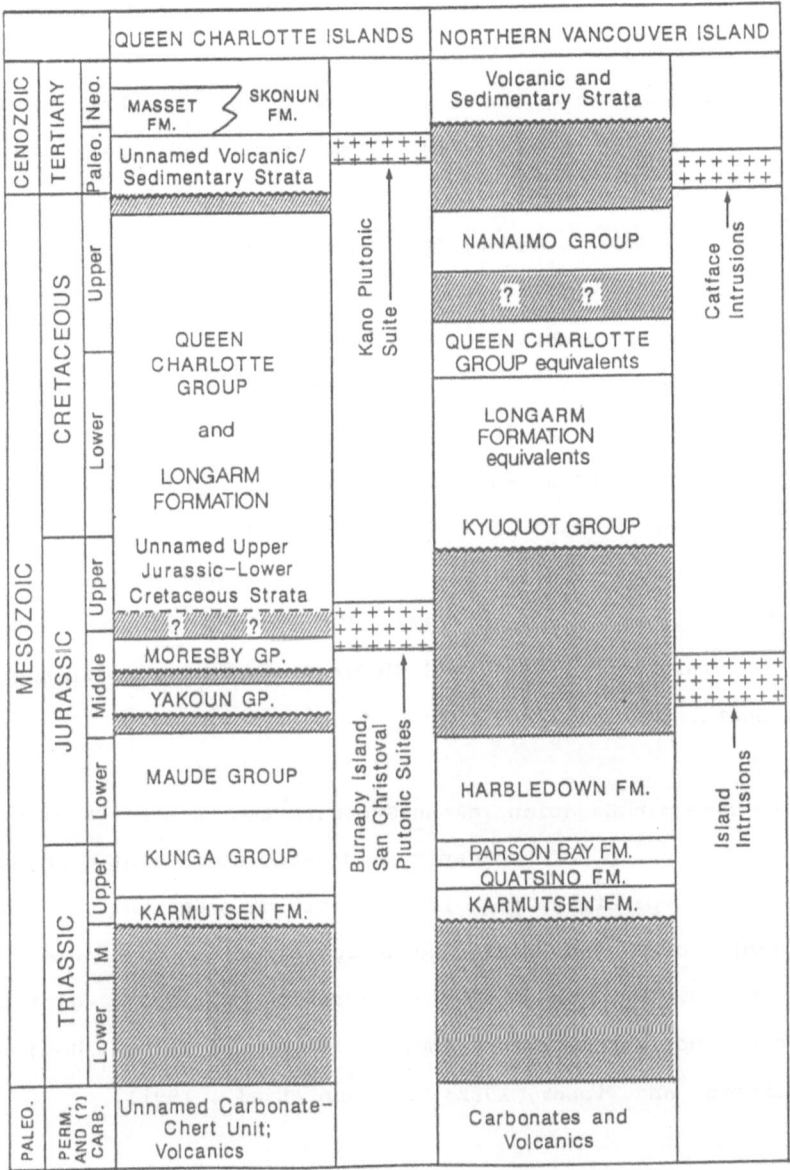

Table I. Generalized stratigraphic columns for the Insular Belt: the Queen Charlotte Islands and northern Vancouver Island (based on Thompson et al., 1991; Lewis et al., 1991a; Lyatsky, 1993a).

fundamental changes in lithofacies, on Vancouver Island. However, Lower Jurassic rocks on Vancouver Island are only partly sedimentary. Abundance of volcanic flows, tuffs and breccias of intermediate composition increases upsection, and the Bonanza Group is largely volcanic (Muller et al., 1974, 1981; Jeletzky, 1976; Andrew and Godwin, 1989a; Desrochers, 1989).

Late Triassic to Early Jurassic sedimentation took place in a single marine basin which covered not only much of the Insular Belt but also extended far into the interior of the northwestern Cordillera. This broad and shallow basin lay in a tectonically quiescent region (Tipper and Richards, 1976). Stratigraphic differences between the Queen Charlotte Islands and Vancouver Island (Jeletzky, 1976; Muller, 1977a; Cameron and Tipper, 1985) attest to migration of Early and Middle Jurassic tectonism along the continental margin.

The Middle Jurassic Bajocian Yakoun Group on the Queen Charlotte Islands is made up of volcanic and associated clastic rocks. Unconformities separate it from from underlying older units. The unconformably overlying Middle Jurassic volcanic-derived clastic rocks of the Moresby Group and Upper Jurassic siliceous clastic strata are poorly preserved. Their thickness is several hundred meters (Cameron and Tipper, 1985; Thompson et al., 1991).

These rocks do not extend on Vancouver Island (Jeletzky, 1976; Muller, 1977a), where tectonic movements and the accompanying erosion were apparently stronger.

Cretaceous rocks, in contrast, are widespread across the Insular

Belt. Their cumulative thickness on the Queen Charlotte Islands is about 3000 m. The Lower Cretaceous Longarm Formation and Albian to Maastrichtian Queen Charlotte Group comprise coarse to fine clastic rocks, primarily lithic wackes, locally including subarkoses and arenites. Volcanics are minor and occur only in places (Yagishita, 1985; Haggart, 1991). Equivalent sedimentary rocks on northern Vancouver Island are less well studied (Muller et al., 1974, 1981; Jeletzky, 1976; Nixon et al., 1995).

The area of sedimentation in the Cretaceous was narrower than in the Early Jurassic. It was elongated, covering the Insular Belt from the present-day Queen Charlotte Islands to northern Vancouver Island (Lyatsky and Haggart, 1993). The generally quiescent tectonic conditions were disturbed in the Turonian and Coniacian, when the Honna conglomerate was laid down, but thereafter deposition of finer-grained sediments resumed. Cretaceous sedimentary rocks contain abundant granitic and volcanic fragments derived from the east, where an Andean-type orogenic belt was developing at that time (Yagishita, 1985).

The long-lived Cretaceous basin evolved along much of the Insular Belt apparently in response to the development of the Mesozoic continental margin (Haggart, 1993). Continuity and paleogeographic uniformity of this basin suggest that a similar tectonic regime existed all along this belt.

Only in Late Cretaceous time did separate Nanaimo and Comox basins (regarded sometimes as sub-basins of the Georgia Basin) develop in the area from eastern Vancouver Island to the Gulf Islands. Formation of these and other small intracontinental depressions was followed by local marine incursions and construction of deltas and submarine fans, but presence of coals (Kenyon and Bickford, 1989) suggests that non-marine paleoenvironments also existed. Up to 4000 m of conglomerate, sandstone and mudstone, intercalated with coal, make up the Nanaimo Group (Mustard, 1991; England and Hiscott, 1992). Clastic sediments derived from surrounding uplifted areas were deposited in fault-bounded depressions (Pacht, 1984). Rocks correlative with the Nanaimo Group are represented only in local grabens on northern Vancouver Island, and the Queen Charlotte Islands.

Development of local horsts and grabens is suggested by the proximal sourcing of Nanaimo Group sediments from the Coast Belt on the mainland to the east and from Vancouver Island to the west. Abrupt thickness variations, lithofacies changes, and variations in coal rank and vitrinite reflectance are consistent with an evolution of these basins in actively subsiding grabens.

Tectonic stages of pre-Tertiary geologic evolution

Paleozoic interval

Several poorly resolved episodes of tectonism occurred before the extrusion of Upper Triassic Karmutsen basalts. In many localities in southeastern Alaska and on Vancouver Island, Paleozoic and older rocks are metamorphosed. Structures in them are considerably more complex than structures in younger rocks (Muller, 1980a; Woodsworth and Orchard, 1985; Massey and Friday, 1989; Gehrels, 1990).

Late Triassic to Early Jurassic interval

The pronounced angular unconformity at the base of the Karmutsen

Formation suggests that an episode of strong tectonism occurred before the eruption of Upper Triassic basalts. However, the Karmutsen and especially the conformably overlying Upper Triassic to Lower Jurassic formations accumulated under conditions of relative tectonic quiescence. A quiescent regime existed at that time in much of the western Canadian Cordillera (Tipper and Richards, 1976; Cameron and Tipper, 1985; Miller et al., 1992).

Mid-Jurassic episode of tectonism

Regional magmatism, metamorphism and deformation occurred in the Jurassic. However, its onset was not simultaneous across the region. Vancouver Island was affected in the Early Jurassic (Muller et al., 1974, 1981; Jeletzky, 1976), but the Queen Charlotte Islands were affected only later, in the Middle Jurassic (Cameron and Tipper, 1985; Thompson et al., 1991).

Voluminous magmatism occurred at that time across the Insular Belt. On Vancouver Island, it is represented by widespread granodioritic Island Intrusions and comagmatic intermediate extrusive rocks of the Bonanza Group (Armstrong, 1988; Andrew and Godwin, 1989a; Andrew et al., 1991). Ar/Ar cooling ages of the Island Intrusions are scattered widely around 170-175 Ma (Archibald and Nixon, 1995). From geological field evidence (Jeletzky, 1976) and magnetic data (Arkani-Hamed and Strangway, 1988), many granitoid plutons probably merge at shallow depths into broad batholiths. Similar Jurassic magmatism on the Queen Charlotte Islands is represented by the Bathonian-Oxfordian San Christoval and Burnaby Island plutonic suites of intermediate composition, and by comagmatic Bajocian Yakoun Group volcanics (Anderson and Reichenbach, 1991; Woodsworth et al., 1991).

No strong regional metamorphism accompanied the Jurassic tectonic episode. In most parts of the Insular Belt, Jurassic and older Mesozoic rocks are metamorphosed not at all or to very low grades usually no higher than prehnite-pumpellyite. A notable exception is a belt of high-grade rocks along almost the entire length of western Vancouver Island. In the narrow Westcoast belt, just 15-30 km across, Triassic-Early Jurassic and older rocks native to the island were buried to depths of 15-25 km and metamorphosed to amphibolite grade around 180 Ma, i.e. near the Early-Middle Jurassic boundary. No high-grade equivalents of this complex have been recognized on the Queen Charlotte Islands, though geophysical data suggest they may lie on trend with the Westcoast complex just west of these islands offshore (see following chapters).

Along the western and partly southern periphery of Vancouver Island, uplift and rapid unroofing in the Early Tertiary exposed the metamorphic rocks of the Westcoast complex in a series of blocks which form a fault-bounded belt (Muller, 1977a; Isachsen, 1987). In this belt, a series of blocks were brought to the surface from mid-crustal depths along steep, deep faults.

The intensity of mid-Jurassic folding and thrusting was much greater to the north. On northern Queen Charlotte Islands, the Middle Jurassic Bajocian compression led to 50% shortening on Graham Island and northern Moresby Island, accommodated by NE-SW-trending folds and thrust faults (Thompson et al., 1991; Lewis et al., 1991a). The amount of shortening declines rapidly to the south. The significance of the Middle Jurassic unconformity diminishes correspondingly: though prominent on Graham Island and northern Moresby Island, it is less apparent on southern Moresby Island (Lewis, 1991). On northern Vancouver Island, the amount and exact age of Jurassic tectonic compression are unclear, but Cretaceous rocks overlie the older units unconformably (Nixon et al., 1995). Vertical block movements on Vancouver Island were large enough to bring protoliths of the Westcoast metamorphic complex to mid-crustal levels.

Late Jurassic to Late Cretaceous interval

Vertical block movements became dominant across most of the Insular Belt in the Late Jurassic. On the Queen Charlotte Islands, high-angle faulting and block movements influenced sedimentation since that time (Lewis et al., 1991a,b; Thompson et al., 1991; Haggart, 1991). The reported offsets on a number of large dip-slip faults in the Late Jurassic are more than 1000 m. As block movements ceased, sedimentation spread from local depocenters into broader areas and continued into the Cretaceous.

Major Late Jurassic faults have a predominant NNW trend. The new prominence of this trend reflected the establishment of a regional structural grain which has predominated since that time. The large Cretaceous sedimentary basin in the Insular Belt follows the same trend. The Coast Belt, rising parallel to it, provided abundant sediments for this basin from the present-day mainland (Yagishita, 1985; Haggart, 1991). Tracing the Cretaceous strandlines of this basin shows that they migrated very slowly eastward (Haggart, 1991, 1993). However, they mostly remained on the Queen Charlotte Islands and northern Vancouver Island throughout the Cretaceous.

A markedly different tectonic regime existed on southernmost Vancouver Island and the neighboring San Juan Islands. Between 100 and 84 Ma, tectonic compression, caused by the development of orogens on the site of the Coast and North Cascade mountains to the east, affected the area of the San Juan Islands, where the Northwest Cascade thrust system was created during that time (Brandon et al., 1988). In contrast, Late Cretaceous extensional tectonism on the Gulf Islands and eastern Vancouver Island created a series of grabens filled with thick Turonian to Maastrichtian sediments of the Nanaimo Group (Pacht, 1984).

Thus, continental-crustal tectonics dominated the geologic evolution of the region during most of the Mesozoic.

Latest Cretaceous(?) to earliest Tertiary tectonism

The next significant episode of tectonism in the Insular Belt took place in the latest Cretaceous(?) and/or earliest Tertiary. Many older high-angle faults were rejuvenated, controlling horst-and-graben movements. That these faults are mostly steep is confirmed by the regional association of Tertiary igneous rocks with faults (Woodsworth et al., 1991).

Offsets of Cretaceous sedimentary rocks on reactivated old faults on the Queen Charlotte Islands indicate earliest Tertiary shortening of ≤10% (Thompson et al., 1991; Lewis et al., 1991a,b). Similar compression, as well as local right-lateral movements, occurred in the Late Cretaceous to early Tertiary on northern Vancouver Island (Nixon et al., 1995).

In contrast, compression on the Gulf Islands and southeastern

Vancouver Island created the west-vergent Cowichan thrust system (England and Calon, 1991) whose east-dipping faults cut sedimentary rocks of the Nanaimo Group. This west-vergent thrust system does not extend on Vancouver Island past the Alberni-Cowichan Lake system of high-angle faults, which divides Vancouver Island into the eastern and western blocks.

Timing of terrane accretion in the western Cordillera

The Cordilleran orogenic system evolved under a twin control of internal and external factors, the latter including plate interactions at the continental margin. The relationship between the internal and external factors, however, remains unclear (Cowan and Bruhn, 1992).

Convergence of oceanic plates with North America during much of the Mesozoic might have produced subduction-related magmatic-arc formations on the site of the western Cordillera. Less clear is the role of "exotic" terranes thought to be accreted to the western edge of the North American continent (Jones et al., 1977; Monger et al., 1982; Saleeby and Busby-Spera, 1992). Most of the Insular Belt in these models is occupied by the Wrangell terrane, or Wrangellia. Its current definition is stratigraphic, based on the presence of Karmutsen basalts and overlying Upper Triassic-Lower Jurassic limestone and clastic rocks, underlain by Paleozoic units (Monger, 1991). In this definition, Wrangellia or its fragments have been identified in southeastern Alaska, the Queen Charlotte Islands and Vancouver Island. Presence of Karmutsen basalts on a small island near the mainland shore of Hecate Strait (Woodsworth, 1988) indicates that Wrangellia probably continues beneath most of the British Columbia interior shelf.

Most terranes recognized in the western Cordillera are commonly regarded as "suspect", because their boundaries, origins and travel paths remain unclear (Cowan and Bruhn, 1992). This is particularly true of the two largest terranes - Wrangell and Alexander - presumed to exist in the Insular Belt. Initially, Wrangellia was regarded as fundamentally distinct from the Alexander terrane, which is recognized as dominant in many parts of southeastern Alaska (Gehrels and Saleeby, 1987b; Brew et al., 1991). The rock assemblage typical of the Alexander terrane is complex and laterally variable, and the understanding of its tectonic evolution is sketchy (Woodsworth and Orchard, 1985; Gehrels and Saleeby, 1987a). A Pennsylvanian granitic pluton in Alaska stitches the Alexander and Wrangell terranes, indicating that at least since that time they have formed a single entity (Gardner et al., 1988).

The docking of Wrangellia to the North American continent was once thought to have taken place as recently as in the Cretaceous (Monger et al., 1982), but this contention has been questioned all along (Brew and Ford, 1983; van der Heyden, 1992). The eastern boundary of the composite Alexander-Wrangellia terrane is still uncertain. Its suture with crustal blocks in the Cordilleran interior was previously thought to lie in the Coast Belt, where it is masked by vast plutons (e.g., Crawford et al., 1987). From new studies, however, a single Alexander-Wrangellia-Stikinia "megaterrane" docked with North America in the Middle Jurassic, causing regional compression and magmatism (van der Heyden, 1992).

For the late Mesozoic and Cenozoic, at any rate, magmatic,

metamorphic and structural evidence offers no indication that accretion of lithospheric blocks of exotic origin took place in this region.

Place of the Coast Belt orogen in the tectonic evolution of western Cordillera

A major formative event in the history of the North American Cordillera was the Late Jurassic Nevadan orogeny. In its tectonotype in the Sierra Nevada in California, it has a complex tectonic expression including regional contractional folding and faulting. To the west, at the continental margin, subduction of oceanic lithosphere at that time and later resulted in emplacement of ophiolitic and magmatic-arc sequences. In particular, the huge ophiolitic Franciscan complex, made up of mostly Cretaceous rocks with oceanic-crustal affinities, was emplaced at California's late Mesozoic continental margin (Saleeby and Busby-Spera, 1992; Cowan and Bruhn, 1992).

North of the Klamath Mountains block on the California-Oregon state boundary, the youngest Mesozoic ophiolitic-type rocks are Middle to Late Jurassic. The Fidalgo complex in the Northwest Cascade fold-and-thrust belt contains an assemblage including mafic and ultramafic rocks but also quartz diorite and tonalite, whose presence complicates the interpretation of this complex as an ophiolite. These rocks lie in a thrust slice on the San Juan Islands (Whetten et al., 1980; Brandon et al., 1988).

Early Jurassic magmatism, in contrast, manifested itself regionally. It affected large parts of both western and central Canadian Cordillera, and intermediate-composition plutons of that age are found from the Intermontane to the Insular Belt (Woodsworth et al., 1991).

Late Early to early Middle Jurassic tectonism in the Insular Belt in British Columbia, according to Thompson et al. (1991), fits into a coherent eastward-younging trend. From Late Jurassic to early Tertiary time, emplacement, cooling and uplift of plutons on the western mainland also shifted gradually eastward (Hutchison, 1982; van der Heyden, 1992).

In the Cretaceous, the narrow Coast Belt orogen was superimposed on the previous grain of the northwestern Cordillera. This event marked the final partitioning of the once-single geologic province into three contrasting tectonic belts observed today: Intermontane, Coast and Insular.

The major orogeny which created the Coast Belt as a distinct tectonic entity took place in the mid-Cretaceous. A new zone of crustal contraction developed, flanked on both sides by outward-verging thrust belts (Rusmore and Woodsworth, 1991, 1994). The Coast Belt orogen stretches from the North Cascades in Washington, through the British Columbia mainland, to Alaska (Brew et al., 1991; Gabrielse et al., 1991; Brown et al., 1994).

Large mid-Cretaceous to early Tertiary granodioritic plutons make up 80% of the exposed Coast Belt rock volume (Crawford et al., 1987; Monger, 1991). Vigorous magmatism took place mostly after the mid-Cretaceous compressional orogenesis, which thickened the crust to the point of melting at its lower levels (Hollister, 1993). Buoyant uplift and unroofing of deep crustal horizons in the Coast Belt occurred during the Late Cretaceous and Tertiary (Parrish, 1983; van der Heyden, 1992). Elongated compensatory depressions developed along the flanks of the rising orogen, from which they were supplied with abundant detritus. Large sedimentary basins formed in Late Cretaceous and Tertiary time in front of the mountainous orogen (the Georgia and Queen Charlotte basins; Monger, 1991; Lyatsky, 1993a).

Local uncommon rock complexes on the western and southern periphery of Vancouver Island

Pacific Rim mélange complex (including Pandora Peak unit)

An unusual assemblage of volcanic and sedimentary rocks of Mesozoic age has been distinguished on western Vancouver Island. It lies in a narrow - a few kilometers across - coastal strip northwest of Barkley Sound, near the towns of Ucluet and Tofino (Muller, 1977a). These rocks, denoted Pacific Rim complex, are different from rocks on the rest of the island and are separated from them by a large steep fault.

From recent detailed geologic mapping, Brandon (1989a,b) divided the Pacific Rim complex into three units. The lower unit comprises unstratified volcanic rocks with subordinate limestone and chert. Fossils in the limestone point to a Late Triassic Carnian-early Norian age, and those in the chert are Early Jurassic. Volcanics of this unit are partly coeval with the Karmutsen Formation, but unlike the Karmutsen basalts, they have andesitic composition.

Next upsection in the Pacific Rim complex is a unit of deformed but unmetamorphosed clastic rocks. A mudstone-sandstone matrix hosts discrete blocks of pillow basalt with ribbon chert and, rarely, of ultramafic material. Some blocks are as big as 300 m. Radiolaria from the chert give Late Jurassic ages.

The top unit is Lower Cretaceous in age. It contains sedimentary rocks - mudstone, sandstone, conglomerate and chert - more than 2000 m thick. Blocks derived from the lower units also occur. This unit is contorted but still partly coherent.

Rocks similar to those in the Pacific Rim complex are also found in isolated localities near the southern tip of Vancouver Island, where they have been named Pandora Peak unit (Rusmore and Cowan, 1985). It comprises black mudstone, greywacke, chert, tuff and greenstone of Upper Jurassic-Lower Cretaceous age. These rocks were metamorphosed to lawsonite grade between 99 and 83 Ma, i.e. in late Albian to Santonian-Campanian time. Like the Pacific Rim complex, the Pandora Peak unit lies in fault-bounded slices; small thrust splays also occur in a complex fault system on the southern tip of Vancouver Island.

Close lithological similarities of the Pacific Rim complex and the Pandora Peak unit with the Fidalgo complex on the San Juan Islands have been noted previously (Brandon et al., 1988; Brandon, 1989a,b). The Pacific Rim complex and the Pandora Peak unit are considered to be pieces of the Northwest Cascade thrust sheets, displaced from their original position on the San Juan Islands by strike-slip movements in Late Cretaceous or early Tertiary time (Brandon, 1989a).

Leech River metamorphic complex

In fault contact with the Pandora Peak unit on the southern tip of Vancouver Island lies another unusual local rock assemblage - the Leech River complex. Once known as Leech River Schist, it was mapped by Muller (1977a) and later re-examined by Fairchild and Cowan (1982) and Rusmore and Cowan (1985). Its protoliths are mostly sandstone and basalt, suspected to be Jurassic-Cretaceous in age. From available descriptions, they seem to be similar to rocks widespread elsewhere on Vancouver Island.

These rocks were buried to mid-crustal depths and, according to K/Ar dates, by 41-39 Ma reached the greenschist to amphibolite grade of metamorphism. At the same time, they were intruded by felsic sills. Metamorphic foliation in the Leech River complex is parallel to the closely spaced bounding San Juan and Survey Mountain faults (see also Mayrand et al., 1987). This suggests that metamorphism was probably synkinematic.

This metamorphic complex is sandwiched between the steep San Juan, Survey Mountain and Leech River faults. The first two faults separate it from Wrangellian rocks, and locally from the Pandora Peak unit, to north; and the Leech River fault juxtaposes it with the Tertiary Metchosin igneous complex to the south. The most probable origin of the Leech River complex is by metamorphism of protoliths of Wrangellian affinity, displaced and uplifted within the South Vancouver Island fault system.

CHAPTER 4 - TERTIARY STRATIGRAPHIC FRAMEWORK
OF COASTAL PROVINCES IN WASHINGTON AND BRITISH COLUMBIA

Early Tertiary paleonvironments

Early Paleogene

By the end of the Mesozoic, continental crust was underlying most of the continental marginal region from Oregon to Alaska. Prominent landmasses existed on the sites of the Klamath Mountains in northern California and southern Oregon and of Vancouver Island and perhaps the Queen Charlotte Islands in British Columbia. The position of the early Tertiary submarine margin between the Klamath Mountains and Vancouver Island is less clear, but sedimentological evidence indicates that during most of the Tertiary, on the site of the Olympic Peninsula lay a large deep-marine embayment (Heller et al., 1992; Niem et al., 1992a-c).

With the possible exception of the Olympic Peninsula (Babcock et al., 1992, 1994), Paleocene sedimentary rocks are virtually absent between western Oregon and western British Columbia. This suggests that broad regional uplift occurred at that time in the westernmost Cordillera (Miller et al., 1992).

Partly because much of the older Pacific oceanic crust bearing the magnetic-stripe record has been subducted, current plate models are unclear about the details of local history of plate interactions (Riddihough, 1982a; Stock and Molnar, 1988). Paleogeographic reconstructions based on geologic mapping are hampered by the scarcity of Paleocene sedimentary rocks. The key to interpretation therefore lies in studies of the volcanic rocks of early Tertiary age, which are widespread across the region.

Early Tertiary basaltic magmatism

Manifestations of Eocene mafic magmatism abound in western Oregon and Washington, where a large Coast Range igneous province is distinguished (Duncan and Kulm, 1989). The nature of these basalts is still controversial. They were once treated as Eocene oceanic crust (MacLeod et al., 1977), and Duncan (1982) interpreted them as a series of seamounts accreted to western North America. In contrast, Brandon and Vance (1992) and Babcock et al. (1992, 1994) viewed these basalts as products of continental rifting. Basalts of the Siletz River Formation in Oregon and of the Crescent Formation in Washington are generally coeval but differ in composition and style of volcanism (cp. Snavely, 1987; Babcock et al., 1992, 1994).

Best-studied is the Crescent Formation, which is exposed in many uplifted crustal blocks in western Washington. It is well represented on the rim of the Olympic Mountains, as well as on the southern tip of Vancouver Island, where it is known as the Metchosin Formation (Muller, 1977a-c). Sparse fossils from carbonate lenses and radiometric dates from basalts suggest diachronous ages of eruption between about 57 and 45 Ma, spanning the Early and Middle Eocene. No regional age progression has been found (Babcock et al., 1994).

Crescent basalts are not geochemically uniform. Submarine and subaerial flows have been mapped, with many lateral variations and facies substitutions. These basalts erupted from many discrete centers (Babcock et al., 1994).

Separate basalt bodies are exposed in topographic highs in the Willapa Hills south of the Olympic Peninsula. Much bigger is the Dosewallips massif, which lies between the Olympic Mountains and Puget Sound. Several narrow basaltic bodies form a WNW-trending belt on northern Olympic Peninsula, parallel to the Strait of Juan de Fuca. From the northwest shore of Puget Sound, this belt includes the Port Ludlow, Marmot Pass, Hurricane Ridge and Crescent Lake massifs.

The Metchosin massif (Muller, 1977a-c, 1980b; Massey, 1986) is parallel to the Hurricane Ridge and Crescent Lake massifs but lies on the north shore of the Strait of Juan de Fuca (Fig. 5). Its exposed thickness is 7000 m. Rocks of this massif are metamorphosed, mostly to prehnite-pumpellyite and in places to greenschist grade. They are cut by dikes and by small plutons known as Sooke Intrusions. The abundance of dikes decreases upsection, as expected in a feeder system. The plutons are mainly made up of gabbro, but some of these stocks contain quartz diorite and tonalite with 10% to 30% quartz (Muller, 1977c).

Links between the Marmot Pass, Hurricane Ridge and Crescent Lake massifs on the south side of the Strait of Juan de Fuca and the Metchosin massif on the north side are indicated by their compositional similarities. Continuity of basalt under the Strait of Juan de Fuca has been suggested from examination of magnetic anomalies (MacLeod et al., 1977) and by gravity modeling (Dehler and Clowes, 1992).

In the scope of the prevailing tectonic models, the Metchosin complex has been interpreted as a piece of obducted oceanic crust

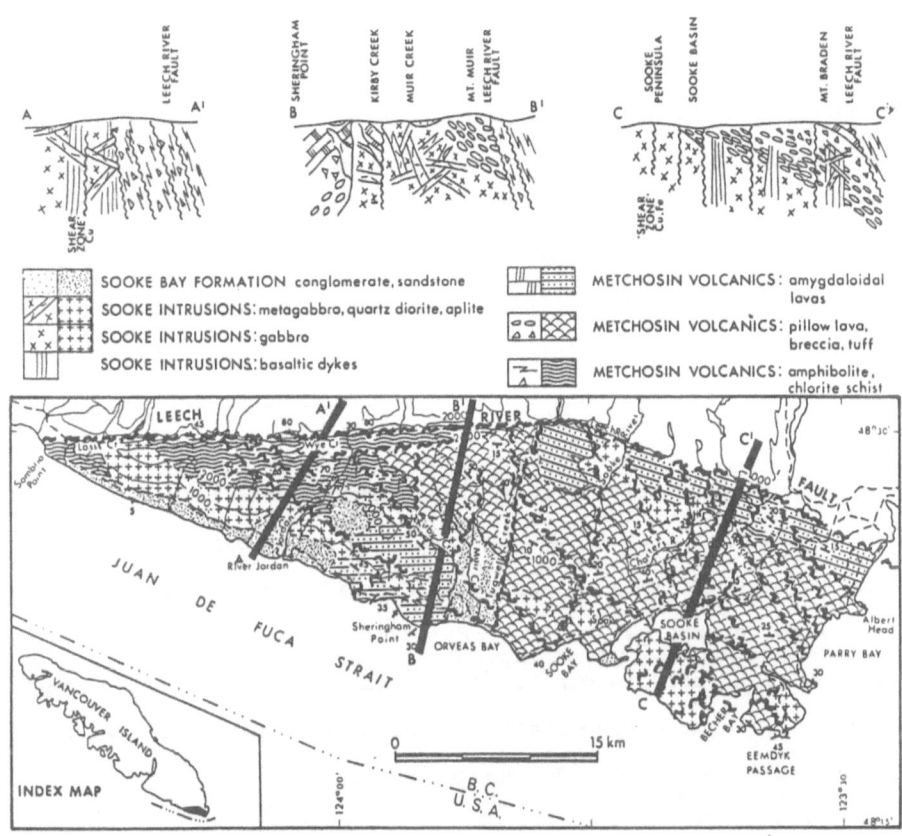

Figure 5. Geologic map of the Eocene Metchosin igneous massif on the southern tip of Vancouver Island (modified from Muller, 1977c). The massif is broken into blocks bounded by steep faults (broken wavy lines). The Leech River fault consists of two straight, steep segments meeting at an angle of about 25°. Such a structural configuration is inconsistent with emplacement of the Metchosin massif along a Leech River "thrust", as suggested in some tectonic models. The Metchosin massif is one of many massifs in the Eocene Crescent Formation, whose basalts erupted from many discrete centers in a rift setting (Babcock et al., 1992, 1994). The Metchosin massif lies within the Olympic-Wallowa structural zone (OWSZ).

(Massey, 1986). This contradicts several lines of field evidence. No ophiolite characteristics are observed in the Metchosin complex, and the interpretation of feeder roots as sheeted dikes is tenuous. Subaerial volcanics 3000 m thick, mapped in the Metchosin massif (Muller, 1977c), are not expected in oceanic crust. Presence of felsic rocks is incompatible with an oceanic-crust interpretation. The fault-block mosaic with variously oriented folds and steep faults mapped in the Metchosin massif (Fig. 5) is not expected in a unit accreted compressionally.

The reported radiometric ages of Metchosin rocks vary depending on the method: 57.8±0.8 Ma by Ar/Ar, 52 Ma by U/Pb (Duncan, 1982; Babcock et al., 1994). K/Ar dates from the Sooke Intrusions are scattered around 45 Ma (Muller, 1977c). The Hurricane Ridge and Crescent Lake massifs can be correlated with the Metchosin massif temporally and geochemically, though the Metchosin massif is eroded to deeper levels due to greater uplift.

Relationship of Crescent Formation massifs with early Tertiary sedimentary sequences

In the Deer Park area on northeastern Olympic Peninsula, basalts of the Crescent Formation are underlain, with a hot contact, by sedimentary rocks of the Blue Mountain unit (Tabor and Cady, 1978; Babcock et al., 1994). This sedimentary unit consists of thinly bedded mudstone and massive sandstone sourced from proximal areas such as the Coast Mountains and the San Juan Islands (Babcock et al., 1994).

The Blue Mountain unit is unfossiliferous. Syn-volcanic age of its upper part is suggested by interfingering of Blue Mountain sedimentary rocks with Crescent basalts through the entire section of the Hurricane Ridge and Dosewallips massifs, up to the Middle Eocene. The Middle Eocene Adwell Formation overlies Blue Mountain and Crescent rocks conformably, in places with a gradational contact. Sedimentation began in the Paleocene and continued simultaneously with Crescent volcanism (Babcock et al., 1994).

The fact that the Crescent Formation has stratigraphic contacts with older, coeval and younger sedimentary sequences negates the idea that these basalts as pieces of oceanic crust. Layered material revealed by seismic data beneath the Siletz River basalts in Oregon may also be sedimentary (Keach et al., 1989).

Structural position of Crescent basalts likewise suggests that eruptions occurred in a continental setting. The grid of steep faults in western Washington is well expressed in gravity and magnetic maps (Finn, 1990), as a rectangular pattern of crustal fractures seems to control the distribution of potential- field anomalies (see also McCrumb et al., 1989a,b).

Stratigraphic record of mid-Eocene to Miocene sedimentary basins
The most complete Tertiary stratigraphic section in the region is found on the south shore of the Strait of Juan de Fuca (see Fig. 6 for basin locations). There, a conformable sedimentary package of the Fuca Basin, about 8 km thick, spans the period from the mid-Eocene to the mid-Miocene (Niem and Snavely, 1991; Niem et al., 1992a). The section begins with the 900-m-thick Adwell Formation of Middle Eocene age. It consists of thin-bedded siltstone with interbedded sandstone, interpreted as products of mud flows on a continental slope.

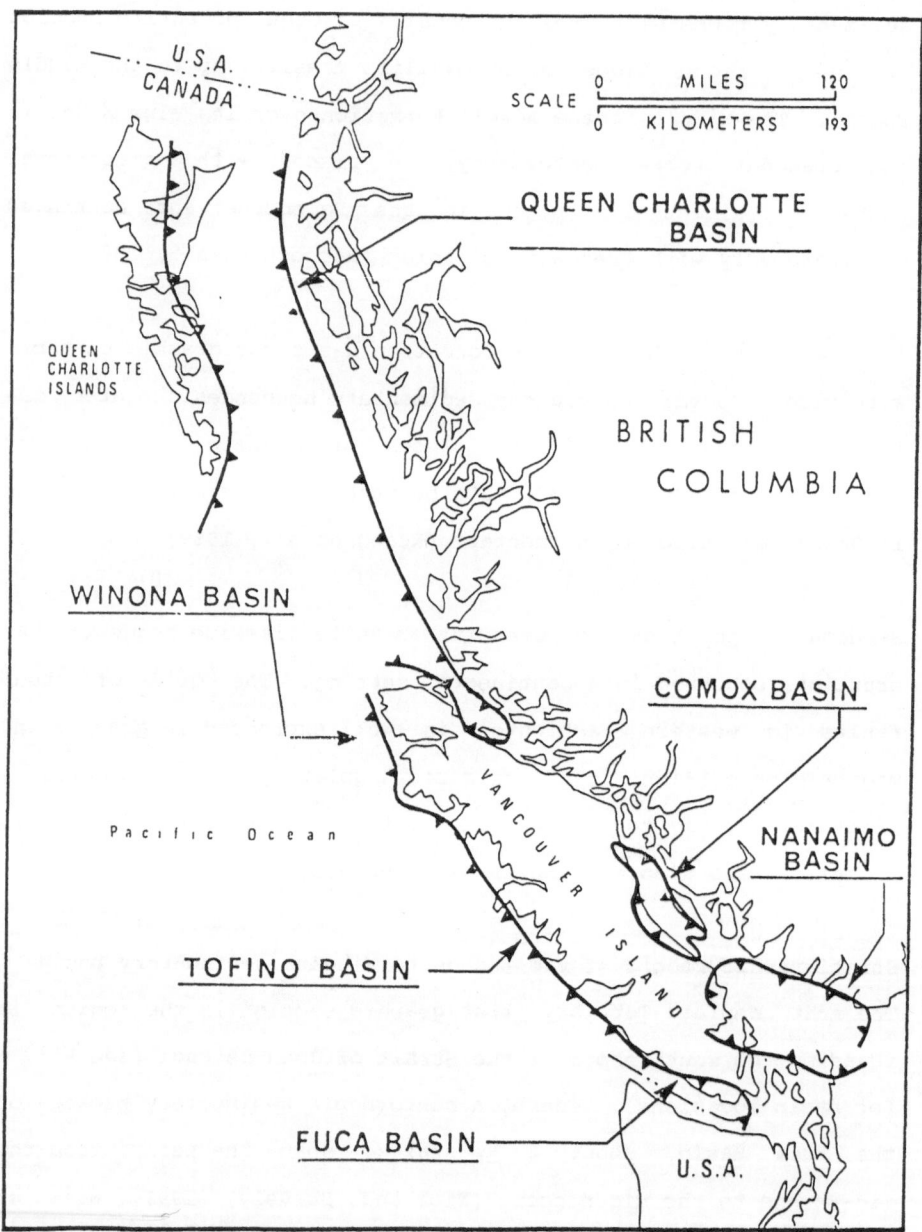

Figure 6. Location of principal Late Cretaceous and Tertiary sedimentary basins along the western Canada continental margin. The Tofino and Fuca basins are regarded together in much of the current literature, but the Winona Basin is distinct. The Nanaimo and Comox basins both lie in the Georgia depression.

The overlying early Late Eocene Lyre Formation, 300 to 600 m thick, is made up of sandstone (lithic arenite and lithic wacke) and breccia. It is thought to have been deposited on a continental slope, in submarine fans or by sediment slumping.

The Late Eocene Hoko River Formation is much thicker - 1600 to 2300 m - and contains massive to thin-bedded siltstone with lesser sandstone and conglomerate rich in phyllite and basalt fragments. The Late Eocene to Late Oligocene Makah Formation is more than 2800 m thick. It is made up of thin-bedded siltstone interbedded with abundant sandstone. The sandstone is mostly turbiditic and occasionally olistostromal.

The 1400-m-thick, Late Oligocene to Early Miocene Pysht Formation was deposited in shallower-water environments. It begins with pebble and boulder conglomerate probably deposited in channels on a shelf and/or upper slope. The rest of this formation comprises mudstone to sandy siltstone with conglomeratic channel deposits. The overlying Lower Miocene Clallam Formation, 800 m thick, is made up of sandstone and subordinate conglomerate; abundant shallow-water marine fauna and wood fragments indicate that basin subsidence was compensated by sedimentation.

The roughly coeval sedimentary rocks of the Carmanah Group lie on the southern and western periphery of Vancouver Island (Tiffin et al., 1972; Muller, 1977a-c; Muller et al., 1981). They occur in just a few partial exposures on the south coast of the island, where their thickness is reduced due to several episodes of erosion. Paleogene sedimentary rocks are absent in this area, and the Miocene to Pliocene Sooke Formation is made up of coarse, shell-bearing sandstone and conglomerate.

The Carmanah Group is exposed in discontinuous outcrops along the western coast of Vancouver Island and is most complete on Hesquiat Peninsula. It dips to the west, and almost 4 km of similar sedimentary rocks have been drilled on the exterior shelf (Shouldice, 1971). On Hesquiat Peninsula, the Carmanah Group is about 1400 m thick. The oldest in it is the Late Eocene Escalante Formation which consists of 150 m of calcareous sandstone with conglomerate at its base. It grades into the Late Eocene to Oligocene Hesquiat Formation, which is some 1200 m thick and also contains siltstone, sandstone and conglomerate.

These rocks are unconformably overlain by the Sooke Formation. Unlike the Escalante and Hesquiat formations, which are throught to have been deposited in a bathyal environment, it was laid down in very shallow water (Muller et al., 1981).

Sediments of the Carmanah Group were derived from elevated parts of Vancouver Island, and throughout the Cenozoic the shelf and the entire continental margin in this area must have been close to their present position.

Stratigraphic record of late Tertiary sedimentary basins
Subsidence between Vancouver Island and the Olympic Peninsula resumed in the Pliocene, when sedimentation began in the Strait of Juan de Fuca and a new WNW-trending graben developed over the old Fuca Basin. The new depocenter is parallel to the old one, but is narrower. Changes in depocenter shape and location occurred also

on the sites of other old Tertiary sedimentary basins on the Olympic Peninsula (Snavely, 1987; Niem et al., 1992a,c). New basins developed in the Neogene on the exterior shelf off southeastern Alaska (Bruns and Carlson, 1987).

On the interior shelf between the mainland and the Queen Charlotte Islands, the Queen Charlotte Basin is up to 6 km thick (Shouldice, 1971; Rohr and Dietrich, 1992; Lyatsky, 1993a). On the islands, Tertiary volcanic rocks as old as Paleocene crop out sporadically, and Early Eocene to Early Oligocene black shale, mudstone, sandstone, conglomerate and coal are found in a few outcrops and drillholes (White, 1990). They seem to have been deposited in fault-bounded structural depressions whose appearance marked the beginning of the Queen Charlotte Basin (Lewis et al., 1991a). Tertiary normal faulting also occurred on northern Vancouver Island (Nixon et al., 1995).

Much more widespread on the Queen Charlotte Islands is the Late Oligocene to Early Pliocene Masset Formation, which consists of mafic and felsic lava flows and pyroclastic deposits (Hickson, 1991). These volcanics, which erupted from several distinct centers, are in places up to 3000 m thick.

The Queen Charlotte Basin lies mostly beneath the interior shelf. It consists principally of Neogene sedimentary rocks of the Skonun Formation, which is up to 6 km thick. In a few exposures on eastern and northern Graham Island, these rocks have sedimentological characteristics suggesting marginal-marine paleoenvironments (Sutherland Brown, 1968; Higgs, 1991).

On the shelf beneath Hecate Strait and Queen Charlotte Sound, the Skonun Formation in eight deep wells is non-marine to marginal-marine in the north and marine in the south (Shouldice, 1971; Higgs, 1991). The Skonun Formation contains marine to continental mudstone and lithofeldspathic sandstone, generally semi-consolidated. Coal seams, volcaniclastic beds and basalt flows are found in it at different stratigraphic levels.

Lithofeldspathic sediments of the Queen Charlotte Basin were derived from neighboring continental areas - the Coast Mountains, the Queen Charlotte Islands, and uplifted parts of southeastern Alaska. Numerous local facies changes and unconformities attest to a tectonically unstable setting, as sedimentation was controlled by differential subsidence of blocks in the pre-Tertiary basement (Lyatsky, 1993a).

Overview of Tertiary geologic evolution of coastal provinces
<u>Two main geologic provinces along the continental margin from Oregon to southeastern Alaska</u>
Two main geologic provinces are usually distinguished on the Pacific continental margin of northwestern North America. One, in western Oregon and Washington, is the Coast Range province. The other, in western British Columbia and southeastern Alaska, is the Insular Belt.

These two provinces differ in many aspects of their geology (Gabrielse and Yorath, eds., 1991; Burchfiel et al., eds., 1992). The Insular Belt is made up mostly of Paleozoic and Mesozoic rocks, the Coast Range province of Cenozoic rocks. The Insular Belt experienced large-amplitude local block movements in the

Tertiary, whereas evolution of the Coast Range province was more uniform: general subsidence during much of the Tertiary, followed by invertsion.

During the Cenozoic, the Coast Range province has continually been affected by subduction along the Cascadia zone (Atwater, 1970, 1989), and a megathrust has existed between the overriding continental plate and the undergoing oceanic slab (Duncan and Kulm, 1989; McCrumb et al., 1989b). In contrast, the nature of plate interactions off British Columbia is still in dispute (Cousens et al., 1984, 1985; Allan et al., 1993; vs. Riddihough and Hyndman, 1989; Hyndman et al., 1990). In contrast to the Oregon continental margin, no manifestations of a subduction-related thrusting have been found by ODP drilling off southern Vancouver Island (Carson et al., 1993; MacKay et al., 1994).

Variations in tectono-magmatic style along the continental margin

Pre-Tertiary rocks of the Insular Belt have been proposed to be exotic, accreted to the North American continent in a chaotic fashion after traveling long distances from remote regions of South America or southeast Asia (e.g., Jones et al., 1977). More recent investigators generally favor a native origin for most western Cordilleran suspect terranes (this does not preclude considerable displacements of some crustal blocks along the margin, but in a regular fashion, as can be inferred from geologic mapping). Paleozoic and Mesozoic rocks, variously metamorphosed, form the basement for Tertiary formations in the Insular Belt.

Scarcity of pre-Tertiary basement exposures in the Coast Range province leaves room for a variety of speculations about the nature of the crystalline basement. Existence of a pre-Tertiary basement in western Oregon and Washington has been inferred from gravity models (Couch and Riddihough, 1989). Deep seismic data in Oregon show under Siletz River basalts a wide zone of reflections interpreted as stratified rocks (Keach et al., 1989).

Crescent basalts are now interpreted as products of Eocene rifting of continental crust or lithosphere (Babcock et al., 1992, 1994). Their radiometric ages correspond to the more regional Challis episode of volcanism, which according to Armstrong (1978) occurred in the Cordillera from 55 to 36 Ma. Once treated as a subduction-related arc, the Challis volcanic suite is now thought to have an intracontinental origin (Cheney, 1994). The connection between Crescent and Challis magmatism is still unclear (op. cit., p. 131, 132), but at any rate, arc-related volcanism is now thought to have begun in western Washigton and Oregon only at 36 Ma, i.e. near the Eocene-Oligocene boundary (Brandon and Vance, 1992).

Patterns of volcanism along the continental margin remained irregular after 36 Ma. Even the Pleistocene to Recent volcanic chain consists, variously, of a cluster of eruption centers in northern California and a belt of volcanoes in western Oregon to west-central Washington (Sherrod and Smith, 1990; Scott, 1990). In northwestern Washington and southwestern British Columbia, a belt of isolated volcanic cones (Read, 1990) follows a linear fracture zone in the Paleozoic and Mesozoic basement and may have an extensional origin (Green, 1990).

In current models, the vigorous Crescent magmatism in the Eocene is linked to the appearance under the marginal part of the continent of a slab window between the subducting Kula and Farallon oceanic plates. Creation of such a window in the upper mantle caused extension in the continental crust, producing rifting and basaltic magmatism. In these models, rifting began around 60 Ma, in the Late Paleocene, allowing the volcanism to reach its peak between 57 and 50 Ma (Babcock et al., 1992, 1994).

Coast Range mafic volcanism in the Eocene was especially voluminous in western Oregon (the Siletz River Formation; Snavely, 1987) and in the area of the Olympic Peninsula. Despite some geochemical differences (Babcock et al., 1992), eruptions in both areas had a common characteristic of occurring from many distinct volcanic centers. Subsequent tectonism caused rotation of many blocks in western Oregon and southwestern Washington (Wells and Coe, 1985), but such rotations on the Olympic Peninsula were minimal (Babcock et al., 1994).

Eocene mafic magmatism also affected the southern part of the Insular Belt, producing the Metchosin igneous massif on southern Vancouver Island and the Prometheus massif on the Vancouver Island exterior shelf (see Snavely, 1987). Thus, presence or absence of Crescent Formation basaltic bodies does not, by itself, allow to distinguish the Coast Range province from the Insular Belt.

Gravity data suggest that some of the igneous bodies in Washington may continue to very deep crustal levels, to depths of up to 30 km (Finn, 1990). The near-surface Crescent and Siletz River volcanic bodies have an average thickness of just 5 km (Duncan, 1982; Keach et al., 1989), though some massifs exceed 7 or 8 km (Muller, 1977c; Babcock et al., 1994). Unexposed deep feeder systems and frozen magma chambers may be present beneath some of the larger volcanic bodies, such as Metchosin.

Importantly, Paleogene magmatism in the coastal provinces was not restricted to mafic eruptions, and both mafic and felsic rocks have been found. Sooke plutons, 47 to 31 Ma in age, in the Metchosin complex, are largely gabbro, but quartz diorite and tonalite stocks with 10% to 30% quartz have been reported among them (Muller, 1977c). Granodiorite, tonalite and granite of the early Tertiary Catface suite, dated at 52 to 35 Ma, are widespread on Vancouver Island. They lie along large faults in the southern and western parts of the island (Muller et al., 1981; Andrew and Godwin, 1989c; Woodsworth et al., 1991). On the west coast of Vancouver Island, one such granitoid stock, dated at 52 Ma, intrudes the Pacific Rim complex (Brandon, 1989a). Felsic sills around 40 Ma in age have been found in the Leech River Complex on southern Vancouver Island (Fairchild and Cowan, 1982).

Felsic Eocene rocks have also been found on northern Olympic Peninsula (Snavely, 1987). Quartz diorite dikes dated at 41 Ma have been mapped in the core of the Olympic Mountains. They cut both volcanic and sedimentary rocks of older Tertiary units. Dikes and small plugs of dacite and andesite, dated at 44 Ma, have been reported from Striped Peak on the northern Olympic coast and from northeastern Olympic Peninsula.

Unlike the predominantly mafic magmatism in the Coast Range province, felsic magmatism was the norm during the Paleogene in the Insular Belt (Woodsworth et al., 1991). Felsic rocks of that age are present in many areas in western British Columbia and

southeastern Alaska (Brew et al., 1991). They have been studied
in detail on the Queen Charlotte Islands, where the Kano plutonic
suite contains quartz monzodiorite and mondodiorite, with some
diorite and granite. U/Pb ages of these rocks suggest that
magmatism occurred in three pulses: 46 to 39 Ma, 36 to 32 Ma, and
28 to 27 Ma (Anderson and Reichenbach, 1991). Intrusive bodies of
the Kano suite are mostly small plutons and dikes, some of which
may have been feeders for comagmatic volcanism.

Thus, the mafic and felsic Paleogene igneous domains correspond
approximately to the two principal geologic provinces of the
Pacific margin of North America. A broad boundary between these
zones lies in the area of southern Vancouver Island and northern
Olympic Peninsula.

Distribution of Tertiary sedimentary basins along the margin in time and space

On the northeastern Olympic Peninsula, in the Deer Park area,
basalts of the Crescent Formation overlap stratigraphically, with
a hot contact, older Tertiary sediments of the Blue Mountain unit.
Throughout the entire Crescent section in that area, basalts and
sediments are intercalated (Babcock et al., 1992, 1994). In this
context, indications from seismic data that equivalent Eocene
basalts in Oregon are underlain by stratified material (Keach et
al., 1989) may suggest presence of sedimentary units beneath these
basalts. Seismic reflections are also present under the Metchosin
complex on southern Vancouver Island (Clowes et al., 1987). Thus,
circumstantial evidence suggests that Tertiary sedimentary basins
began to develop in the continental-margin region as early as in
the Paleocene.

After the early Middle Eocene, two subparallel belts of sedimentary basins formed along the margin (Fig. 6). In the inner belt, abundant arkosic sediments were deposited in western Oregon north of the Klamath Mountains (the Tyee Formation; Snavely, 1987) as well as in several basins in Washington (Niem et al., 1992a). Since the Oligocene, along the Cascade magmatic arc, Willamette and southern Puget basins have developed in a fore-arc setting (see also Dickinson, 1976). To the north, the northern Puget, Georgia and Queen Charlotte basins evolved in compensatory depressions in front of the mountains of the Coast Belt orogen (Lyatsky, 1993a). Neogene subsidence along parts of these mountain ranges created the western Canada interior shelf.

Outboard of these basins lies a series of islands stretching from northern Olympic Peninsula to southeastern Alaska. A single island chain along the coast of British Columbia, including Vancouver and Queen Charlotte islands, is jointly referred to hereinafter as Western Canada Archipelago. This archipelago separates the interior and exterior continental shelves.

Eocene to mid-Miocene marine clastic deposits on central and western Olympic Peninsula (Central Olympic and Hoh basins) were laid down in a continental-slope and trench setting (Tabor and Cady, 1978; Niem et al., 1992b). Such conditions existed in the entire Tertiary Olympic embayment, which was at least 100 km wide and lay between the Dosewallips igneous massif and the western end of the Hoh Basin on the present-day shelf. Such broad areas with fairly uniform slope or trench conditions of sedimentation are uncommon at continental margins. Tectonic evolution of the Central

Olympic and Hoh basins is also unusual: these basins are slightly metamorphosed, compressionally shortened, and inverted (see next chapter). All other Tertiary basins in the region are unmetamoprphosed and considerably less deformed.

In the Tertiary depressions which developed along the rising Coast Belt, subsidence was considerable, but non-marine conditions nonetheless predominated throughout the Tertiary. In the Georgia depression between the Coast Mountains and Vancouver Island, a succession of conglomerate, sandstone, mudstone and coal is 6 km thick. Marine incursions were minor, and fluvial sediments derived from the surrounding mountains accumulated in large delta complexes (Mustard, 1991).

To the south, clastic sediments of the Middle to Late Eocene Puget Group, some 3500 m thick, were also laid down in a deltaic setting by streams draining granite-rich areas in the Cordilleran interior. These deposits were then covered by submarine fan complexes containing Oligocene volcanic-rich sediments shed from subduction-related arcs in the South Cascades. The overlying Miocene sediments are non-marine, containing detritus from the Olympic Mountains to the west (Niem et al., 1992a,c).

On the exterior shelf, a series of Tertiary sedimentary basins has been identified from gravity and seismic data, and in places confirmed by drilling, along the submerged Pacific margin of North America from Oregon to southeastern Alaska. Off the coast of Washington and Oregon, such basins are 3-4 km deep but some are 6-8 km deep (Snavely, 1987; Couch and Riddihough, 1989). The Tofino Basin on the exterior shelf off Vancouver Island has been drilled to almost 4 km depth (Shouldice, 1971). Several undrilled fault-bounded depressions are also found on the shelf off southeastern Alaska (Bruns and Carlson, 1987). These basins, which according to seismic and gravity data contain at least 3000 m of stratified sediments, are several tens of kilometers wide and elongated parallel to the continental margin. Farther outboard, on the continental slope, several other deep basins lie between strands of the plate-boundary fault system.

Most of these basins were formed by differential subsidence of crustal blocks bounded by steep faults. Where deformation has occurred, its causes are varied - contractional folding and faulting (as in western Oregon and in the Central Olympic Basin; Tabor and Cady, 1978; Snavely, 1987; Niem et al., 1992a-c), flowage of overpressured sediments (as in the Tofino and Hoh basins; Tiffin et al., 1972; Snavely, 1987) or fault movements (as in the Puget and Queen Charlotte basins: Finn, 1990; Rohr and Dietrich, 1992; Lyatsky, 1993a).

Most of the Tertiary basins are not metamorphosed, and the Hoh Basin is metamorphosed only slightly. Only the Central Olympic Basin is penetratively sheared and metamorphosed (Tabor and Cady, 1978; Brandon and Vance, 1992).

Anomalous in the structural sense is the Fuca Basin. Its WNW orientation, discordant with the general trend of the continental margin and the geologic grain of coastal provinces, has not yet been satisfactorily explained. The Fuca Basin lies on trend with the trans-Cordilleran Olympic-Wallowa zone of crustal weakness.

CHAPTER 5 - SIGNIFICANCE OF THE TRANS-CORDILLERAN OLYMPIC-WALLOWA

ZONE IN GEOLOGIC EVOLUTION OF THE WASHINGTON AND BRITISH COLUMBIA

COASTAL REGIONS

Recognition of the Olympic-Wallowa Zone of crustal weakness

Observed in regional topographic maps is a long lineament across

the Cordillera. It runs from the Pacific coast into Washington

and northwestern Oregon (Fig. 7). First noted by Raisz (1945), it

was traced from northern Olympic Peninsula, across the Columbia

Plateau, into the Wallowa Mountains near the Oregon-Idaho state

line. Raisz (1945) named this topographic feature Olympic-Wallowa

Lineament (OWL) and speculated that it may be coincident with a

large fault.

Geological, geophysical and geographical surveys have shown that

the OWL is not unique, but rather one of many large lineaments

that trend across the general N-S grain of the Cordillera (e.g.,

King, 1969; Reidel et al., 1994). Other WNW-ESE lineaments, to

name just a few, are Lewis and Clark zone in western Montana and

northern Idaho, Brothers zone in central Oregon, Eugene-Denio zone

in southern Oregon and northern Nevada. The western branch of the

curved Snake River plain in southern Idaho also has a WNW-ESE

orientation, as do many valleys and mountain ranges (Mann and

Meyer, 1993).

The scale of the structural zone along the topographic OWL became

apparent in the 1970s and 1980s, as mapping revealed its spatial

and temporal continuity and internal structural complexity (Reidel

and Hooper, eds., 1989). A series of WNW-trending faults and

folds run parallel to the OWL, in places coinciding with it

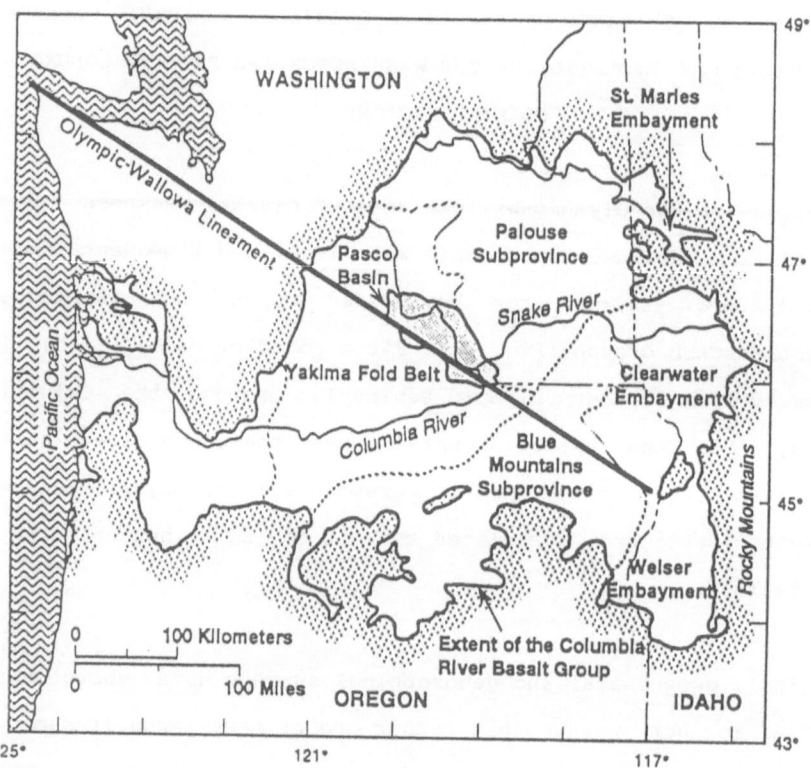

Figure 7. Regional sketch of the Olympic-Wallowa structural zone (OWSZ) approximate location and extent (Reidel et al., 1994). The Olympic-Wallowa Lineament is the topographic expression of the OWSZ. The OWSZ is in fact tens of kilometers wide, much broader than sketched by these workers.

exactly. The OWL was proposed to be a part of a much longer Olympic-Wichita Lineament presumably crossing the Cordillera from Oklahoma to the Pacific margin (Baars, 1978).

More evidently, the interregional Olympic-Wallowa structural zone (OWSZ), as designated herein, runs from the Late Cretaceous Idaho batholith, across the Columbia River volcanic province and the Cascade Ranges, to the Pacific coast. Along its trend, the OWSZ changes its width and surface expression, but its elements are everywhere connected with one another to form a continuous, distinct structural zone. In places, it is just a 15-km-wide system of closely spaced faults. Elsewhere, it widens to 50-70 km and includes, besides faults and anastomosing folds, raised blocks (Wallowa and Cuddy Mountains uplifts) and depressions (Pine Valley and Grande Ronde grabens, as well as the Fuca Basin). In some areas, the OWSZ controlled volcanic eruptions, such as those of Eocene Crescent basalts on northern Olympic Peninsula and southern Vancouver Island, and of the Miocene Columbia River basaltic province in central Washington.

The OWSZ is now recognized to be one of the major shear zones in the western part of the North American continent. Geophysical data show it to be deep-seated (Saltus, 1993), possibly reaching into the upper mantle (Catchings and Mooney, 1988). Structural evidence suggests that different parts of the OWSZ were zones of high tectonic mobility throughout at least the Tertiary. What remains unnoted is that this zone of crustal weakness not only transects the Cordillera but also controls the position of some parts of the Pacific continental margin.

The OWSZ in eastern Oregon and Washington

West of the huge, N-S-trending Late Cretaceous Idaho batholith, the OWSZ is expressed mainly as a series of horsts, grabens and closely spaced faults whose orientation is persistently WNW. Some of the faults bound coherent crustal blocks, variously uplifted and downdropped.

One uplifted block in northeastern Oregon, topographically expressed as the Wallowa Mountains, is made up largely of Mesozoic rocks. It was raised in the Neogene and so was bypassed by lava flows of the Columbia River province (Hooper and Conrey, 1989). Large faults have been mapped on both flanks of the Wallowa Mountains horst: the Wallowa fault on the northeast, the Eagle fault on the southwest (Fig. 8). These structures, as well as the Grande Ronde graben and Baker Valley to the south, are on trend with the OWSZ. On the ESE, the OWSZ continues at least as far as the N-S-oriented Long Valley fault system in western Idaho (Mann and Meyer, 1993), and on the WNW it runs into Washington.

Neogene right-lateral movements on the OWSZ have been proposed by Hooper and Conrey (1989) and Mann and Meyer (1993). This idea is supported by focal mechanisms of some earthquakes. However, transverse fault systems - Long Valley in western Idaho, Hite on the Oregon-Washington boundary - cross the OWSZ at high angles. To the west, a Miocene pluton that intrudes the OWSZ is not disrupted, and other Miocene rocks lie on both side of the OWZ without lateral offset (Frizzell et al., 1984; Reidel et al., 1989, 1994; Price and Watkinson, 1989). This suggests strike-slip movements were local in scope and limited in magnitude.

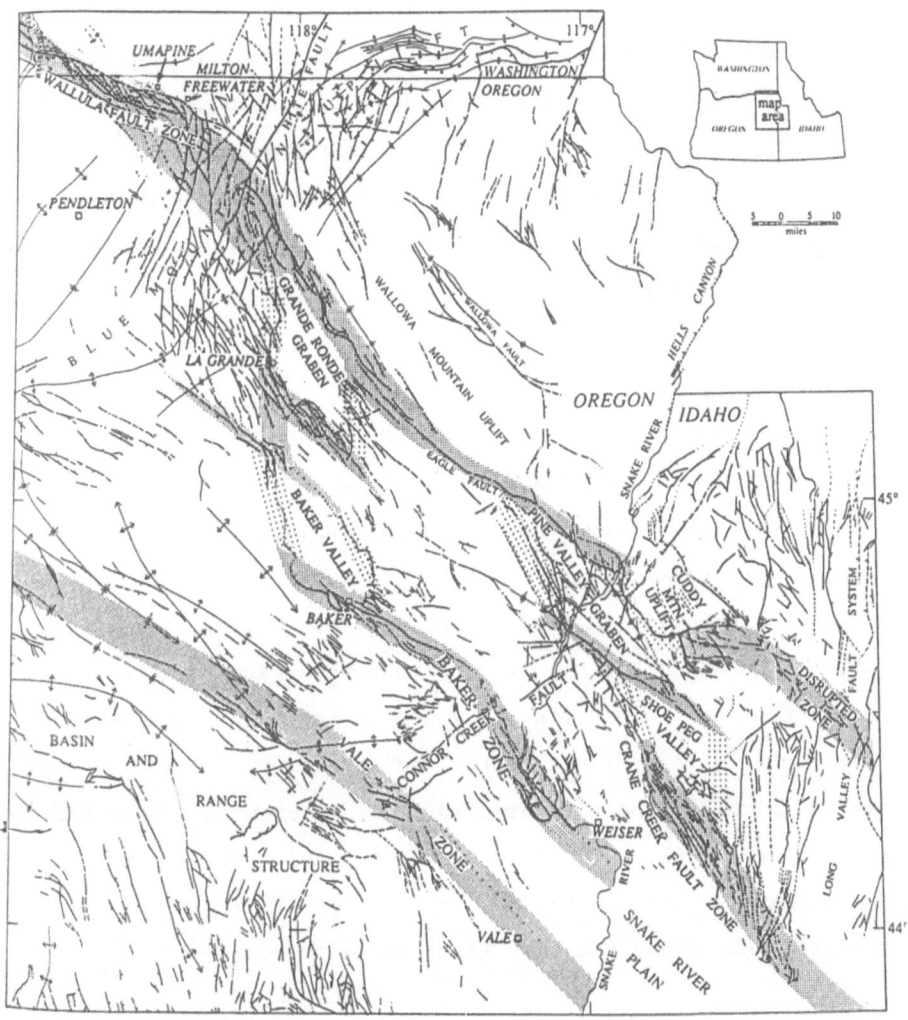

Figure 8. Structural map of the area near the southeastern end of the OWSZ, showing the Wallowa Mountains horst and bounding faults, as well as the Long Valley and Hite fault zones which transect the OWSZ (modified from Mann and Meyer, 1993).

The OWSZ truncates the Blue Mountain block of Permian and Triassic rocks and the NE-trending Klamath-Blue Mountain Lineament in Oregon (Hooper and Conrey, 1989). In Washington, the OWSZ was apparently active in the Miocene, providing conduits for the abundant lavas of the Columbia River flood basalt province; these basalts are estimated to be up to 4500 m thick (Reidel and Hooper, eds., 1989; Saltus, 1993).

The OWSZ in central Washington

North of the Oregon-Washington boundary, the OWSZ is a series of closely spaced, WNW-trending fault zones and long anticlines. The Wallupa Gap fault zone, Umtanum and Manashtash ridges, Rattlesnake Mountain anticline, White River-Naches River fault zone, all have a WNW orientation (Reidel and Hooper, eds., 1989; Reidel et al., 1994). The Pasco Basin in the northwestern part of the Columbia River province owes its structure to the OWSZ. The Yakima fold belt contains long, fault-controlled anticlinal folds separating stable blocks, and its structural grain changes across the OWSZ (Reidel and Campbell, 1989). In the central part of the Yakima Belt, faults and folds are spaced particularly closely and oriented WNW, along the OWSZ (Fig. 9).

Anticlines in the OWSZ are asymmetric and commonly cored by small reverse faults. Rocks between them, though tilted, are mostly undeformed (Watters, 1989; Reidel et al., 1994). Shallow near-surface splays of some of the reverse faults dip at low angles (Price and Watkinson, 1989). Yet, evidence of strike-slip displacements of a few kilometers is found in several localities (Reidel et al., 1994). Analysis of gravity data suggests that many of the faults reside only in the basalt succession (Saltus,

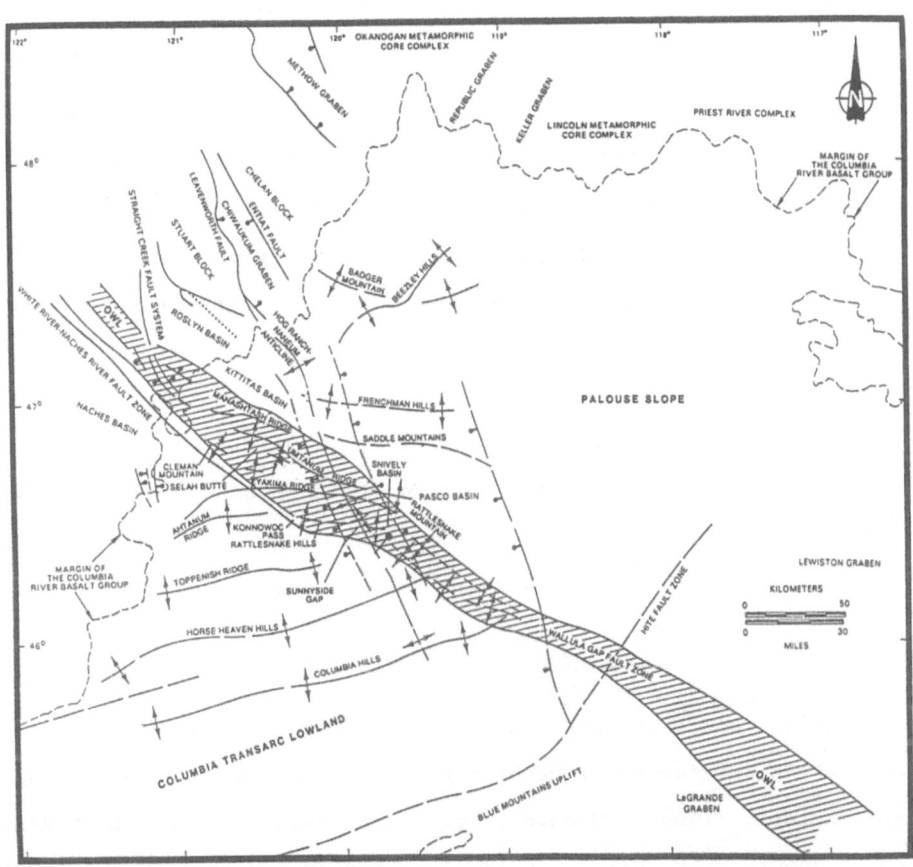

Figure 9. Distribution and orientation of faults and folds in the Yakima Belt in southern Washington (modified from Reidel et al., 1994). OWL - Olympic-Wallowa Lineament (topographic manifestation of the OWSZ). Structures along the OWSZ trend WNW.

1993). Seismic reflection data show them to dip 15°-45° (Jarchow et al., 1994; Lutter et al., 1994).

However, bigger faults are crustal in scale and penetrate the pre-basalt basement. Magnetic maps (Johnson et al., 1990) show a linear anomaly crossing the Columbia River province along the OWSZ. Seismic refraction (Catchings and Mooney, 1988) and gravity (Saltus, 1993) surveys have revealed large perturbations across the OWSZ in the velocity and density structure of the upper and lower crust and even of the uppper mantle. Geophysical and drillhole data show that thickness of both Miocene basalts and underlying Paleogene sedimentary rocks increases dramatically within the OWSZ. The post-Miocene Pasco sedimentary basin also trends WNW and lies in the OWSZ tract.

Seismicity on the OWSZ occurs in clusters, and some earthquakes have NW-SE dextral focal mechanisms (Ludwin et al., 1989; Tolan and Reidel, 1989). However, structural evidence from the Yakima belt shows that N-S compressive stresses were predominant throughout the Neogene. Microseismicity suggests this area is still experiencing N-S compression (Reidel et al., 1994).

At the western edge of the Columbia River province, the Straight Creek-Fraser fault system merges with the OWSZ. A system of N-S-oriented Straight Creek and Fraser faults extends hundreds of kilometers across Washington into British Columbia. In the early Tertiary it accommodated tens of kilometers of right-lateral motion (Monger, 1991). No continuation of the Straight Creek-Fraser fault system is found south of the OWSZ, and Paleogene dextral motions seem to have been dissipated at least partly through a set of large NNW-oriented faults on this system's east side (Evans, 1994).

At its junction with the OWSZ, the N-S structural trends of the Straight Creek-Fraser fault system switch their trend to ESE over a distance of only 10 km (Campbell, 1989; Reidel et al., 1994). Whereas the Straight Creek-Fraser fault system does not extend south of the junction, the OWSZ continues WNW across the Cascade Ranges and beyond.

The OWSZ as a boundary between North and South Washington Cascades
The OWSZ is clearly expressed as structural and topographic breaks in the N-S-trending mountain ranges of the Cascades. The narrow White River-Naches River zone of severe faulting, mapped for a distance of several tens of kilometers (Frizzell et al., 1984; Campbell, 1989), marks the physiographic boundary between the North and South Cascades in Washington. These two topographic domains are separated by straight segments of the Cedar and Snoqualmie rivers, which run along the OWSZ trend. To the south, the mountains are less than 2000 m high, but they are more than 2500 m high in the north (Galster et al., 1989).

The North and South Cascades are distinguished also by their contrasting rock types and structures. The South Cascades throughout the Tertiary were characterized by broad crustal blocks movements which were slow and had a relatively low amplitude. The Mesozoic crystalline basement is exposed only in a few places, and no deep crustal rocks were brought to the surface in the Tertiary (Evarts, 1990). Cenozoic magmatic suites (Armstrong, 1978) and unconformity-bounded stratigraphic sequences (Cheney, 1994) have been correlated south of the OWSZ across much of Washington, over hundreds of kilometers.

The North Cascades, in contrast, are a direct continuation, from southeastern Alaska and British Columbia, of the Coast Belt orogen (Haugerud, 1989; Brown et al., 1994). As elsewhere in this orogen, the North Cascades are cored by polygenetic metamorphic and plutonic rocks exhumed from middle- to lower-crustal depths. Igneous and sedimentary rocks of mostly Mesozoic ages were remobilized at pressures >9 kbar (depth ≈ 30 km) in the Late Cretaceous. In the latest Cretaceous and early Tertiary they were intruded by large granitic plutons, whose buoyancy caused them to rise back to the surface.

The style, rate and volume of Late Cenozoic magmatism also differ sharply across the OWSZ. Estimated Quaternary eruption rates are several times lower north of Mt. Rainier than south. Typical arc magmatism related to subduction of the Juan de Fuca plate is occurring in western Oregon and southwestern Washington (Taylor, 1990). In contrast, volcanoes in northwestern Washington and southwestern British Columbia are isolated cones that have produced lavas of intermediate composition, contaminated with material from continental crust (Sherrod and Smith, 1990; Read, 1990; Green, 1990).

At 36 Ma, arc magmatism began along the western North America continental margin from California to southern British Columbia (Brandon and Vance, 1992). Later, however, the arc became strongly segmented (Guffanti and Weaver, 1988), and contrasts developed south and north of the OWSZ (cp. Sherrod and Smith,

1990). This segmentation of the arc probably reflects ongoing fragmentation of the Juan de Fuca oceanic slab, which has been inferred from teleseismic data (Michaelson and Weaver, 1986; VanDecar et al., 1990; Dueker and Humphreys, 1994a,b).

From field mapping, geochemical analyses and interpretation of potential-field data, several igneous domains have been distinguished along the South Cascades (Guffanti and Weaver, 1988; Sherrod and Smith, 1990; Blakely and Jachens, 1990; Scott, 1990). These domains are characterized by diachronous eruptions and igneous rocks with different compositions. The northern and southern ends of the Cascade arc, in northern Washington and northern California, are characterized by intermediate to silicic lavas which erupted from distinct composite volcanoes. In contrast, the middle part of the arc in Oregon and southern Washington contains overlapping fields of mafic (basaltic) lava.

North of the OWSZ, the volume of Pleistocene-Quaternary volcanism decreases substantially (Sherrod and Smith, 1990). Intermediate to silicic lavas are contaminated with material from continental crust (Green, 1990). Unlike in areas to the south, volcanoes north of the OWSZ are considered dormant.

Several separate Quaternary volcanoes in the Washington North Cascades (Glacier Peak, Mt. Baker) and the Garibaldi volcanic belt in southwestern British Columbia (Fig. 10) form a linear zone which runs NNW (Read, 1990; Green, 1990; Smith, 1990). This probably indicates control by a long crustal fracture. The Garibaldi belt, which contains andesite, dacite, rhyolite and basalt, is made up of stratovolcanoes, volcanic domes and isolated

84

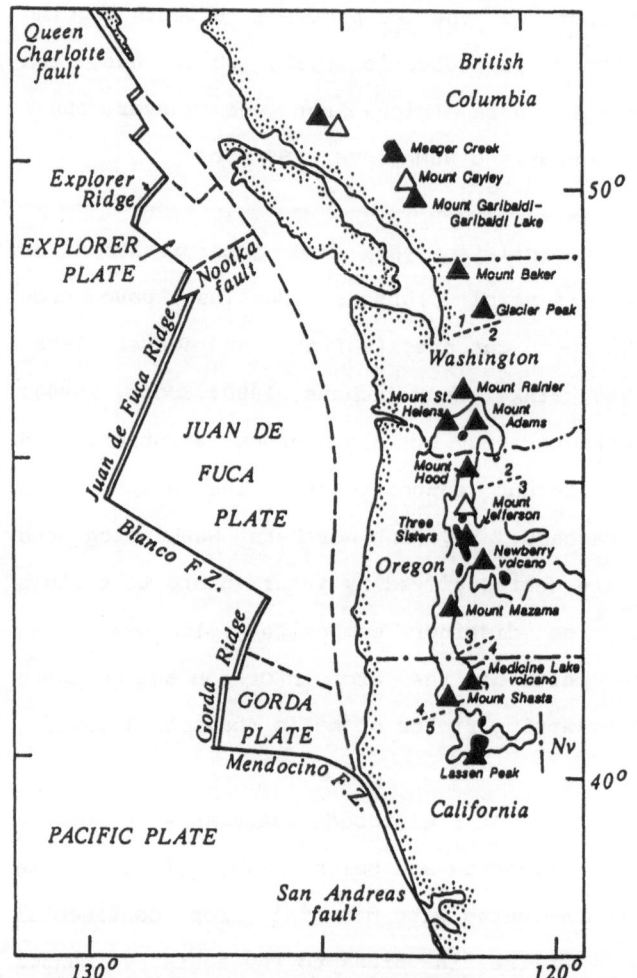

Figure 10. Distribution of Quaternary volcanoes along the western North America continental margin (mofified from Scott, 1990; cp. Fig. 3a). Segmentation of the High Cascade arc and the Garibaldi belt (segments 1 to 5) according to Guffanti and Weaver (1988). Recent evidence (Green, 1990) suggests the Garibaldi belt (segment 1) formed in an extensional tectonic regime: if so, it is not part of the current subduction-related arc. The conventionally assumed plate boundaries are shown offshore.

lava flows. Continuing tectono-magmatic activity at depth is suggested by the presence of hot springs.

Geochemical and petrological evidence suggests the Garibaldi belt was created in an intracontinental extensional tectonic regime. Even the basaltic rocks in this belt "most closely resemble magmatic associations considered to characterize regions of recent uplift, extensional tectonism, and high heat flow" (Green, 1990, p. 173). These features are similar to those in other Cenozoic volcanic belts across the Cordillera in British Columbia and the Yukon, which were related to deep fracturing of the crust in an extensional tectonic regime (Souther, 1990).

The OWSZ west of the Washington Cascades

The Puget and Georgia sedimentary basins partly cover up the continuation of the OWSZ west of the Washington Cascades. Faults trending WNW have been described on the eastern shore of Puget Sound, near Everett (Adair et al., 1989). From geophysical data, a buried basement horst between sediment-filled depocenters lies on trend with the OWSZ near Seattle (Lees and Crosson, 1990; Finn, 1990). The San Juan Rise (Galster et al., 1989) separates the Puget and Georgia basins (Figs. 6, 11). Faults trending WNW, some accommodating Late Cretaceous movements (e.g., the Lopez thrust), have been described along its southern flank (see Whetten et al., 1980; Brandon et al., 1988).

Running SE from the city of Victoria, across eastern Strait of Juan de Fuca, is a series of strong, linear, short-wavelength magnetic anomalies up to +800 nT in amplitude (Fig. 12). This band of anomalies forms a magnetic domain boundary: negative

86

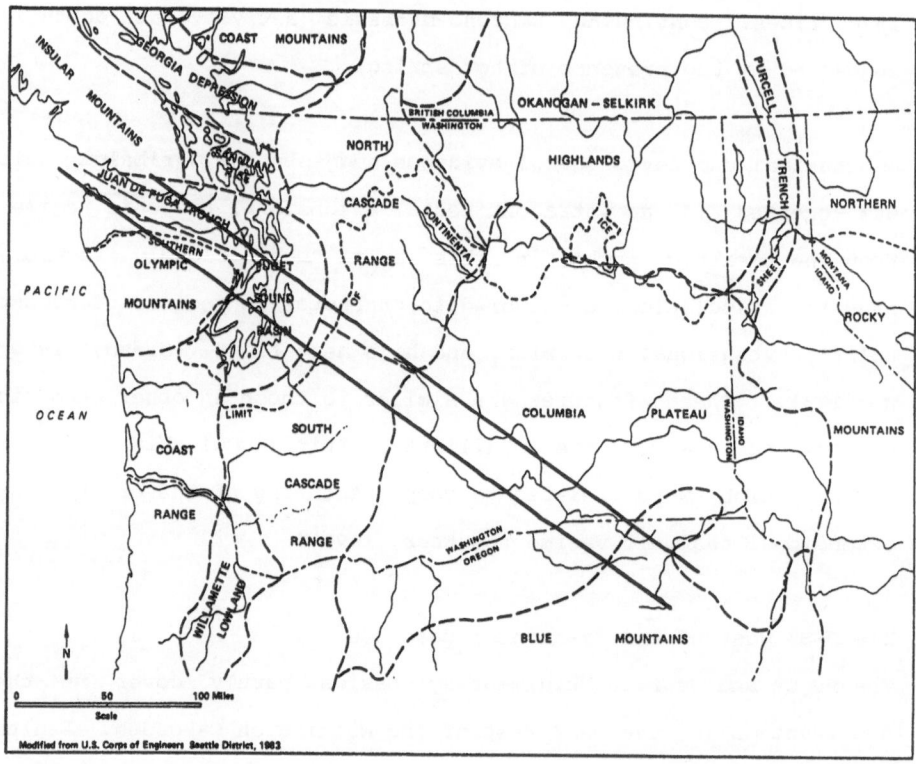

Figure 11. Geologic provinces of Washington and adjacent regions (modified from Galster et al., 1989). The OWSZ (whose approximate position is marked by the two parallel, solid lines) separates the North and South Cascades; on trend with it lies the Juan de Fuca Trough. The southern boundary of the San Juan Rise lies on trend with the OWSZ. Detailed mapping (Muller et al., 1977a; Monger, 1991) shows the northern boundary of the San Juan Rise is a NE-trending system of faults beginning on southern Vancouver Island, so unlike in this map, the rise is actually a trianglar block.

values as low as -300 nT .are observed to the north, whereas to the south the values are strongly positive. On northeastern Olympic Peninsula, short-wavelength anomalies of this band are sourced by volcanics, which have been drilled at a depth of 1660 m in the Dungeness Spit well. Gravity anomalies in the Strait of Juan de Fuca (Fig. 13) are also oriented WNW, as anomaly magnitides decline from Vancouver Island towards the Olympic Peninsula (MacLeod et al., 1977).

Marine surveys off Whidbey Island in eastern Strait of Juan de Fuca have revealed several subrarallel zones of structural disturbance in sediments as young as Quaternary (Atwater, 1994).

The long Strait of Juan de Fuca and the Fuca sedimentary basin lie on trend with the OWSZ. This basin is a long WNW-trending graben filled with 8 km of Tertiary sediments (Niem and Snavely, 1991). Gravity anomaly values decrease over the basin from >+60 mGal on southern Vancouver Island to <-20 mGal on the northern Olympic coast, and anomaly contours trend overwhelmingly WNW (Fig. 13).

Magnetic anomalies in the basin trend WNW and are generally >+400 nT (Fig. 12). Presence of these positive magnetic anomalies as far as Cape Flattery suggests continuity of Eocene basalts under the Strait of Juan de Fuca (MacLeod et al., 1977; Dehler and Clowes, 1992). Locally, as between the cities of Victoria and Port Angeles on the opposite shores of the Strait, short-wavelength anomalies are similar to those in eastern Strait of Juan de Fuca, though without the strong linear fabric. This similarity suggests that the Fuca Basin may also contain shallow volcanic rocks.

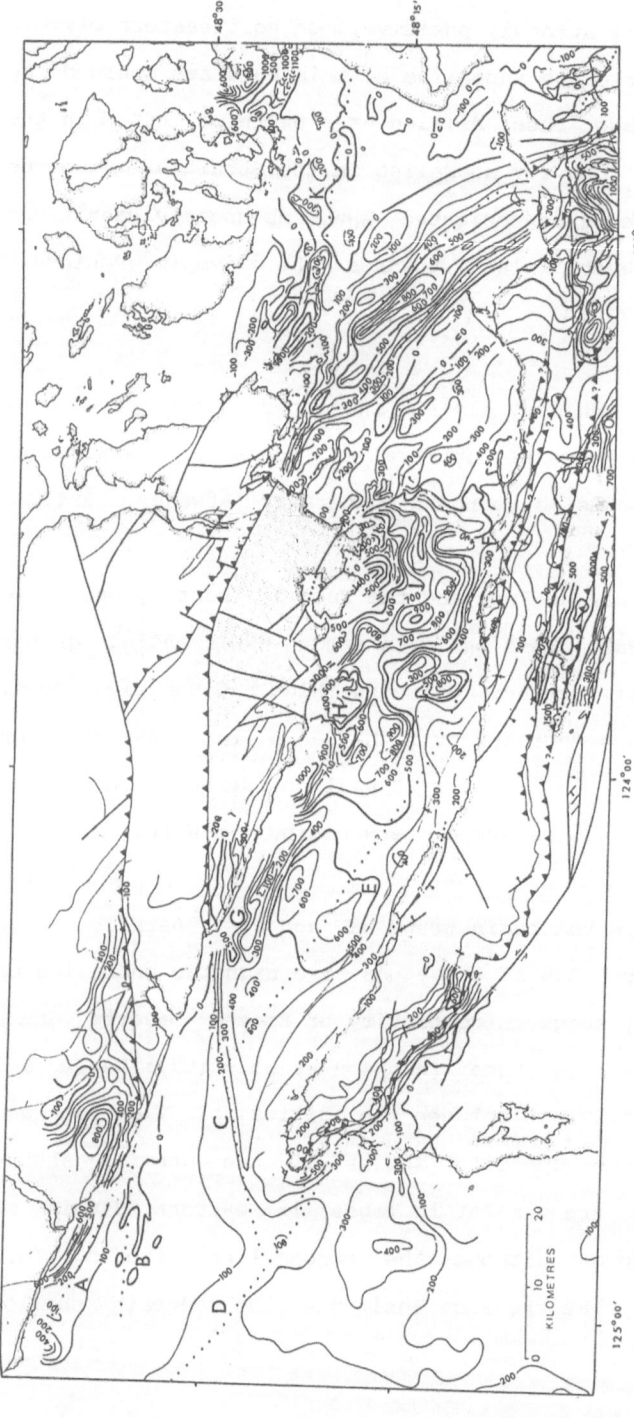

Figure 12. Magnetic anomaly map of the Strait of Juan de Fuca and vicinity, superimposed on a map of faults of the northern and southern strands of the western OWSZ (modified from MacLeod et al., 1977). Anomalies are in nT. The predominant WNW orientation of anomalies reflects the orientation of the OWSZ. Letters A to L represent anomalies discussed in detail by MacLeod et al. (1977).

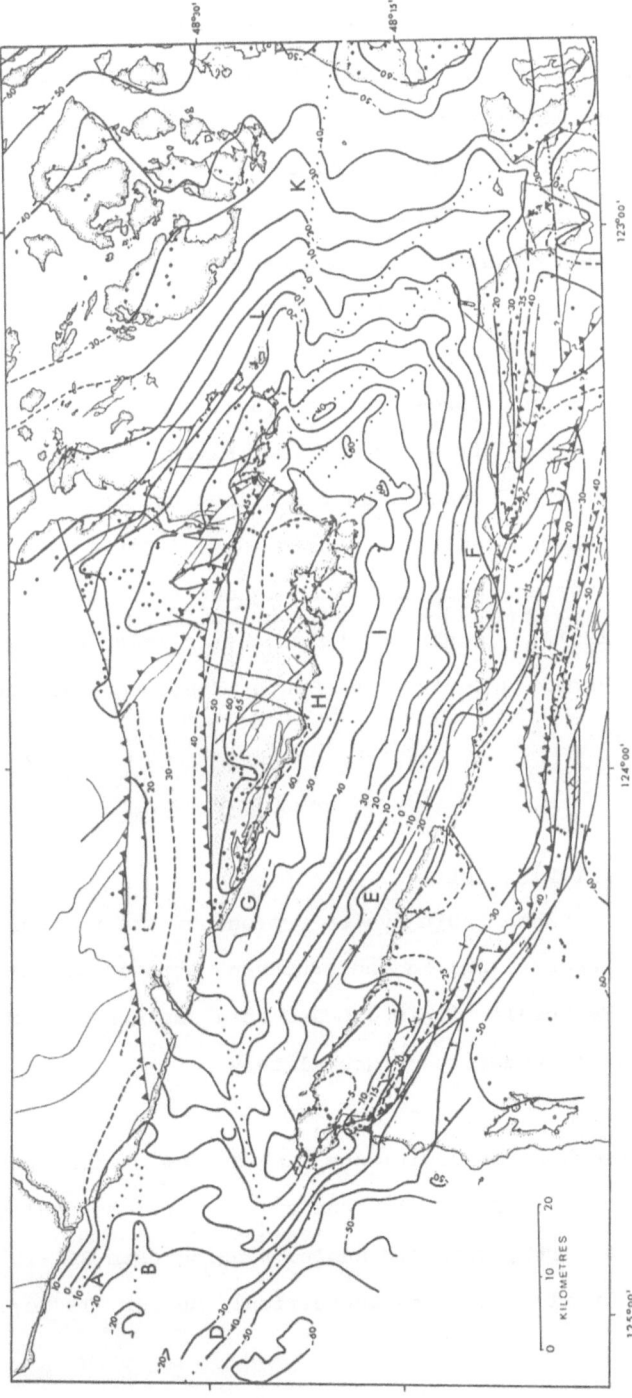

Figure 13. Bouguer gravity anomaly map of the Strait of Juan de Fuca and vicinity, superimposed on a map of faults of the northern and southern strands of the western OWSZ (modified from MacLeod et al., 1977). Anomalies are in mGal. The main WNW orientation of anomalies reflects the orientation of the OWSZ. Letters A to L represent anomalies discussed in detail by MacLeod et al. (1977).

The Fuca graben is bounded by long, straight faults. On northern Olympic Peninsula, the boundary fault system is marked by WNW-trending volcanic massifs of the Crescent Formation and by outcrop belts of sedimentary rocks of the Fuca Basin. Several big WNW-trending faults have been mapped along the Hurricane Ridge and Crescent Lake basaltic massifs (Tabor and Cady, 1978). One of them, the Calawah fault, runs WNW through Cape Flattery onto the exterior continental shelf (MacLeod et al., 1977; Snavely, 1987).

On southern Vancouver Island, large, steep faults separate the Metchosin basaltic massif from the Leech River metamorphic complex, and the latter from Wrangellian formations to the north (Muller, 1977a-c). These faults - Leech River, San Juan and Survey Mountain - interacted complexly throughout the Tertiary (see also Monger, 1991). To stress their genetic relationship, they are herein designated jointly, as South Vancouver Island fault system (Fig. 14).

Boundary fault systems of western OWSZ
<u>South Vancouver Island fault system</u>
The South Vancouver Island fault system represents the northern strand of the OWSZ. Because, unlike the southern strand of the OWSZ on the Olympic Peninsula, it is exposed in a small area, its tectonic history is still understood incompletely.

At the surface, the Leech River fault zone is a few kilometers wide (Fairchild and Cowan, 1982). It consists of straight faults dipping steeply to the north, with dip angles ranging from 36°-70° northward (MacLeod et al., 1977) to subvertical (Muller, 1977c).

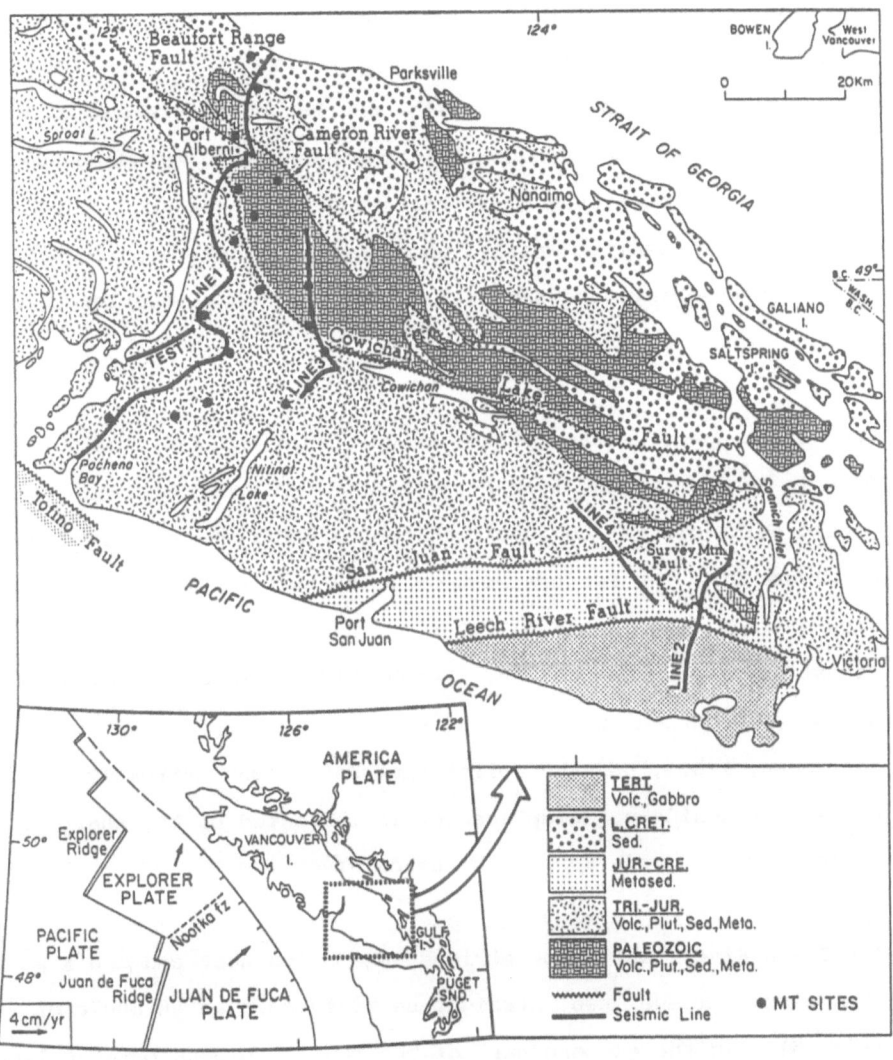

Figure 14. Location of LITHOPROBE seismic reflection profiles 84-01 to 84-04 (lines 1 to 4) and magnetotelluric (MT) sites on Vancouver Island, superimposed on a simplified geologic map of that area (modified from Green, ed., 1990). A more detailed geologic map of the entire Vancouver Island is presented in Fig. 16. Inset shows the conventionally assumed plate boundaries and rates of motion.

Movements on these faults might have begun even before the Tertiary, if protoliths of the Leech River metamorphic complex were related to Wrangellian units on Vancouver Island. These rocks foundered to mid-crustal depth and were metamorphosed to amphibolite grade around 41-38 Ma. They were juxtaposed against the Metchosin basaltic massif, whose grade of metamorphism did not exceed greenschist, and together these two rock complexes were brought to the surface.

The exact age of metamorphism and uplift is unclear. On northern Olympic Peninsula, the Lyre Formation of the Fuca Basin contains metamorphic clasts derived from the Leech River complex, as well as clasts from the Pandora Peak unit (Babcock et al., 1994; see also Garver and Brandon, 1994). Because the age of the Lyre Formation is 42 to 40 Ma, Leech River complex metamorphism probably occurred before 41-38 Ma (the age proposed by Fairchild and Cowan, 1982), and the uplift must have been extremely rapid. Large north-side-up displacement is indicated by the sharp break in metamorphic grade across the Leech River fault.

The Leech River fault zone strikes WNW on its east end and E-W in the west. These two straight segments meet at an angle of 25° (Fig. 5). In the eastern part of the massif, blocks between local faults are tilted 10° to 30° NE towards the massif-bounding Leech River fault and are subparallel to it, but just 10 km from the fault the tilts are in the opposite direction. This indicates a broad antiformal fold. The Leech River fault truncates these structures in the Metchosin massif, but is slightly offset by several transverse, mostly NNE-oriented faults (Muller, 1977c).

The Leech River metamorphic complex is strongly disrupted and sheared. Foliation in it, being synkinematic with uplift, is steep and parallel to the bounding faults. The Leech River, Survey Mountain and San Juan faults restrict the outcrops of the Leech River complex to an area between just 2 and 15 km across. The narrowest part occurs in the east, where the Leech River and Survey Mountain faults approach each other, but the Leech River complex widens to the west.

The amplitude of fault-controlled uplift of these amphibolite-grade, foliated rocks is similar to or greater than the width of the Leech River complex itself. This complex should therefore be regarded as part of the deep structural zone on southern Vancouver Island. The exact history of fault movements in this zone is so far unresolved. Based on field mapping, various ideas invoke dip-slip (Monger, 1991) or strike-slip (Fairchild and Cowan, 1982; Johnson, 1984; Rusmore and Cowan, 1985) movements. From seismic data, another idea holds that in the Late Eocene or Oligocene(?), these faults also acted as thrusts (Clowes et al., 1987). Several low-angle events in seismic sections have been interpreted as downdip continuations of the Leech River and San Juan faults (Yorath et al., 1985a; Green et al., 1985, 1987). Because of these differences, no comprehensive interpretation is available.

Steep dips of these faults are indicated by field mapping (Muller, 1977a; Fairchild and Cowan, 1982; Rusmore and Cowan, 1985). Locally, thrust-like splays are associated with steep master faults. Detailed reinterpretation of seismic data has shown that faults postulated to be thrusts are in fact steep (Mayrand et al., 1987; vs. Clowes et al., 1987).

Magnetic and gravity anomalies suggest the eastern segment of the Leech River fault continues across the Strait of Juan de Fuca (MacLeod et al., 1977) and meets faults that form the southern boundary of the San Juan Rise (Brandon et al., 1988). On the Washington mainland, in the vicinity of Everett, parallel faults were active in the Late Oligocene (Adair et al., 1989). Last movements on the western segment of the Leech River fault occurred in the Neogene: flat-lying Miocene sediments overlap the Metchosin massif to the south, whereas Oligocene sediments overlap the Leech River complex to the north (Muller, 1977a).

West of Vancouver Island offshore (Figs. 12, 15), the north-side-down magnetic gradient zone associated with the Leech River fault turns slightly to the SW (anomaly C of MacLeod et al., 1977). It is truncated off Cape Flattery by a magnetic anomaly coincident with the WNW-oriented Calawah fault (Snavely, 1987). The western offshore extension of the San Juan fault is unknown, but it is commonly assumed to swing to the NW and follow the Vancouver Island coast line (MacLeod et al., 1977; Snavely, 1987).

The WNW-oriented Survey Mountain fault is a link between the Leech River and San Juan faults. Though short, it may be the oldest in the South Vancouver Island system. It truncates from the south the gneisses known locally as Wark complex. These rocks are similar to the high-grade rocks of the Westcoast complex (Fig. 16), which were metamorphosed in the Early Jurassic (Muller, 1977a; Isachsen, 1987) and uplifted sometime before or during the Middle Eocene. Between these episodes of downward and upward movement, large strike-slip displacements might have occurred on

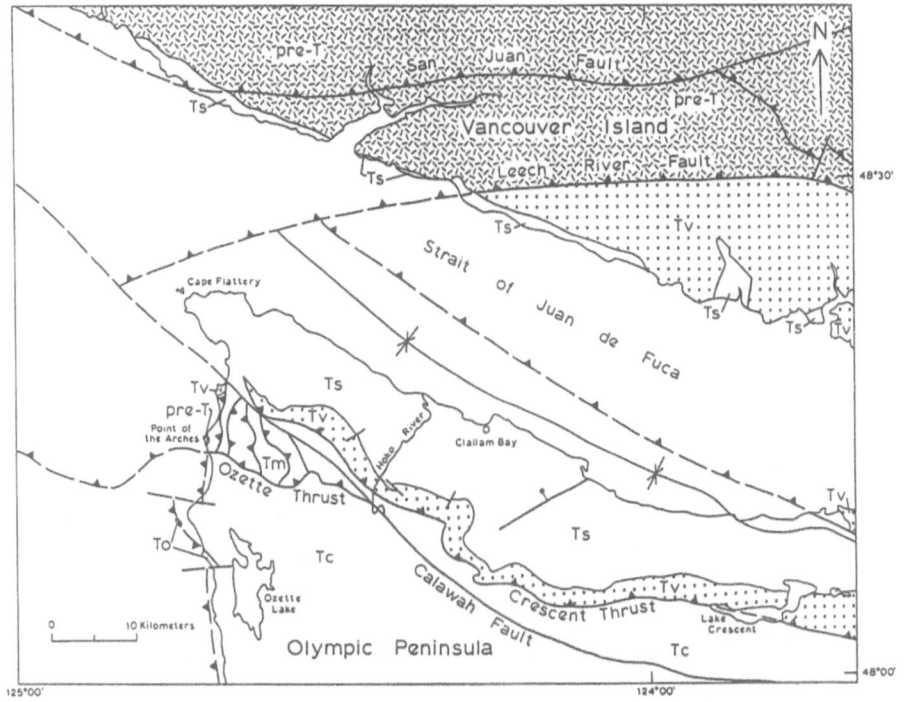

Figure 15. Fault map of northwestern Olympic Peninsula and southwestern Vancouver Island, according to Snavely (1987). Offshore, the continuation of the Calawah fault had been inferred previously, from gravity and magnetic data, by MacLeod et al. (1977). Details of fault configuration in the Point of the Arches area are probably unjustified, as this interpretation assumes the Mesozoic granitic rocks in that area are part of an accreted terrane rather than basement (see text). Oceanward dip of the thrust fault along the western Olympic Peninsula coast suggests that obduction, as well as subduction, has occurred. Tc - highly deformed Tertiary sedimentary rocks in the Olympic Mountains core; Tm - thrust-faulted pre-Tertiary to Oligocene broken formation; To - Upper Oligocene sandstone and conglomerate; Ts - Tertiary sedimentary rocks (undifferentiated); Tv - Tertiary volcanics.

96

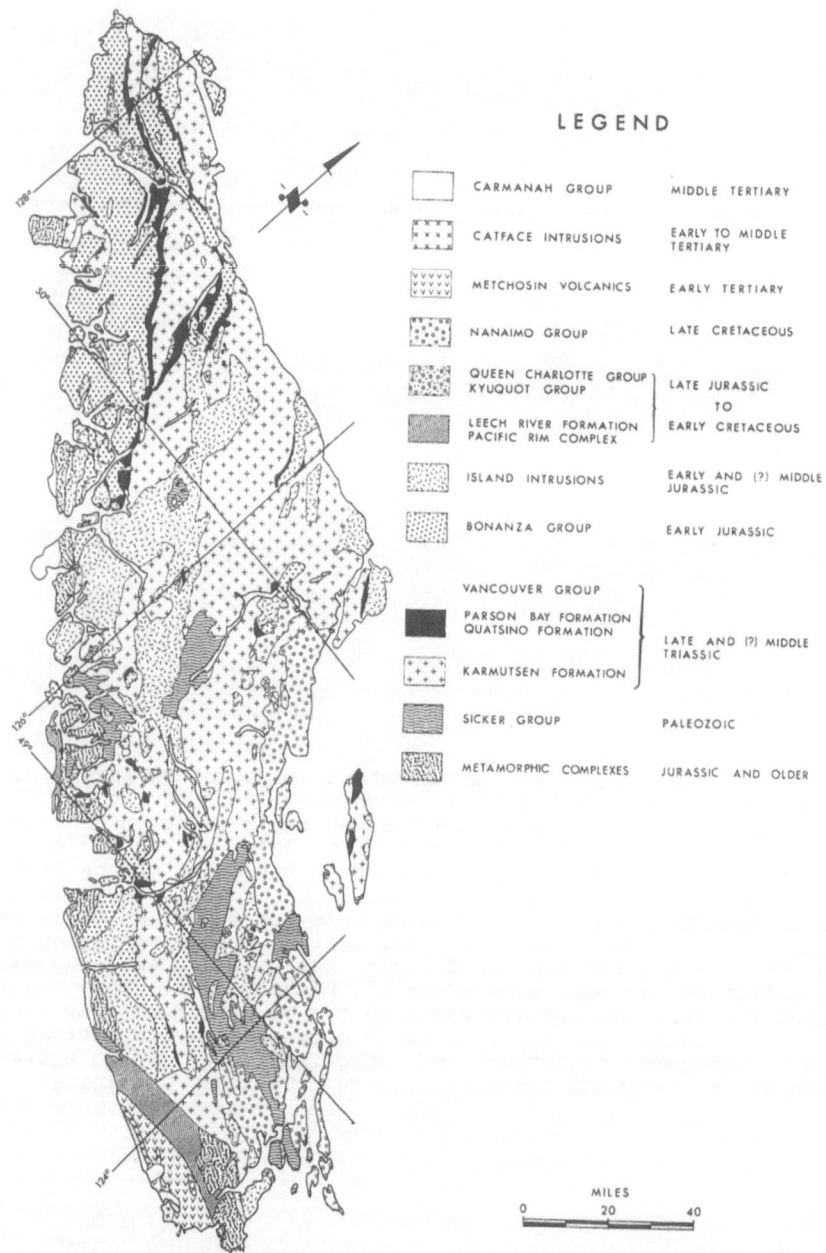

LEGEND

	CARMANAH GROUP	MIDDLE TERTIARY
	CATFACE INTRUSIONS	EARLY TO MIDDLE TERTIARY
	METCHOSIN VOLCANICS	EARLY TERTIARY
	NANAIMO GROUP	LATE CRETACEOUS
	QUEEN CHARLOTTE GROUP KYUQUOT GROUP	LATE JURASSIC TO
	LEECH RIVER FORMATION PACIFIC RIM COMPLEX	EARLY CRETACEOUS
	ISLAND INTRUSIONS	EARLY AND (?) MIDDLE JURASSIC
	BONANZA GROUP	EARLY JURASSIC

VANCOUVER GROUP

	PARSON BAY FORMATION QUATSINO FORMATION	LATE AND (?) MIDDLE TRIASSIC
	KARMUTSEN FORMATION	
	SICKER GROUP	PALEOZOIC
	METAMORPHIC COMPLEXES	JURASSIC AND OLDER

MILES
0   20   40

Figure 16. Geologic map of Vancouver Island and the Gulf Islands (after Muller et al., 1981). Location of LITHOPROBE seismic lines in relation to the Vancouver Island geology is shown in Fig. 14.

the Survey Mountain fault, as a part of a regional fault system, in the Late Cretaceous and early Tertiary (Johnson, 1984). Such movements apparently resulted in the dispersal of Pandora Peak tectonic slices from the Northwest Cascade thrust belt on the San Juan Islands to their current positions along the San Juan fault (Rusmore and Cowan, 1985). Presence of Pandora Peak clasts in the western lithofacies zone of the Lyre Formation on northern Olympic Peninsula (Babcock et al., 1994) suggests these offsets occurred prior to 40-42 Ma.

Steep foliation in the Leech River metamorphic complex is parallel to the Survey Mountain, San Juan and Leech River faults. This indicates the three faults acted in tandem during the uplift of the Leech River complex.

The eastern part of the San Juan fault, from its merger point with the Survey Mountain fault, runs NE. Muller (1977a) linked it with the Orcas fault belt which runs across the Strait of Georgia and forms the northern boundary of the San Juan Rise. On the British Columbia mainland, the Vedder fault on trend with this NE-oriented fault belt is thought to have accommodated normal movements in the Neogene (Monger, 1991).

Presence of laterally displaced slices of the Pandora Peak unit in the San Juan fault zone (Rusmore and Cowan, 1985) suggests that this fault is steep. This evidence contradicts the interpretation of this fault as a low-angle thrust in a Tertiary subduction complex (Clowes et al., 1987). Mayrand et al. (1987) have shown that this fault remains steep in the subsurface, as far as it is resolved with seismic data.

From the description in this section, the northern strand of the OWSZ extends from the interior of the Cordillera, across eastern Strait of Juan de Fuca, through southern Vancouver Island, onto the exterior continental shelf.

North Olympic fault system

The southern strand of the OWSZ is well exposed on northern Olympic Peninsula. It consists of several major WNW-oriented faults which were complexly interrelated during their evolution. The most prominent of them are the Hurricane Ridge, Crescent and Calawah faults (Figs. 12, 13, 15, 17).

These faults lie between the Hurricane Ridge and Crescent Lake basaltic massifs and the sedimentary successions to the south. Near the eastern end of the Hurricane Ridge fault, a local gap occurs in the basaltic buttress. Sediments on both sides of the fault in this area are deformed similarly, and no distinct structural-domain boundary is observed across this part of the fault (Tabor and Cady, 1978; Babcock et al., 1994).

The Hurricane Ridge fault zone strikes about N70°W. It contains several subparallel strands dipping about 60° to the north. As along the OWSZ in the Cordilleran interior (see Reidel et al., 1994), small thrust splays are also found along the master fault, but the extent of thrusting is insignificant. The recently established stratigraphic relationships of the Crescent Formation basalts with sediments of the Olympic Mountains core to the south and of the Fuca Basin to the north (see Babcock et al., 1994) reveal continuity that allows only small displacements across the Hurricane Ridge fault.

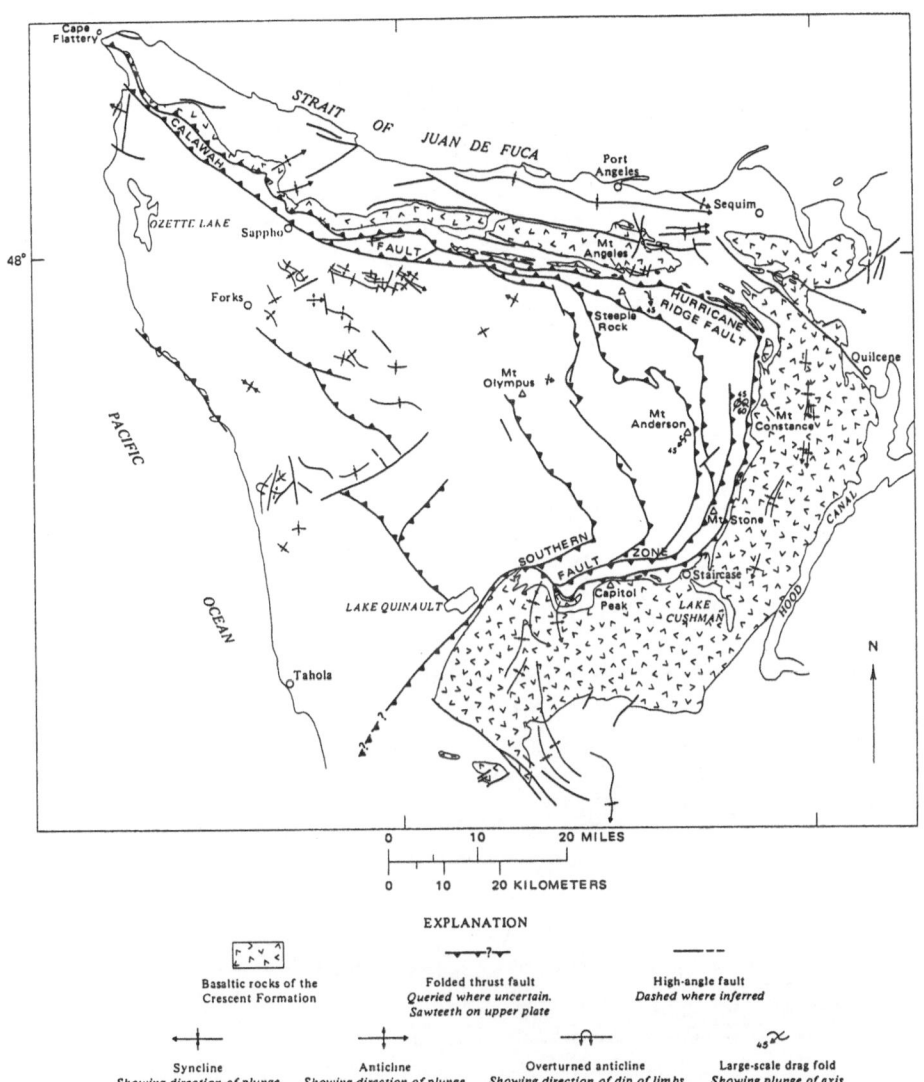

Figure 17a. Distribution of faults and Crescent Formation basaltic massifs in the southern strand of the OWSZ and elsewhere on the Olympic Peninsula (modified from Tabor and Cady, 1978).

100

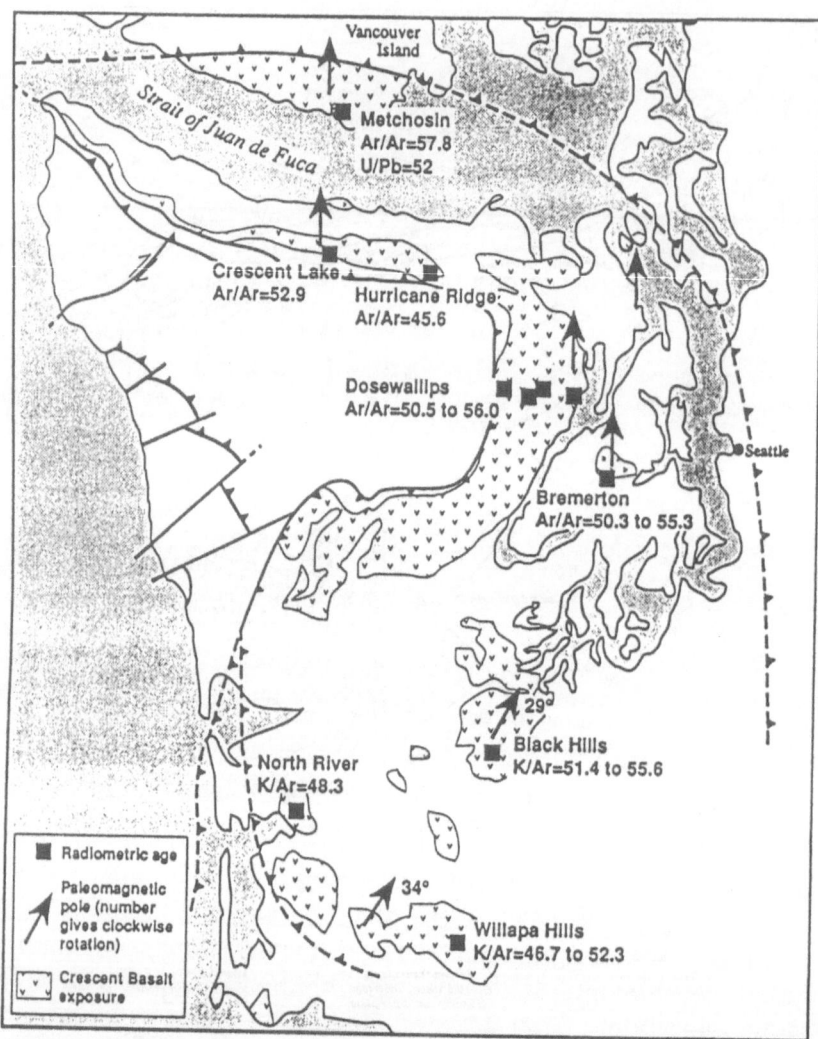

Figure 17b. Distribution and age of Crescent Formation massifs in western Washington and British Columbia (modified from Babcock et al., 1994). The northward swing of the Leech River fault along Puget Sound is not supported by the data presented in this volume: rather, the Leech River fault appears to be part of the northern strand of the OWSZ.

The Crescent fault has a more northwesterly trend than the Hurricane Ridge fault, which it meets near longitude 124°W at a an angle of some 25° (Tabor and Cady, 1978; Babcock et al., 1994). The Crescent fault runs parallel to the closely adjacent Calawah fault, with which it merges at some localities. Between the Crescent and Calawah faults lies a narrow strip of Eocene clastic rocks with minor basalt.

Like in the Hurricane Ridge fault zone, the master Crescent fault plane dips 50°-60° to the north, but thrust sheets have also been described along its trend in the west (MacLeod et al. 1977; Snavely, 1987). Thrust movements in the late Middle Eocene have been interpreted on the Crescent fault (Snavely, 1987).

The Calawah fault is straight and near-vertical along its entire length from north-central Olympic Peninsula to Cape Flattery and beyond. Motion on was once interpreted as reverse (Tabor and Cady, 1978) but is now usually thought to have been sinistral strike-slip (Snavely, 1987). Post-Pleistocene movements on this fault are indicated by deformation of young deposits onshore and offshore, and by details of shelf bathymetry (MacLeod et al., 1977; Snavely, 1987).

The Hurricane Ridge, Crescent and Calawah faults, together named herein North Olympic fault system to emphasize their genetic relationship, have existed since at least the Early Eocene, when they served as magma conduits. Several basaltic massifs - Marmot Pass, Hurricane Ridge, Crescent Lake - lie along these faults.

A long history of the southern strand of the OWSZ on northern
Olympic Peninsula is apparent also from the geologic contrasts
across this fault system. The North Olympic fault system
controlled the position of different sedimentary basins, one on
the site of the Olympic Mountains, the other on the site of the
Strait of Juan de Fuca. Their beginnings may lie in the Paleocene
(the Blue Mountain unit), but the well-preserved stratigraphic
record of the Fuca Basin spans the Middle Eocene to the Miocene.
In the late Tertiary, the North Olympic fault system served as an
important structural boundary. To the south, rocks are highly
deformed (Central Olympic and Hoh basins) and metamorphosed
slightly (Central Olympic Basin). To the north, in the Fuca
Basin, they are deformed only mildly and unmetamorphosed.

Central Olympic Basin

A large sedimentary basin, designated herein Central Olympic
Basin, existed on the site of the present-day Olympic Mountains
from Paleocene to Miocene. Fairly stable paleogeological settings
resulted in accumulation of thick, monotonous mudstone with
subordinate sandstone and conglomerate, occasionally interbedded
with basalt (Tabor and Cady, 1978). Sedimentation took place
dominantly in continental-slope environments, which during the
early Tertiary existed in a large area from the site of the
Dosewallips basaltic massif outboard.

Despite the generally monotonous lithology and scarce fossil
control, sediments of the Central Olympic Basin have been grouped
into four assemblages based on relative abundance of coarse
material and fossil ages (Tabor and Cady, 1978). They were later

interpreted in terms of two main successions, distinguished principally by their fission-track ages (Brandon and Vance, 1992). Lithological differences between these two successions lie chiefly in the abundance of stringers and blocks of basalt, which occur mostly in older rocks. The older succession has ages from 48 to 32 Ma (Middle Eocene to Early Oligocene), and the younger one from 27 to 19 Ma (Late Oligocene to Early Miocene).

Sediments of the Central Olympic Basin have been traced to source areas in the interior of southern Canadian Cordillera (Heller et al., 1992). Rocks of the older succession were metamorphosed slightly in the mid-Oligocene, at 30-29 Ma (Tabor, 1972). Later, both successions were affected by more intense metamorphism, reaching the prehnite-pumpellyite grade. This event occurred in the late Early Miocene, around 14-13 Ma (Brandon and Vance, 1992).

The Central Olympic Basin is now inverted, and its exhumed sedimentary rocks are well exposed in the core of the Olympic Mountains. Even though these rocks are rather soft and penetratively sheared, which makes them fissile and easily eroded, the mountains rise as high as 2427 m at Mt. Olympus.

Large NW-trending faults divide sedimentary rocks in the Olympic Mountains into several tectonic slices (Fig. 18). These slices are characterized by different rock assemblages. Distinguished by relative proportion of sand and shale and abundance of basalt, these rocks reacted slightly differently to compressional tectonic stresses in late Tertiary time.

The oldest, Lower-Middle Eocene rocks make up the Elwha assemblage

104

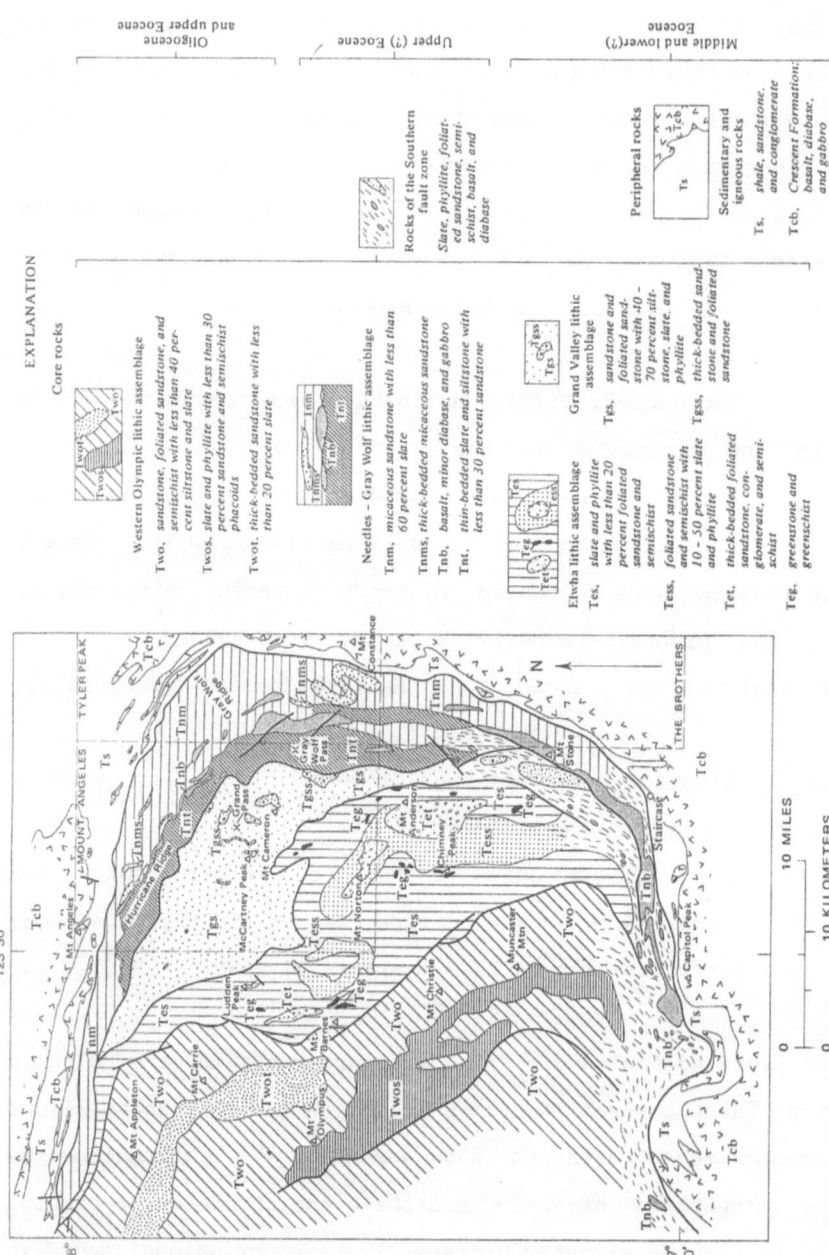

EXPLANATION

Core rocks

Western Olympic lithic assemblage

Two, sandstone, foliated sandstone, and semischist with less than 40 percent siltstone and slate

Twos, slate and phyllite with less than 30 percent sandstone and semischist phacoids

Twot, thick-bedded sandstone with less than 20 percent slate

Needles – Gray Wolf lithic assemblage

Tnm, micaceous sandstone with less than 60 percent slate

Tnms, thick-bedded micaceous sandstone

Tnb, basalt, minor diabase, and gabbro

Tnt, thin-bedded slate and siltstone with less than 30 percent sandstone

Grand Valley lithic assemblage

Tes, sandstone and foliated sandstone with 40-70 percent siltstone, slate, and phyllite

Tgss, thick-bedded sandstone and foliated sandstone

Elwha lithic assemblage

Tes, slate and phyllite with less than 20 percent foliated sandstone and semischist

Tess, foliated sandstone and semischist with 10-50 percent slate and phyllite

Tet, thick-bedded foliated sandstone, conglomerate, and semischist

Teg, greenstone and greenschist

Rocks of the Southern fault zone

Slate, phyllite, foliated sandstone, semischist, basalt, and diabase

Peripheral rocks

Sedimentary and igneous rocks

Ts, shale, sandstone, and conglomerate

Tcb, Crescent Formation; basalt, diabase, and gabbro

Oligocene and upper Eocene

Upper (?) Eocene

Middle and lower(?) Eocene

Figure 18. Geologic map of the Olympic Mountains area, showing faults and tectonic slices in the Central Olympic Basin (modified from Tabor and Cady, 1978).

in the fault-bounded slice in the center of the mountains. In a
slice next to it on the east, Middle and Upper Eocene as well as
Lower Miocene rocks are contained in the Grand Valley assemblage.
In a slice still further east, Upper Eocene and Lower Oligocene
are found in the Needles-Gray Wolf assemblage. To the west of the
older core lies the Western Olympic slice, which contains Upper
Oligocene to Lower Miocene sediments near Mt. Olympus and older,
Middle Eocene to Lower Oligocene sediments farther west. Thus, no
regular E-W age progression of sedimentary rocks exists across the
northern Olympic Peninsula, and rocks of different successions may
be found in the four tectonic slices.

These tectonic slices are partly thrusted over one another,
usually with a westerly vergency (Tabor and Cady, 1978; Brandon
and Vance, 1992). The most deformed are the Elwha and Grand
Valley slices, whereas to the west and east, the Western Olympic
and Needles-Gray Wolf tectonic slices are relatively less
deformed. These less-deformed rocks lie on both sides of the
more-deformed rocks in the middle.

The coincidence of rock assemblages having slightly variable
lithologies and degrees of deformation with identifiable tectonic
slices suggests that the original Central Olympic Basin might have
contained several depocenters. They were later telescoped into a
compressional thrust-sheet complex. Evidence of polyphase
deformation found in the Olympic Mountains indicates variability
of tectonic stresses (Snavely, 1987), though the strongest was E-W
compression probably induced by Juan de Fuca plate subduction.
However, the lack of a systematic E-W progression in deformation
intensity also suggests other influences.

Hoh Basin

A different stratigraphy characterizes the rock assemblages exposed west of the Olympic Mountains and of the Central Olympic Basin. Common rocks in outcrops on the western coastal lowlands of the Olympic Peninsula and in drillholes on the shelf offshore are turbiditic sandstone and organic-rich siltstone and mudstone (Snavely, 1987; Orange et al., 1993). They are thought to have been deposited in continental-slope or trench environments (Niem et al., 1992b).

Snavely (1987) tentatively distinguished in the Hoh succession two units separated by a large stratigraphic hiatus: Middle to Late Eocene, and Late Oligocene to Middle Miocene. Orange et al. (1993) were later unable to confirm the existence of these two units, but the age span of the succession was not challenged. The Hoh Basin was largely coeval with the Central Olympic Basin, and paleoenvironments were strikingly similar from the Dosewallips massif to at least the edge of the present-day shelf. This would be possible if a large embayment existed on the site of these two basins, forming a deep re-entrant in the continental margin (cp. Niem et al., 1992a-c).

Unlike the Central Olympic Basin, where sedimentation ended in the Miocene, the Hoh Basin continued to develop much later (Niem et al., 1992c). Its structural evolution was also considerably different. In the late Middle Miocene, the Hoh Basin was affected by thrust faulting, induced perhaps by interactions of the continental and oceanic plates. This tectonic episode resulted in partial underthrusting of pre-Late Miocene rocks (Niem et al.,

1992b), but in many places the same rocks have been mapped as thrusted over other rocks (Snavely, 1987, p. 320; Orange et al., 1993). Both faults and sedimentary strata are dipping towards both the continent and the ocean. Along the coast, a N-S-oriented thrust is dipping oceanward (Fig. 15).

The entire Hoh assemblage has been described as consisting mostly of mélange and broken rocks (e.g., Snavely, 1987). Detailed mapping in several areas has revealed large variations in structural styles along the west coast of the Olympic Peninsula. Mélanges are of different origins including diapiric, and the distribution of broken and sheared rocks is uneven (Orange, 1990; Orange et al., 1993). Many zones of shearing are correlated with steep faults. Deformation in the Hoh Basin was not uniform, and between sheared zones lie broad areas where sediments are coherent. Not all the deformation had tectonic origins. Sediment mobility in a continental-slope setting and post-depositional mud diapirism were important factors in rock deformation.

Accumulation of the upper Hoh assemblage was interrupted in the early Middle Miocene by tectonic uplift and erosion. This event resulted in a large stratigraphic gap spanning eight million years between about 15 and 7 Ma. When sedimentation resumed, the Quinault Formation was laid down in the latest Miocene and Pliocene, over a pronounced angular unconformity.

The Quinault Formation contains up to 2000 m of shallow marine deposits, as conditions of sedimentation were comparable to those at present. On the shelf, 25 km from the coast, the Pan-American P-0141 well encountered a similar thickness of Pliocene and Upper

Miocene clastic rocks, underlain by Middle Miocene to Upper
Oligocene(?) sediments which continued to the bottom of the well
at 3160 m (Snavely, 1987).

Therefore, the Hoh Basin has evolved since Middle Eocene time.
Bathyal environments on a very wide continental slope existed in
the early Tertiary on the site of the present-day Olympic
Peninsula. By Late Miocene to Pliocene time, however, the basin
shallowed, and shelf environments were established in large parts
of western Olympic Peninsula and adjacent submerged margin.

The boundary between the Hoh and Central Olympic basins is
probably a basement high inferred on the Olympic Peninsula from
gravity data. Structure in this area is complex and poorly
understood, with a variety of thrust faults dipping in opposite
directions as is common in fold-and-thrust belts of orogens.
Gravity data suggest that the Hoh Basin consists of two distinct
depocenters which lie in large depressions under the coastal plain
and the continental shelf. These two depressions are separated by
a basement high in the Point of the Arches area, which contains
the only known exposure of Mesozoic crystalline rocks on the
Olympic Peninsula (MacLeod et al., 1977; Niem et al., 1992a).

Deep structure of the Olympic Peninsula area from gravity data
New gravity data for western Washington and southwestern British
Columbia make it possible to bring geophysical interpretations up
to date and in line with recent geological information. To
examine large-scale crustal structure, these data were upward
continued to a range of nominal elevations from 5 to 100 km.
Final interpretations were constrained by seismic refraction and
reflection profiles and geological facts.

In the Bouguer anomaly map (Fig. 19), prominent highs correspond to large basaltic massifs. The most pronounced is the positive anomaly of >+60 mGal over the Metchosin massif. Similar but smaller gravity highs lie over basaltic bodies in southwestern Washington. Northwest of the Metchosin massif, local gravity highs north of Barkley Sound are probably related to mafic igneous rocks in the subsurface. In the Strait of Juan de Fuca, anomaly values decrease southward from the Metchosin high across several steep gradient zones parallel to the OWSZ, and become negative over the Olympic Mountains. This southward decrease of anomaly values is consistent with the asymmetry of the Fuca Basin, whose axis lies along the northern Olympic coast (Niem and Snavely, 1991). The gravity low over the Olympic Mountains may reflect either thick sediments or a low-density crystalline basement.

Despite a strong density contrast between the Tertiary sediments and Crescent basalts (the latter in this area have a density of 2,700 kg/m3, considerably heavier than the surrounding sediments; Finn, 1990), not all basaltic massifs are associated with strong gravity anomalies. Unlike the Metchosin massif, the Hurricane Ridge and Crescent Lake bodies are not strongly expressed in the gravity map, probably due to lack of deep roots.

Over the Olympic Peninsula and the adjacent continental shelf, three local minima can be distinguished. Isometric in shape, they are separated by elongated relative highs. The gravity low over the Olympic Mountains is about -90 mGal in magnitude. On the west, it is flanked by a relative high (-50 to -70 mGal) bounded by gradient zones trending NE and NW.

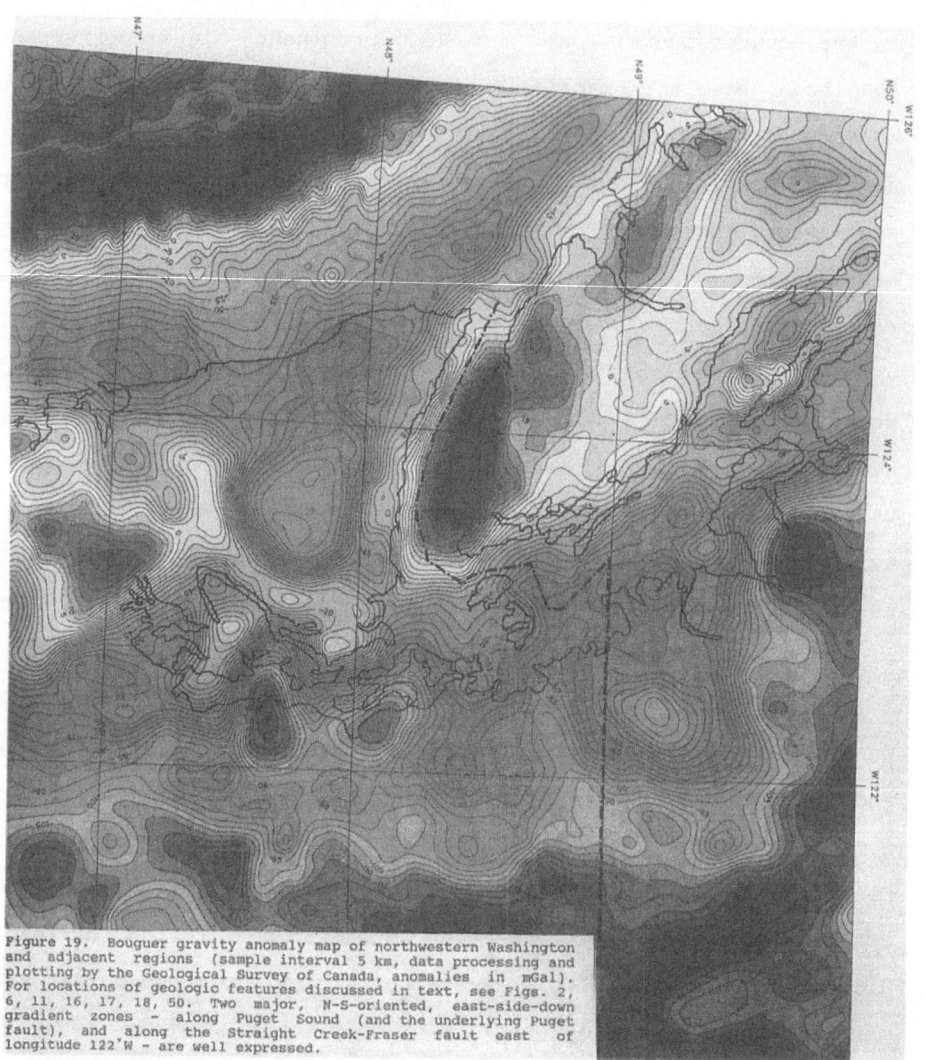

Figure 19. Bouguer gravity anomaly map of northwestern Washington and adjacent regions (sample interval 5 km, data processing and plotting by the Geological Survey of Canada, anomalies in mGal). For locations of geologic features discussed in text, see Figs. 2, 6, 11, 16, 17, 18, 50. Two major, N-S-oriented, east-side-down gradient zones - along Puget Sound (and the underlying Puget fault), and along the Straight Creek-Fraser fault east of longitude 122°W - are well expressed.

The Hoh Basin farther west is marked by two isometric lows, probably corresponding to distinct depocenters. One low, nearly -75 mGal in amplitude, lies over the central part of the western Olympic coast. Another low, also around -70 mGal in the Bouguer map but <-80 mGal in the free-air reduction (Finn et al., 1991; Dehler and Clowes, 1992), lies just west of Cape Flattery offshore. Between these two negative gravity anomalies, a relative positive anomaly, with values higher than -50 mGal, lies on the northern west coast of the Olympic Peninsula. It corresponds to the Point of the Arches structural high, where Mesozoic crystalline rocks are exposed at the surface. By analogy, the gravity high between the Central Olympic and Hoh basins may represent a buried basement arch.

Straight gravity gradient zones separating the southern Hoh and Central Olympic lows trend NW and NE. In outcrop, faults between these basins have the same trends (Fig. 17). The southern boundary of the Central Olympic gravity low and the eastern boundary of the southern Hoh low are in part oriented NW and lie on trend with the Nisqually lineament at the southern end of Puget Sound (McCrumb et al., 1989b). This NW-trending gravity gradient zone is part of a larger set of anomalies suggested by regional isostatic maps to begin in northern Oregon (Blakely and Jachens, 1990). The Nisqually fault zone thus extends from the Washington Cascades into the Olympic Peninsula and even beyond, towards the submerged Pacific margin.

Many straight gradient zones with various orientations, bounding polygonal gravity highs, are probably associated with faults. Many of those gravity highs lose their sharply polygonal shape, and the gradient zones become considerably less distinct, when the gravity data are upward continued to just 5 km. Prominent and persistent gradient zones, however, mark the Nisqually fault system as well as the northern and southern strands of the OWSZ - the Leech River fault on Vancouver Island, the Calawah and Crescent faults on northern Olympic Peninsula.

To investigate large-scale structure of this area, the gravity data were upward continued to 20 and 100 km. In the 100-km map (Fig. 20), only the largest regional features are revealed. Gradual eastward decline of Bouguer anomaly values across the continental margin reflects the oceanward attenuation of the continental crust. The large negative anomaly in the northeastern corner of the map lies over the granite-rich, thick crust of the North Cascades. The subducted Juan de Fuca slab is not discernible from gravity data (see also Finn, 1990).

Regional structure of the crust is well expressed in the map of gravity data upward continued to 20 km (Fig. 21). The low over the North Cascades is flanked on the west by an east-side-down gradient zone related to the large Straight Creek-Fraser fault system. The Metchosin high is strong and centrally symmetrical, as it is in the low-level map in Fig. 19. This indicates that the causative high-density body extends directly beneath the exposed Metchosin massif, without significant lateral offset at depth. A frozen magma chamber may lie at mid-crustal levels beneath this massif, whose mapped thickness is only 7 km. From similar gravity highs in southern Washington, Finn (1990) has modeled bodies of Coast Range basalt to be up to 30 km thick.

113

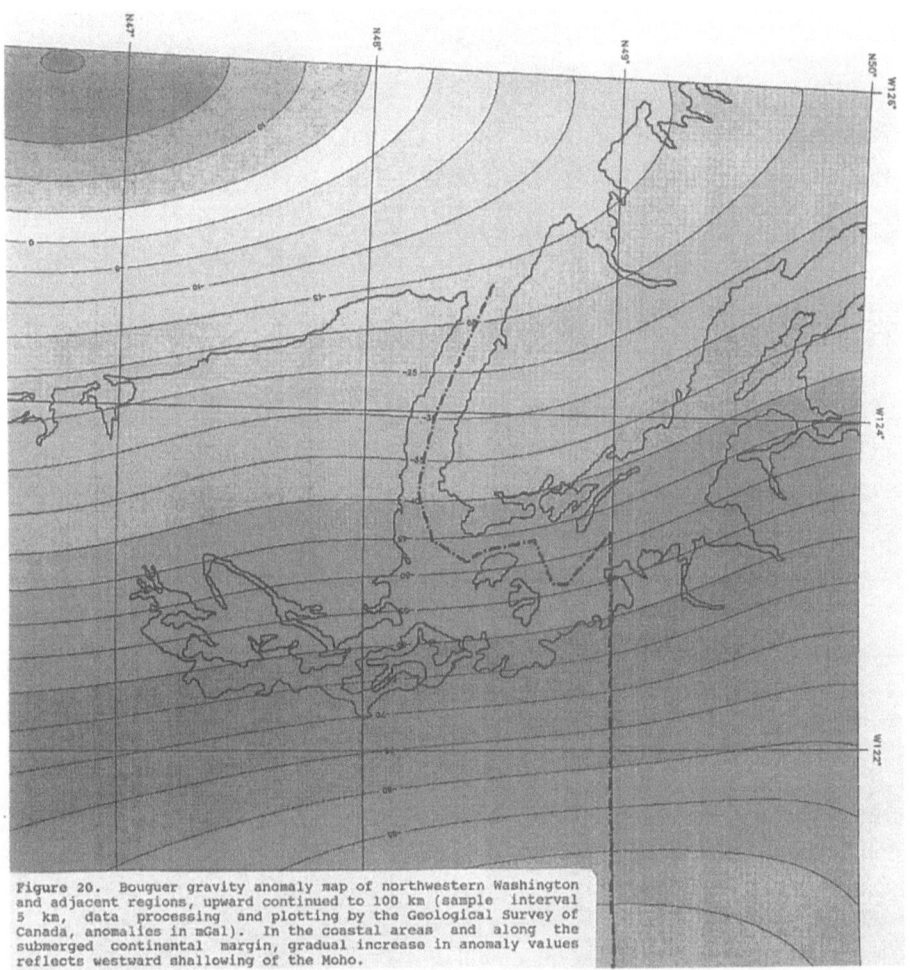

Figure 20. Bouguer gravity anomaly map of northwestern Washington
and adjacent regions, upward continued to 100 km (sample interval
5 km, data processing and plotting by the Geological Survey of
Canada, anomalies in mGal). In the coastal areas and along the
submerged continental margin, gradual increase in anomaly values
reflects westward shallowing of the Moho.

114

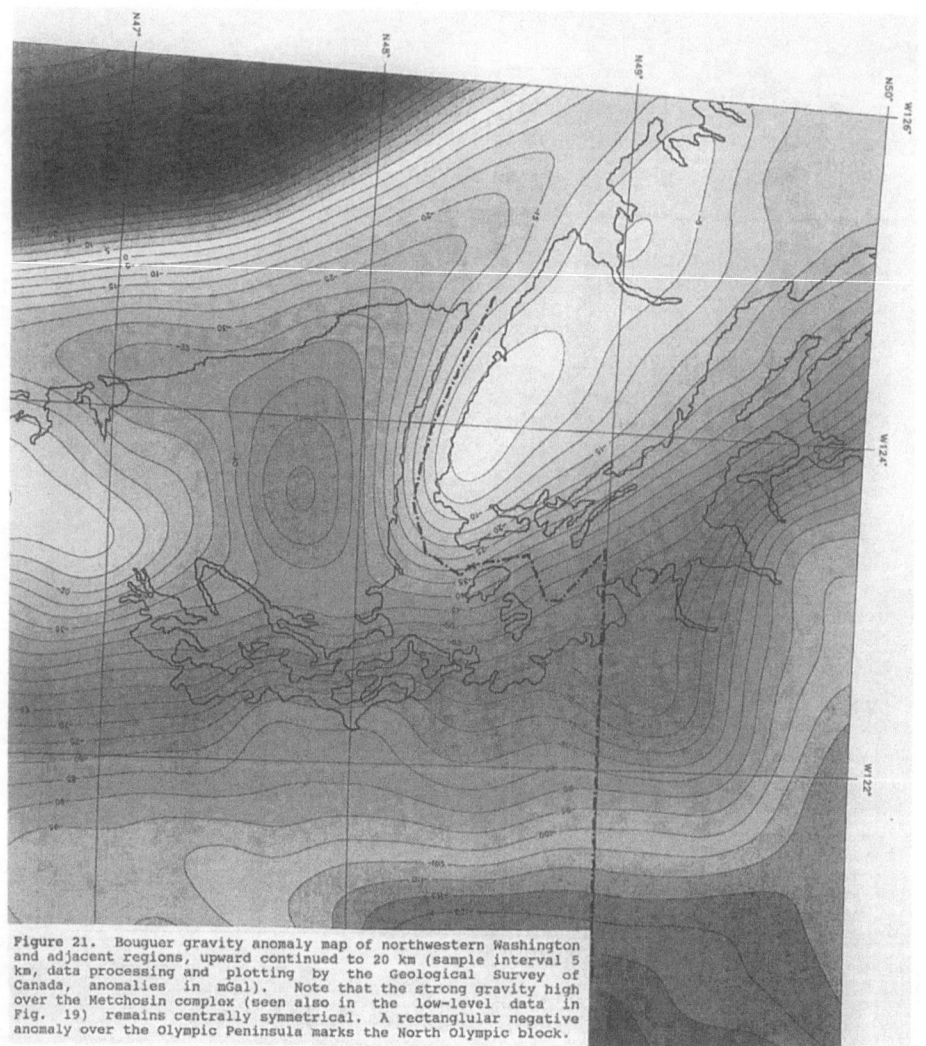

Figure 21. Bouguer gravity anomaly map of northwestern Washington and adjacent regions, upward continued to 20 km (sample interval 5 km, data processing and plotting by the Geological Survey of Canada, anomalies in mGal). Note that the strong gravity high over the Metchosin complex (seen also in the low-level data in Fig. 19) remains centrally symmetrical. A rectanglular negative anomaly over the Olympic Peninsula marks the North Olympic block.

Of particular interest in the 20-km map is the elongated low of -55 mGal over the Olympic Peninsula. This anomaly outlines a rectangular feature bounded by known major fault zones on the Olympic Peninsula (Fig. 17). The northern boundary of this block is the North Olympic fault system. Its eastern boundary is the N-S-trending fault which separates it from the Dosewallips massif. Its southern boundary is the Southern Fault Zone, which lies on trend with an E-W-oriented, north-side-down gravity gradient zone corresponding to the Seattle fault that crosses Puget Sound (Finn, 1990). This fault is shown by the 20-km gravity map to continue across the Olympic Peninsula as a boundary of the crustal block delineated by the rectangular negative anomaly.

This North Olympic crustal block, though large, is not apparent in the 100-km gravity map, which reflects structures on the scale of the entire crust. This block is therefore thought to be rooted above the base of the crust. It has been rigid enough to resist tectonic reworking in the OWSZ, though the distribution of earthquake hypocenters suggests in may be arching upward slightly (cp. Crosson and Owens, 1987). Deflection of the OWSZ around this block's northeastern corner is consistent with the change in orientation between parts of the North Olympic fault system: the Hurricane Ridge fault has a more westerly strike than the Crescent and Calawah faults and the general trend of the OWSZ. Interaction of the OWSZ and the North Olympic block may be causing stress-field anomalies at depth, which are reflected in a strong concentration of seismicity in Puget Sound whose origins have long been considered puzzling (e.g., McCrumb et al., 1989b).

On the nature of crystalline basement of the Olympic Peninsula

In the 1970s, Crescent Formation basalts, presumed to be Eocene oceanic crust, were regarded as the crystalline basement for thick Tertiary sedimentary successions in this region (e.g., Muller, 1977a). Though it remained popular in the 1980s (Duncan, 1982; Clowes et al., 1987), this interpretation has been disproved by recent findings.

The Crescent Formation does not have an oceanic-crustal origin and cannot be regarded as the basement. Many discontinuous Eocene basaltic massifs were produced by eruption from different volcanic centers. These basalts overlie and interfinger with sedimentary rocks of the Blue Mountain unit, and this package is in turn overlain conformably by sediments of the Adwell Formation of the Fuca Basin (Babcock et el., 1992, 1994).

Mesozoic crystalline rocks are exposed on the northern west coast of the Olympic Peninsula, at Point of the Arches. Though these old granitoids are only known from a small area several kilometers across, their presence requires an explanation.

An outcrop of crystalline rocks at Point of the Arches contains gneissic diorite and gabbro of Late Jurassic age, with reported K/Ar radiometric dates on the diorite of 144 Ma (MacLeod et al., 1977). Also exposed in that locality are greywacke, pillow basalt and minor chert believed to be Lower Cretaceous. This Mesozoic assemblage is covered, apparently with a depositional contact, by Tertiary sediments and basalts (Snavely, 1987; Niem and Snavely, 1991; Niem et al., 1991b).

In the context of the oceanic-crust model, these Mesozoic rocks were presumed to be an olistostromal block derived from Vancouver Island and incorporated into the sedimentary matrix by unspecified processes in the Middle Miocene (Niem et al., 1992b). Another interpretation (Snavely, 1987) regarded Point of the Arches as a separate terrane juxtaposed against other small terranes in a very complex manner: numerous, variously oriented and vergent thrust sheets separated by mutually truncating local faults were proposed in this small area (Fig. 15). Brandon and Vance (1992, their Figs. 3, 15) linked all these "terranes" on northwestern Olympic Peninsula, including the one containing the Mesozoic rocks, with the rock assemblage in the core of the Olympic Mountains.

A layer of crystalline crust under the Olympic Peninsula was included in several gravity models (R.W. Couch in Muller, 1977a; MacLeod et al., 1977; Riddihough, 1979). The density assigned to this crystalline crust was 2,900-2,920 kg/m3. The crust-mantle boundary was shown to lie variously at 20 or 27 km. The totality of such characteristics is best explained if the crust in this area is continental (see chapter 1). Though no seismic reflection data are available in the rugged Olympic Mountains, modern deep profiles on southern Vancouver Island help elucidate the stucture of the continental crust.

Deep structure of southern Vancouver Island from seismic data
Two reflection profiles cross the OWSZ on southern Vancouver Island (Fig. 14). Though short and lacking a tie, they provide a partial insight into the crustal structure in this area. Line

84-02 is 24.4 km long (Fig. 22). It begins in the Metchosin basalts, crosses the Leech River metamorphic complex between the Leech River and Survey Mountain faults, and ends in Wark gneisses of the Westcoast complex. Line 84-04 is 20.8 km long (Fig. 23). It begins in the Leech River complex, runs parallel to the Survey Mountain fault, crosses the San Juan fault near its junction with the Survey Mountain fault, and ends in Karmutsen basalts.

Clowes et al. (1987) interpreted these seismic lines in terms of subduction-related thrust structure. All faults in these profiles were postulated to be low-angle thrusts. The Leech River complex and the Metchosin massif were treated as exotic "terranes" underthrusted beneath Wrangellia in the Eocene. Broad bands of seismic events at long traveltimes were interpreted as sediments equivalent to those on the Olympic Peninsula, underthrusted beneath Vancouver Island.

This model (see also Dehler and Clowes, 1992) requires a fossil subduction megathrust at the base of the Metchosin massif and Wrangellia. Such a megathrust was hypothesized to be subhorizontal and undulating, at a depth of 6 to 13 km under southern Vancouver Island and 8-10 km under the Fuca Basin. Farther south, it was correlated with the Hurricane Ridge fault.

This interpretation is rather speculative, insufficiently constrained by the geological information discussed above. Both strands of the OWSZ - the South Vancouver Island and North Olympic fault systems - are deep-rooted, steep, long-lived crustal discontinuities that acted as conduits for Crescent Formation magmas in the Early and Middle Eocene. From the Middle Eocene

LINE 2

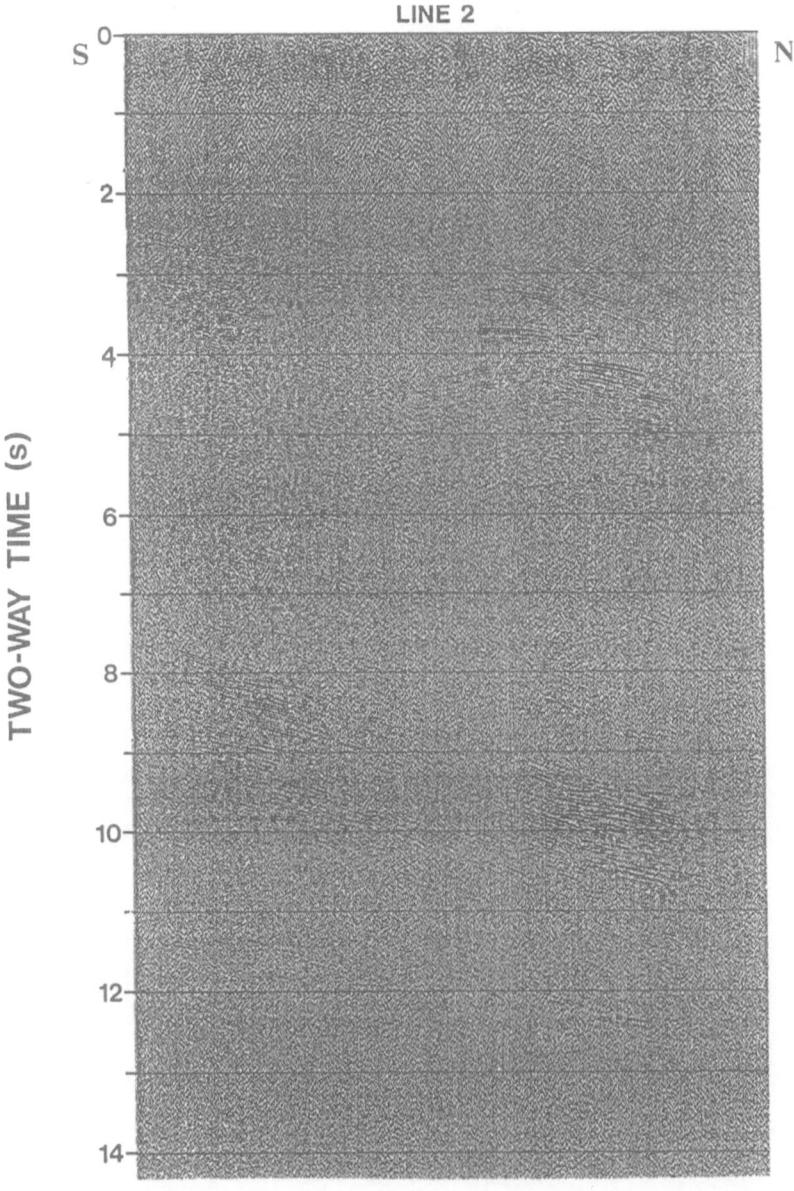

Figure 22. Deep structure of the northern strand of the OWSZ on southern Vancouver Island, imaged in the seismic reflection line 84-02 (modified from Clowes et al., 1987). Note the arcuate events around 4 s and the basal north-dipping events below 8 s. Profile location is given in Fig. 14; profile length is 24.4 km.

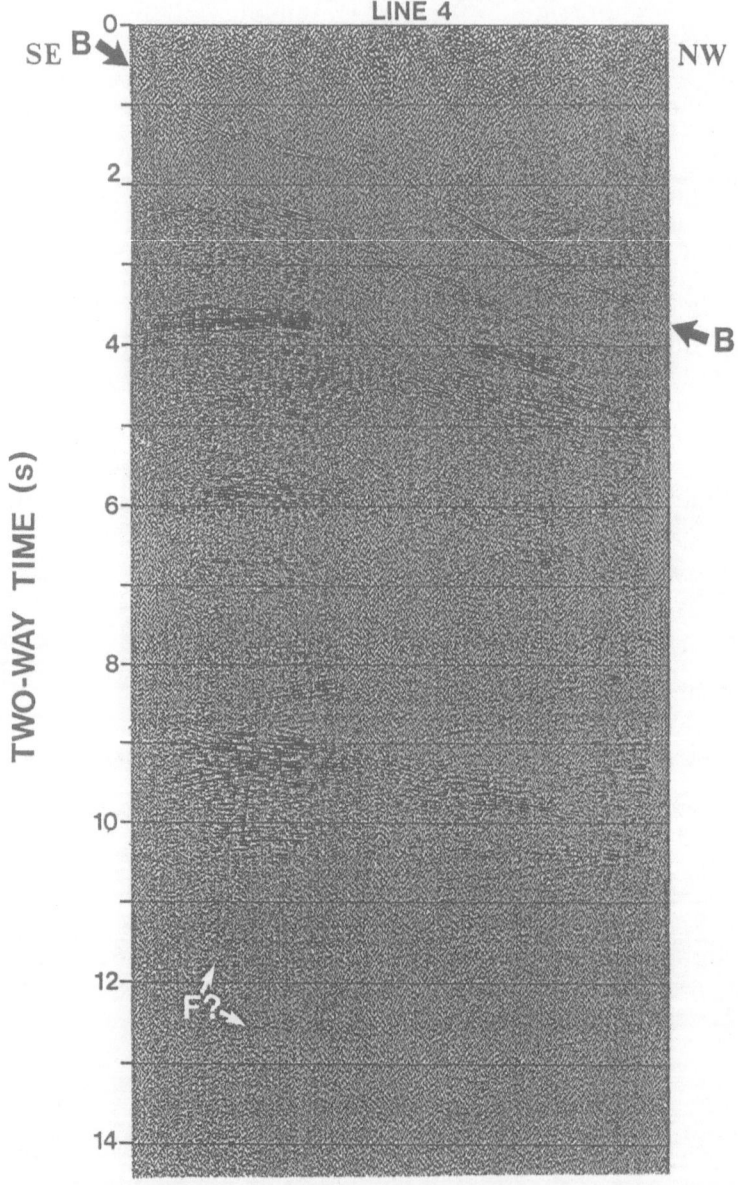

Figure 23. Deep structure of the northern strand of the OWSZ on southern Vancouve Island, imaged in the seismic reflection line 84-04 (modified from Clowes et al., 1987). The events labeled "B" and "F" are discussed separately in Chapter 9. Profile location is given in Fig. 14; profile length is 20.8 km.

until the Early Miocene, the Fuca graben subsided between these fault systems, creating the 8-km-thick Fuca Basin. Faults on the northern and southern strands of the OWSZ are steep. The South Vancouver Island fault system, in particular, has accommodated uplift from mid-crustal depths of the Leech River metamorphic complex, and perhaps also strike-slip movements.

Protoliths of the Leech River complex were buried, metamorphosed in the Eocene, and uplifted along faults of the South Vancouver Island system. During the uplift, the Leech River and Metchosin complexes were juxtaposed. They provided sediments for the late Middle Eocene Lyre Formation and younger units of the Fuca Basin (Babcock et al., 1994; Garver and Brandon, 1994).

Contrary to the thrust model, the San Juan and Leech River faults are subvertical (Muller, 1977c; Fairchild and Cowan, 1982; Rusmore and Cowan, 1985; Mayrand et al., 1987). For detecting high-angle structures, seismic reflection profiling is generally a poor tool. The idea that these faults flatten out at depth relies on a structural interpretation of low-angle events in lines 84-02 and 84-04. However, these events may have other origins. Volcano-sedimentary successions on southern Vancouver Island are diverse, variously metamorphosed and deformed. Reflections may arise from stratigraphic and structural contacts, metamorphic fronts, tops of plutons, sills. Steep faults and flanks of plutons may cause diffractions and sideswipe. Many events are observed in Line 84-04, whose SE part runs along the steep Survey Mountain fault.

Great structural and stratigraphic complexity in Paleozoic and Mesozoic rocks is revealed by geologic mapping (see Chapter 3).

The Paleozoic Sicker Group, made up of many units with contrasting lithologies (basalt, sandstone, siltstone, argillite, chert, limestone) is heavily faulted, folded and in places metamorphosed. These rocks are at least 5.5 km thick. Triassic Karmutsen basalts are up to 6 km thick, and together these two successions form a broad anticlinorium in the eastern half of Vancouver Island (Muller, 1977a,b, 1980a; Massey and Friday, 1989). On the west flank of this positive structure, north of the San Juan fault, lies a synclinorium where the Sicker Group and the Karmutsen Formation are covered by up to 3.5 km of younger sedimentary and volcanic rocks. Upper Triassic limestone and shale of the Quatsino and Parson Bay formations, with a total thickness of about 1 km, are exposed in places. More widespread is the Lower Jurassic Bonanza Group, which contains up to 2.5 km of volcanics, conglomerate, sandstone and shale.

Because deformation on Vancouver Island occurred in many pulses, the Bonanza Group is deformed less than the underlying older rocks, and in many areas its bedding remains subhorizontal. Whereas the Bonanza Group is unmetamorphosed, metamorphic grade in Karmutsen basalts is usually no higher than prehnite-pumpellyite, and in the Sicker Group as high as greenschist.

Further complexity is provided by abundant plutons (Muller, 1977a-c; Woodsworth et al., 1991) and sills (Fairchild and Cowan, 1982). Jurassic Island Intrusions are widespread across Vancouver Island. From geologic field mapping and analysis of magnetic anomalies, many of the exposed pluton are thought to be apophyses of large batholiths at depth (Jeletzky, 1976; Arkani-Hamed and Strangway, 1988). Another pulse of intrusive magmatism occurred in the

Tertiary, producing Catface and Sooke plutons as well as felsic sills in the Leech River complex. Tops of intrusive igneous bodies and sills may serve as low-angle seismic reflectors at different levels in the crust. Concave-upward, arcuate events at about 4 s, which may be correlated between the north end of Line 82-02 and the SE end of Line 84-04, might indicate a pluton-like domal feature at a depth of 10-15 km in the Leech River and Wark complexes, on both sides of the Survey Mountain fault.

The Moho under the southern tip of Vancouver Island lies at about 35 km. Subhorizontal, discontinuous events lie on both ends of Line 84-04 at 8.5 to 10.5 s (28 to 35 km; depth conversions according to Clowes et al., 1987). On the north end of Line 84-02, a deep band of high-amplitude events, dipping northward, lie at 9.5 to 10.7 s. Weak events seem to connect this band with north-dipping events at 7.5 to 10 s on the south end of that profile. Disappearance of reflections below about 10 s in both profiles probably marks the reflection Moho. Such a Moho depth is an indication that the crust is continental.

The continental nature of the crust is also suggested by felsic magmatism which occurred in the Tertiary on both southern Vancouver Island and northern Olympic Peninsula (see Chapter 4). On Vancouver Island, stocks and sills from that magmatic episode have been mapped in various country rocks: gabbro and quartz diorite of the Sooke suite intrude the Metchosin massif, felsic dikes occur in the Leech River complex, granitoid plutons of the Catface suite intrude pre-Tertiary rocks throughout the rest of Vancouver Island. These plutons intrude deformed rocks but bear no sign of compressional deformation themselves (cp. Massey and

Friday, 1989). Rather, they are cut by high-angle normal faults (Muller et al., 1981).

Timing of inversion of the Central Olympic Basin and uplift of the Olympic Mountains

Though soft, sheared clastic rocks make up the Olympic Mountains, the topographic relief is high and rugged, rising at Mt. Olympus to as much as 2427 m. To sustain such a high relief despite rapid erosion, the uplift must be rapid.

The work of Tabor and Cady (1978), supported by Brandon and Calderwood (1990), showed the Olympic Mountains to be a dome centered on Mt. Olympus. This rising dome and the mountain terrain it creates are fairly isometric in map view (Finn et al., 1991). In contrast, metamorphic isograds outline a smaller oval, 30 km long and 15 km wide, elongated E-W (Fig. 24), discordant with topographic contours. Rocks metamorphosed to prehnite-pumpellyite grade are found in this oval (Brandon and Vance, 1992). This metamorphic pattern in the Central Olympic Basin is superimposed on the structural pattern of NW-trending tectonic slices (Fig. 17).

The tectono-metamorphic and tectono-magmatic history of this basin can be restored from the available data with a fair degree of confidence. An early episode of widespread rock recrystallization took place at 30-29 Ma (Tabor, 1972; Brandon and Vance, 1992). This rock-alteration event seems to have accompanied a pulse of deformation which affected the entire northern Olympic Peninsula. General restructuring led to the creation of distinct depocenters where the younger sedimentary succession of the Central Olympic Basin was laid down.

125

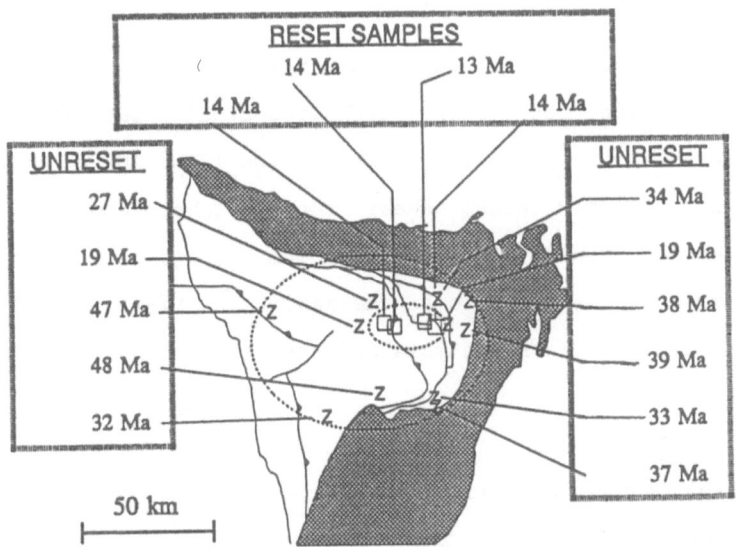

Figure 24. Metamorphic aureole in the Olympic Mountains, based on fission-track data from detrital zircons (modified from Brandon and Vance, 1992). The numbers indicate fission-track ages of zircons. Rocks within the small oval aureole are metamorphosed to prehnite-pumpellyite grade, and zircon ages within it are reset due to metamorphism at 13-14 Ma.

Sedimentation ended with a new pulse of heating and deformation at about 19 Ma, when isoclinal folding and west-vergent thrusting, which affected all the rocks, formed the present-day pattern of tectonic slices (Fig. 17). Quartz veins, age 17 Ma, penetrated rocks of all ages, mostly along shear zones (Tabor and Cady, 1978). Prehnite-pumpellyite metamorphism (Fig. 24) was superimposed on the new structure of the basin at 14-13 Ma (as determined by fission-track analysis of detrital zircons; Brandon and Vance, 1992).

Inversion of the Central Olympic Basin and exhumation of buried rocks began immediately after the mid-Miocene tectonic episode. Uplift in the core of the present-day Olympic Mountains is indicated by the 12 Ma fission-track cooling age of these rocks (Brandon and Calderwood, 1990; Brandon and Vance, 1992). Erosion in that area provided detritus for the Grays Harbor basin to the south (the Late Miocene Montesano Formation; Brandon and Calderwood, 1990; Babcock et al., 1994) and in the Puget Basin to the east (Niem et al., 1992c). This rapid, domal uplift, centered on Mt. Olympus, is continuing to the present day. Its deep roots are suggested by the concave-upward zone of earthquake hypocenters at a depth of some 40 km under northern Olympic Peninsula (cp. Crosson and Owens, 1987).

Possible causes of Olympic Mountains uplift

Buoyant rise of extremely thick young sediments has been proposed as a possible cause of the Olympic Mountains uplift (Tabor and Cady, 1978). Another hypothesis ascribed the uplift to upward

arching of the subducting Juan de Fuca plate. An eastward-
plunging arch in the slab under the peninsula is supposedly caused
by curvature of the continental margin in this area (Crosson and
Owens, 1987). Brandon and Calderwood (1990) speculated that the
margin curvature and the arch in the slab had resulted from
outward push of the western U.S. continental margin due to Basin
and Range extension in the Cordilleran interior in the Miocene.

Field mapping in the western Cordillera, however, does not support
such a model. No structures accommodating Neogene oroclinal
bending have yet been identified (cp. Monger, 1991). The domal
uplift of the Olympic Mountains is hard to explain by the rise of
an elongated arch underneath. An alternative cause of the domal
uplift may be the rise of a buoyant granitic massif under the
Olympic Mountains. Such an explanation is consistent with all the
available geological and geophysical data.

A broad, negative magnetic anomaly with amplitudes of -300 to -600
nT lies over the Olympic Mountains (Finn, 1990). Because Tertiary
sediments are not magnetic, this anomaly is probably caused by a
deeper reversely magnetized source. Elsewhere in the coastal
provinces of western U.S. and Canada, negative anomalies are
occasionally associated with reversely magnetized igneous rocks
which cooled in a reverse-polarity ambient geomagnetic field
(Arkani-Hamed and Strangway, 1988; Finn, 1990).

Stark, negative gravity anomalies also mark the Olympic Mountains.
Appearance of these anomalies in both Bouguer and isostatic maps
(Blakely and Jachens, 1990; Finn et al., 1991) suggests that low-
density material is abundant in this area and/or that the North

Olympic crustal block is below its isostatic equilibrium. In the latter case, a negative imbalance gives this block a natural tendency to rise.

An isostatic imbalance might have been created by emplacement of a large amount of light material, reducing the density of the continental crust above the level of isostatic compensation. Two mechanisms may be envisaged for such emplacement. One is the formation of a subduction-related sedimentary accretionary prism piled up against a rigid continental buttress (e.g., Cloos, 1993). The other is emplacement of a large body of granitoids.

The first mechanism is appealing since the Olympic Peninsula area lies in a continental-margin setting characterized by subduction of the Juan de Fuca plate (e.g., Couch and Riddihough, 1989). However, the continental-rift origin of Crescent basalts and the presence of felsic igneous rocks of various Tertiary ages preclude the possibility that oceanic crystalline crust underlies this area. Besides, the 30-km thickness of accreted sediments required by this scenario (Brandon and Calderwood, 1990) is unusual at other subduction zones.

More conservative estimates of sediment thickness in this area were made in gravity models. MacLeod et al. (1977) modeled a 15-km-thick sediment layer with an average density of 2,700 kg/m3, over a 12-km-thick layer of crystalline rocks with a density of 2,900 kg/m3. This implies an average density for the whole crustal section under the Olympic Mountains to be only about 2,790 kg/m3. By comparison, adjacent areas where the sedimentary cover is known to be much thinner and lying over continental crust, have in that model an average density of about 2,850 kg/m3. A gravity model of Riddihough (1979) showed a sediment layer with a density of 2,600 kg/m3, up to 10 km thick, over crystalline material with a density of 2,920 kg/m3, at least 10 km thick. A subducting oceanic slab was modeled below the base of the continental crust, and a wedge of continental upper mantle was modeled between the oceanic slab and the continental Moho. Finn (1990), however, cautioned that a subducting slab cannot be resolved from gravity data in western Washington.

Largely following Brandon and Calderwood (1990), a recent gravity model for the Olympic Mountains (Dehler and Clowes, 1992, their Fig. 14) showed up to 20 km of sediments on top of a slab of oceanic crust. However, rock parameters assumed in that model are unrealistic. Sedimentary rocks were assigned a high density of 2,700 kg/m3. A Crescent "terrane" was assumed to be a panel about 5 km thick, with an extremely high density of 3,200 kg/m3.

By comparison, measured surface densities of Crescent basalts are 2,200 to 2,950 kg/m3 in Washington (Finn, 1990) and 2,950±60 kg/m3 on Vancouver Island (Currie and Muller, 1976). Rather being a uniform panel, the Crescent Formation is recognized to consist of discontinuous basaltic massifs (Babcock et al., 1992, 1994). Measured densities of the Central Olympic and Hoh basin sediments are 2,400 to 2,600 kg/m3 (see chapter 2).

Intrusion of granitic magma is a simpler way to put a large volume of low-density material into the continental crust. Geology of this region offers many examples of such intrusion. In the North Cascade and Coast mountains, Mesozoic and Cenozoic granitoid rocks occur in abundance (Hutchison, 1982; Brown et al., 1994). Their measured densities are mostly around 2,700 kg/m3.

Felsic intrusive rocks of early Tertiary age are widespread on Vancouver Island. On the west flank of the North Cascades, near Puget Sound, lies the felsic Snoqualmie batholith. It is elongated N-S and intrudes the OWSZ. Its radiometric ages are between 19.7 and 17 Ma (late Early Miocene; Frizzell et al., 1984; Reidel et al., 1994). Similar plutons have been inferred from potential-field data beneath sedimentary basins on the west flank of the Coast Belt orogen, e.g., the Puget (Finn, 1990) and Queen Charlotte (Lyatsky, 1991a) basins.

The conspicuous hydrothermal veins of quartz in the Olympic Mountains may have developed over the apex of another such pluton. The age of these veins is around 17 Ma. A thermal episode related to intrusive magmatism may have also been responsible for metamorphism of basin sediments at 13-14 Ma.

Buyoant rise of granitoid massifs is known from many mountainous regions in the Cordillera (cp. Parrish, 1983; Monger, 1991). In the Olympic Mountains, such rise has produced a topographic dome. Coincidence of the center of this dome with the center of the metamorphic aureole (at Mt. Olympus) suggests a genetic connection. If the oval metamorphic aureole reflects the shape of the apophysis closest to the surface, the more isometric mountain domain may represent the overall shape of the pluton.

The continental crust in this area, as suggested by teleseismic and refraction data, is attenuated somewhat but still about 30 km thick (Taber and Lewis, 1986; Owens et al., 1988; Lapp et al., 1990). This provides a sufficient crustal thickness to produce granitic magmas. Such magmas, commonly formed by partial melting at depths of 20-30 km, may rise to as little as 7 km before solidifying (Hollister, 1993; Petford et al., 1993; Grocott et al., 1994). Because felsic rocks are light, their buoyancy causes uplift involving both the plutons and the country rocks of the upper crust.

Hot granitic magmas cause metamorphism of country rocks, including roof rocks over large intrusive bodies. The prehnite-pumpellyite grade of metamorphism is usually reached at pressures of up to 2.5 kbar (Digel and Ghent, 1994) and temperatures no less than 240°-245° (Brandon and Vance, 1992). Assuming a very low Tertiary geothermal gradient of 19.4±1.7°/km for the Olympic Peninsula, Brandon and Vance (1992) concluded that the peak of metamorphism of the exposed rocks in the core of the Olympic Mountains was reached at a depth of about 12 km. However, sufficiently high temperatures might have existed at shallower depth if a magmatic massif supplied local heat. Besides, pressure-depth relationships might have been complicated if the adjacent subduction zone provided additional compressive stress.

In these circumstances, rocks of the Central Olympic Basin may have reached the prehnite-pumpellyite metamorphic conditions at a depth of 7-10 km. Such thickness is not unreasonable for this basin in Middle Miocene time, when both the Middle Eocene to Early Oligocene and Late Oligocene to Early Miocene successions had already been deposited. After the emplacement of a buoyant granitic pluton and prehnite-pumpellyite metamorphism in the late

Early and Middle Miocene, inversion of the Central Olympic Basin and uplift of the Olympic Mountains provided sediments for the surrounding basins in the Late Miocene. If 7-10 km of material has been eroded from the Olympic Mountains to unroof the metamorphosed rocks, the crystalline basement containing the felsic pluton must now be close to the surface.

Such a scenario explains all the diverse geological peculiarities of the Olympic Mountains. It is also consistent with Bouguer and isostatic gravity anomaly data, as it explains the pronounced gravity low over the Olympic Peninsula without an unrealistic sediment thickness. The strong negative magnetic anomaly over the peninsula may also be accouted for by a pluton with a negative remanent magnetization, under a thin cover of non-magnetic sediments. This would also explain the lack of correlation of metamorphic isograds with the fault-bounded slices or the shape of the mountain terrain. No arbitrary external tectonic forces are needed to drive the uplift of the Olympic Mountains.

CHAPTER 6 – CONTINENTAL MARGIN OFF SOUTHEASTERN ALASKA, THE QUEEN CHARLOTTE ISLANDS, AND NORTHERN VANCOUVER ISLAND

Scope of ideas regarding tectonic nature of the North America-Pacific plate boundary

Since the inception of the plate tectonics theory, the boundary between the continental North America plate and the oceanic Pacific plate has been placed along the northwestern North American continental margin. This boundary has been classified as predominantly transform, dominated by strike-slip motion of the Pacific plate past the western edge of the continent (Atwater, 1970, 1989). By definition, such a margin must be associated with a system of faults of lithospheric scale, across which the plates are juxtaposed. Such faults have been detected by marine seismic surveys along the western North America margin all in southeastern Alaska and farther SSE (Figs. 25, 26). These faults control the geologic structure of the continental margin (von Huene, 1989).

Off southeastern Alaska, the plate-boundary fault zone, named by von Huene et al. (1979) Chichagof-Baranof, has been traced along the lower continental slope from Cross Sound to the Alaska-British Columbia border in Dixon Entrance (Bruns and Carlson, 1987). This fault system has been noted to continue northward, as the Fairweather fault, into continental-crust regions inside Alaska, where it separates large continental terranes. A single Chichagof-Baranof-Fairweather fault system can thus be inferred.

South of Dixon Entrance, off the Queen Charlotte Islands, this plate-boundary fault system has been named Queen Charlotte fault zone (Sutherland Brown, 1968; Chase and Tiffin, 1972). With

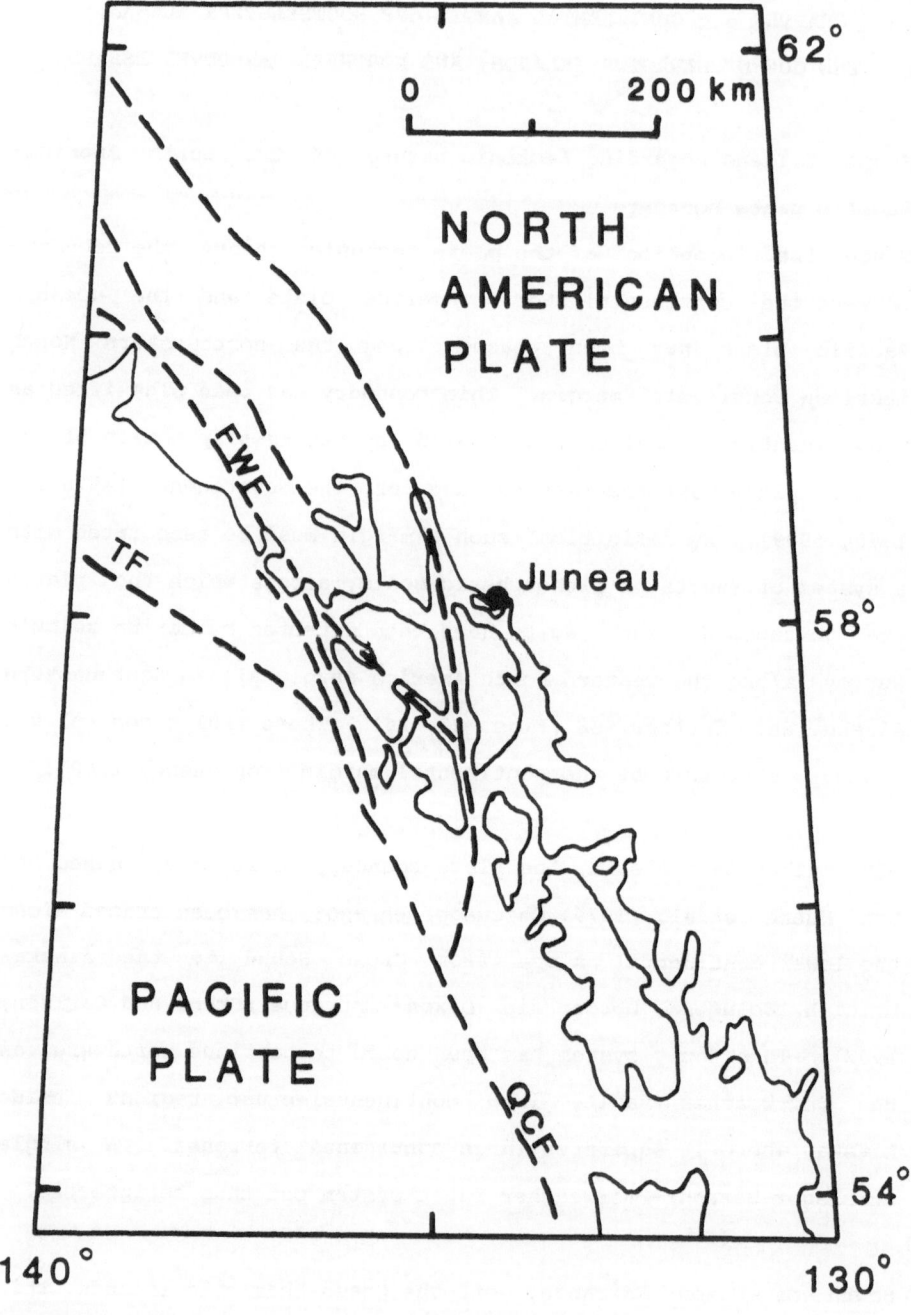

Figure 25. Position of biggest faults in southern and southeastern Alaska onshore and offshore (modified from Bruns and Carlson, 1987). QCF - Queen Charlotte fault; FWF - Fairweather fault; TF - Transition fault.

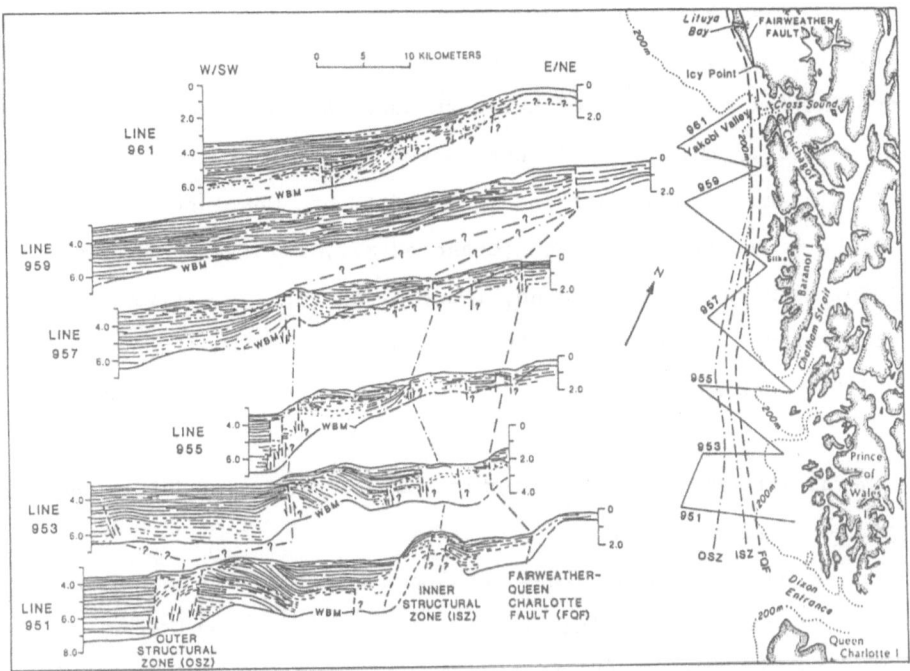

Figure 26. Strands of the plate-boundary fault system off southeastern Alaska. Structure of this system is illustrated in the line drawings of marine seismic reflection profiles identified in the map (modified from Bruns and Carlson, 1987).

modern data, a single Fairweather-Queen Charlotte fault system (Figs. 25, 26) is known to control the position of the submerged continental margin of northwestern North America (von Huene, 1989; Brew et al., 1991). Still unresolved is the southern continuation of this fault system off the mouth of Queen Charlotte Sound and northern Vancouver Island (Chase et al., 1975; Keen and Hyndman, 1979; Riddihough and Hyndman, 1989; Allan et al., 1993). Many alternative ideas about the tectonic evolution of the Canadian Pacific continental margin involve conflicting interpretations.

General structural characteristics of the plate boundary along the southeastern Alaska margin
The Fairweather fault has long been known as a boundary between distinct geologic domains (King, 1969), now considered to be accreted terranes, which compose Alaska. The mostly Mesozoic Chugach terrane is separated by this fault from the Tertiary Yakutat terrane to the west (Plafker, 1987; Brew et al., 1991).

From the Alaskan interior, the Fairweather fault runs SSE into the northeastern Pacific. In the Cross Sound area, it dives beneath the ocean waters onto the continental shelf and slope. There it meets the Chichagof-Baranof fault. This composite fault system, in current models, accommodated the northward shift of the Yakutat terrane, which in the Miocene lay off the Queen Charlotte Islands and later moved to its present location off southern Alaska (Plafker, 1987; Bruns and Carlson, 1987).

Seismic profiles show several major faults on the continental slope off southeastern Alaska (Fig. 26). Two main parallel faults are recognized off Cross Sound, three south of Sitka Sound. In

many places they are covered by thick sediments, but in a number
of seismic sections these faults have broken through to the ocean
floor. Though seismic images of the stratified packages suffer
from limited signal penetration, the resolved sediment thickness
reaches several kilometers. These sediments are undrilled, but
their age is often assumed to be no older then Miocene. The
nature of the underlying crystalline basement is obscure (von
Huene et al., 1979; Bruns and Carlson, 1987; von Huene, 1989).

North of Sitka Sound, sediments along the margin are largely
undisturbed. The intensity of deformation increases to the south,
where the Chichagof-Baranof fault system also becomes wider.
Though massive sedimentation shaped the southeastern Alaska
continental shelf and slope, many of the faults are expressed in a
series of sea-floor scarps and ridges south of Sitka Sound.

The outer and inner fault strands of the Chichagof-Baranof fault
system are only about 8 km apart at Cross Sound, and remain close
as far south as Sitka Sound. There this fault system widens to
20-30 km, and off Dixon Entrance it becomes up to 40 km wide.
Subsidence of coherent crustal slivers bounded by fault strands in
this broad structural zone caused the development of sedimentary
basins, and the basin off Dixon Entrance may be many kilometers
deep (von Huene et al., 1979; Bruns and Carlson, 1987).

The orientation of the continental margin changes slightly south
of Dixon Entrance, from NNW-SSE to NW-SE. Off the Queen Charlotte
Islands, a small amount of oblique convergence between the Pacific
and North America plates has been suggested from estimates of
plate motion (e.g., DeMets et al., 1990), but seismic refraction

surveys have failed to identify a subducted oceanic slab (Mackie et al., 1989; Spence and Asudeh, 1993).

Concerns about fidelity of geophysical models along the western Canada continental margin

In the conventional geophysical models relying primarily on large-scale plate-tectonic reconstructions in the northeastern Pacific, oceanic crust is presumed to underlie the continental slope off the Queen Charlotte Islands as well as the slope and exterior shelf off Vancouver Island (Riddihough and Hyndman, 1989). However, from local geological and geophysical data, the transition between continental and oceanic crust along the Canadian segment of the western North America continental margin may be much more complex.

The striped pattern of linear magnetic anomalies, associated with oceanic crust worldwide, is diagnostic of this type of crust. At continental margins, it is a prime indicator of how far the oceanic crust extends towards the continent. Deep structure of continental margins is commonly interpreted based on the extent of magnetic stripes (e.g., Johnson et al., 1990).

Off southeastern Alaska, magnetic stripes end at the outer strands of the plate-boundary fault system and do not reach the continental slope. At the Washington margin, they dissipate towards the continent gradually, and their faded extensions have been noted even on the upper continental slope and shelf (Finn, 1990, 1991). Reasons for such gradual dissipation of stripes may be two-fold: subduction of the oceanic plate increases the depth to the anomaly source, and thermal alteration of magnetite-bearing rocks at depth destroys the source altogether.

At the British Columbia continental margin, stripes are terminated very abruptly west of the continental slope off northern Vancouver Island (Fig. 27). To the north, they disappear gradually some 40 km off the Queen Charlotte Islands, without reaching the continental slope (Currie et al., 1983a,b).

The absence of oceanic-crustal magnetic signatures in the blank magnetic zone along the western Canada continental margin, in areas supposedly underlain by oceanic crust, has led to complex and disparate explanations. Usually it is assumed that magnetic stripes existed there originally but were destroyed by secondary processes. An early explanation invoked intense structural reworking of the oceanic crust by faulting (Chase and Tiffin, 1972; Srivastava, 1973). Another idea held that original magnetization of oceanic-crust rocks was destroyed by unusually strong hydrothermal alteration under a sediment blanket on the ocean floor (Levi and Riddihough, 1986).

These explanations are not borne out by facts. Off northern Vancouver Island, where the blank magnetic zone is particularly well defined, no severe faulting is revealed by seismic data. Alteration of oceanic-crust basalts by ocean water and hydrothermal fluids is a common phenomenon worldwide, but not to the extent where magnetic stripes are completely destroyed. On the ocean floor off western North America, high heat flow has been recorded in many places, usually where warm fluids are vented near faults (Davis and Riddihough, 1982; Moran and Lister, 1987). Away from these local thermal anomalies, however, the measured heat

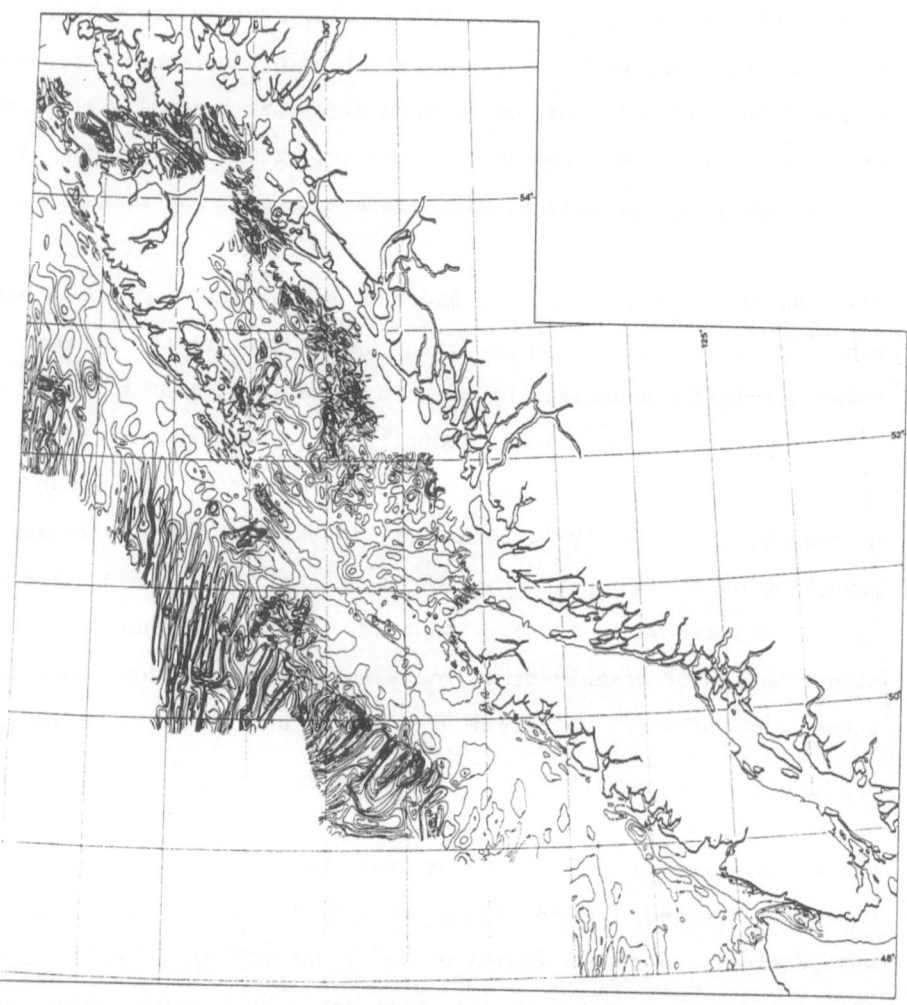

Figure 27. Magnetic anomalies offshore British Columbia (after Currie et al., 1983a,b). Magnetic stripes are absent in a zone some 60 km wide along the submerged margin. Farther west, the stripes are clearly expressed, but broken due to internal deformation of the northern Juan de Fuca plate (compare with the regional magnetic anomaly map in Fig. 3b). The stripes stop abruptly at the Revere-Dellwood fault.

flow is regionally lower than predicted by theory for young oceanic plates such as Juan de Fuca (Riddihough et al., 1983). If sediments are indeed acting as a partial thermal blanket, their effect is insufficient to destroy magnetic stripes elsewhere in the Juan de Fuca plate.

Off western Canada, other evidence also shows that sediment blanketing did not create the blank magnetic zone. The outboard boundary of this zone off northern Vancouver Island is very abrupt, whereas sediments thicken towards the continent gradually. The abrupt termination of magnetic stripes in this area coincides with a major NW-trending regional fault (Revere-Dellwood) which is well expressed in bathymetry and in seismic profiles.

The Revere-Dellwood fault separates blocks of crust with different composition and of different nature. The blank magnetic zone off British Columbia is interpreted to reflect presence along the continental margin of foundered blocks of continental crust, juxtaposed by faults against the oceanic crust of the Pacific and Juan de Fuca plates. The transition from continental to oceanic crust therefore lies at the outer strands of the broad plate-boundary fault system.

Models of western Canada continental margin based on gravity data
An early interpretation of gravity data in the Insular Belt (Stacey and Stephens, 1969) involved correlating some of the prominent anomalies with known geologic features. Later, analysing the overall Bouguer gravity signature of the Insular Belt in comparison with that of other parts of southern Canadian Cordillera, Stacey (1973) concluded that Vancouver Island is

characterized by abnormally dense crust and/or light upper mantle. Riddihough (1979) considered two explanations for such a density distribution: abnormally mafic crust on Vancouver Island or presence of a subducted oceanic slab underneath.

The first explanation is appealing due to subsequent findings from seismic refraction surveys that the crust in the Insular Belt contains mid-crustal layers of P-wave velocity up to and exceeding 7 km/s, whose thickness is up to 10 km (Spence et al., 1985; Yuan et al., 1992). Because these layers were not inferred till the mid-1980s and their nature remains unclear to this day, the conventional interpretation has come to hold that a subducted slab of oceanic lithosphere is present beneath Vancouver Island.

Riddihough (1979) used this idea as a basis for his gravity models of deep structure of the Insular Belt and adjacent oceanic regions. He postulated broad geological similarities between western British Columbia, Washington and Oregon and assumed that subduction of the Juan de Fuca plate is occurring along the continental margin as far north as Vancouver Island. The Cascadia subduction zone was later extended to the mouth of Queen Charlotte Sound, and these models influenced subsequent interpretations of seismic prifiles across the continental margin (Yorath et al., 1985a,b; Spence et al., 1985; Clowes et al., 1987; Hyndman et al., 1990; Yuan et al., 1992).

Non-uniqueness, which is normal in geophysical interpretations (Pakiser and Mooney, eds., 1989), makes these models ambiguous, and new geological data cast doubt on their validity. It was recently shown by modeling that gravity data by themselves cannot resolve a subducted oceanic slab under northwestern North America, and constraints from other sources are required (Finn, 1990). Geological control is of particular importance in testing geophysical models, and below the geological and geophysical information is discussed together.

Bathymetry of the British Columbia continental margin

The Queen Charlotte Islands, which extend from Dixon Entrance to Queen Charlotte Sound, rise no higher than 1000 m above sea level. Along their western shore, they are truncated abruptly by steep faults of the Queen Charlotte fault zone, which cut all Mesozoic and Cenozoic rocks on the islands (Sutherland Brown, 1968; Hickson, 1991). Leaving almost no room for a shelf, these faults form a huge bathymetric scarp next to the shoreline, where the ocean floor deepens abruptly to 1000-2000 m (Fig. 28).

At the bottom of the scarp lies the Queen Charlotte Terrace, a step-like feature on the lower continental slope. Just 20-30 km wide, it runs the length of the islands and is inclined slightly towards the ocean. A smaller scarp flanks it on the west, where water depth increases abruptly from some 2000 m on the terrace edge to about 2800 m in the adjacent Queen Charlotte Trough (Chase et al., 1975; Seemann, 1982; Currie et al., 1983b). This trough, 20-30 km wide, is bounded on its outboard side by another, still smaller scarp across which the ocean floor rises about 200 m.

Farther west, in the pelagic area, lies a broad, gentle, positive ocean-floor feature known as Oshawa Rise. Its slightly elevated oceanic crustal basement is crowned by Tertiary seamounts of the Kodiak-Bowie chain (Chase, 1977). The broad ocean-floor uplift

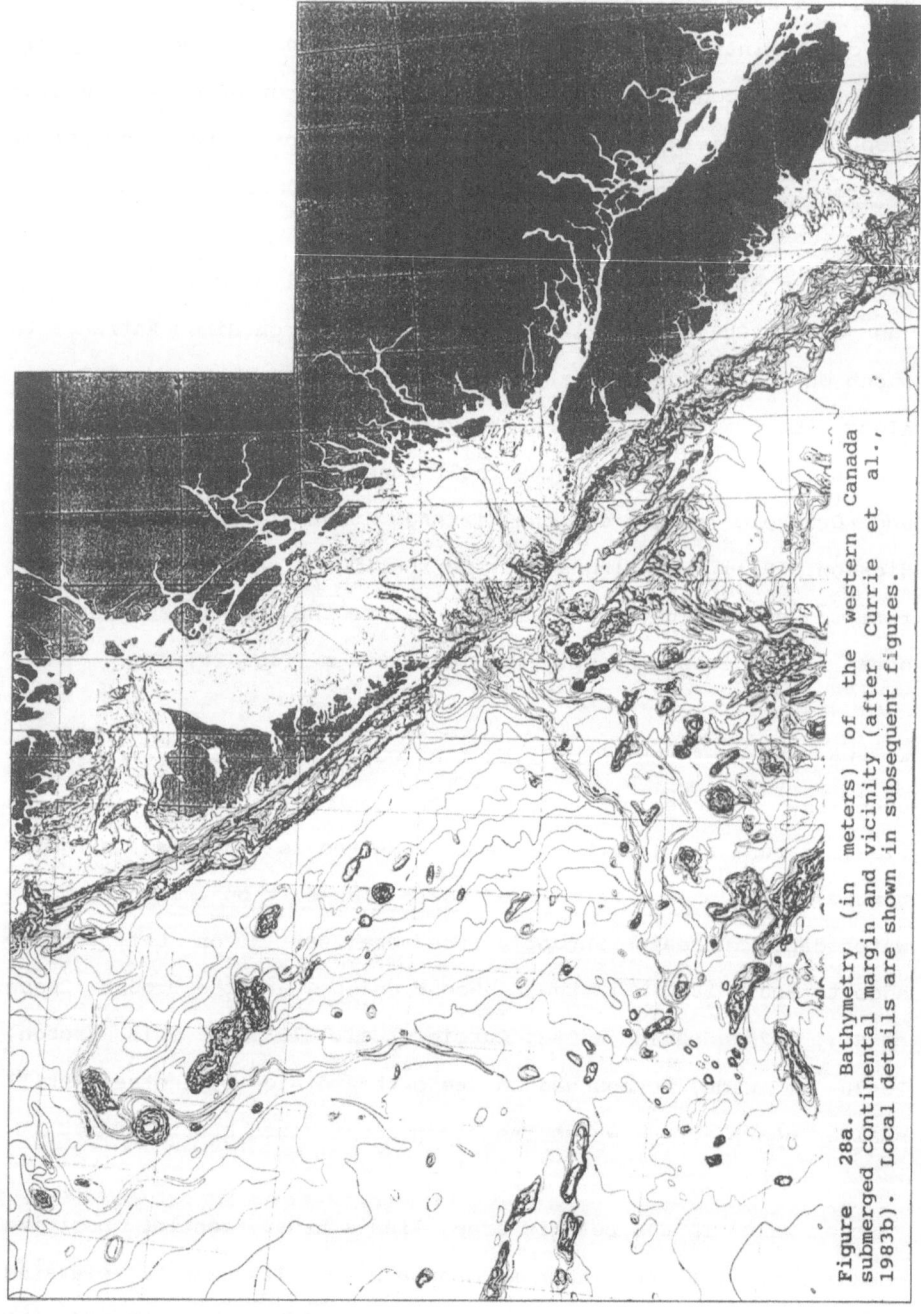

Figure 28a. Bathymetry (in meters) of the western Canada submerged continental margin and vicinity (after Currie et al., 1983b). Local details are shown in subsequent figures.

145

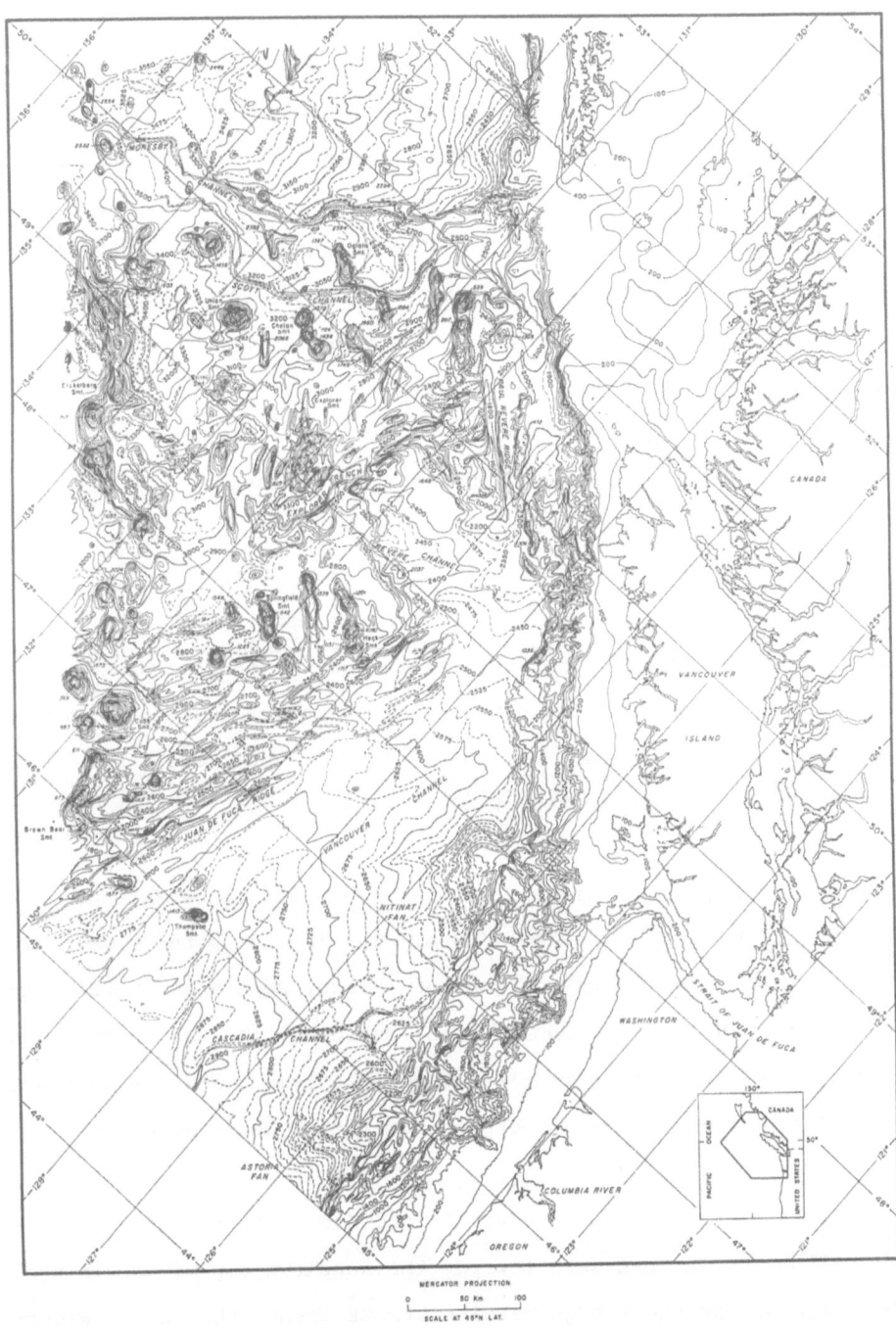

Figure 28b. Bathymetry (in meters) of the Vancouver Island and Washington submerged continental margin and vicinity (from J. Mammerickx and I.L. Taylor, 1971, Scripps Institute of Oceanography, Geological Data Center, Special Chart #1).

and the localized seamounts on it are products of tectono-magmatic activity in the oceanic crust, unrelated to the continent (Cousens et al., 1985). The basement rise and the seamount chain run NW across the Gulf of Alaska, obliquely to the continental margin.

No extension of the Queen Charlotte Terrace is observed north of Dixon Entrance, where the continental slope is very gentle due to blanketing by thick sediments of the Baranof fan (von Huene et al., 1979). No such fan exists off the Queen Charlotte Islands, because there the subsiding interior shelf provided an inboard catchment for detritus shed from the Coast Mountains.

Off the mouth of Queen Charlotte Sound, the terrace ends. The continental slope here is dissected by many turbidite channels which probably follow local, auxiliary faults that transect the slope. On the abyssal plain, small, isolated mounds, known as Dellwood Knolls and Tuzo Wilson Seamounts, protrude above the sedimentary cover (Carbotte et al., 1989).

Vancouver Island is higher than the Queen Charlotte Islands, its elevations locally exceeding 1500 m. The island is the widest in the middle, where only a narrow seaway separates it from the mainland. Vancouver Island becomes narrower in the north and south, and the interior shelf between the island and the mainland (the Queen Charlotte Strait and the Strait of Georgia) widens.

The exterior shelf off Vancouver Island widens to the south. The coastline curves eastward towards the southern tip of the island, but because the shelf edge retains its SE trend, the shelf widens in the south to as much as 80 km. Off northern Vancouver Island, however, the exterior shelf is only about 20 km wide. Like off the Queen Charlotte Islands, the upper slope is straight and steep, apparently controlled by a major fault (Tiffin et al., 1972). On the other hand, the lower slope is broad and gentle.

Width of the continental slope varies irregularly. The large Brooks-Estevan embayment (new name) disrupts it off central Vancouver Island. The slope is only 35-40 km wide in the embayment, but it is 60-70 km wide to the north and 50-60 km wide to the south. The upper slope in the south is gentler than in the north and contains many small terraces and sediment ponds. Several large underwater canyons dissect it and continue to the base of the lower slope.

The Juan de Fuca canyon, originating at the mouth of the Strait of Juan de Fuca, separates the Vancouver Island and Olympic Peninsula margins. Both are broad, up to 120-140 km, but the width of shelf and slope in them varies. Off southern Vancouver Island, the shelf is 80 km wide and the slope about 60 km wide, whereas the Olympic Peninsula margin has a 90-km-wide slope and a 30-40-km-wide shelf.

Deep structure of the continent-ocean plate boundary off Queen Charlotte Islands

The Chichagof-Baranof fault system off southeastern Alaska continues, as the Queen Charlotte fault zone, into the continental margin off British Columbia (Bruns and Carlson, 1987; von Huene, 1989), where deep structure of the plate-boundary fault system is revealed by modern seismic surveys.

The deep levels of the Queen Charlotte Trough have been modeled
from seismic refraction and gravity data as containing 6-km-thick
crystalline oceanic crust tilted towards the continent (Mackie et
al., 1989). Seismic reflection profiles across the trough (e.g.,
Fig. 29) reveal above the crystalline basement a 1.5-km-thick
sedimentary wedge with a fan-shaped pattern of reflections whose
landward dips increase with depth (Chase and Tiffin, 1972; Snavely
et al., 1980; Davis and Seemann, 1981).

Gradual eastward dissipation of oceanic magnetic stripes over the
trough (Fig. 27) may be a result of increasing depth to anomaly
source below the sedimentary wedge blanketing the oceanic crust.
The outboard fault bounding the trough disturbs the magnetic
stripes only slightly. Free-air gravity values decrease gradually
across the trough, from near zero in outboard areas to near -50
mGal at the foot of the slope (Fig. 30).

Geophysical signatures on the Queen Charlotte Terrace are notably
different. Seismic reflection data show the sediments to be
strongly deformed, with local undisturbed sediment ponds between
uplifted blocks (Fig. 29). Srivastava (1973) and Horn et al.
(1984) interpreted the terrace as a set of horsts and grabens.

From a deep refraction survey, the Moho has been modeled at 13 km
depth on the west side of the terrace and at 20 km on the east
side, deepening towards the continent (Fig. 31). The crystalline
crust has variable velocities: 5.3 km/s at the top, increasing to
more than 7.3 km/s below the depth of 10 km. A bipartite
structure has been modeled in the sedimentary cover: the upper
layer, with velocities of 2-3 km/s, is some 2 km thick; the lower

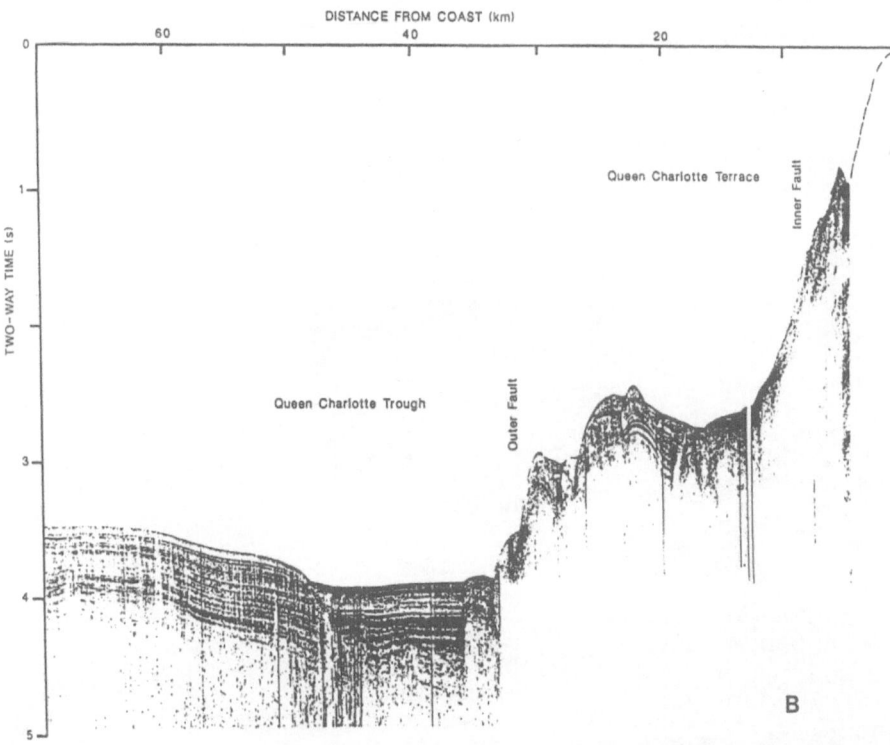

Figure 29. Structure of the Queen Charlotte Terrace and Trough imaged in an old seismic reflection profile (after Davis and Seemann, 1981; Riddihough and Hyndman, 1989): (a) geographical index map with profile location; (b) seismic data. Two steep scarps bound the Queen Charlotte Terrace on the east and west.

150

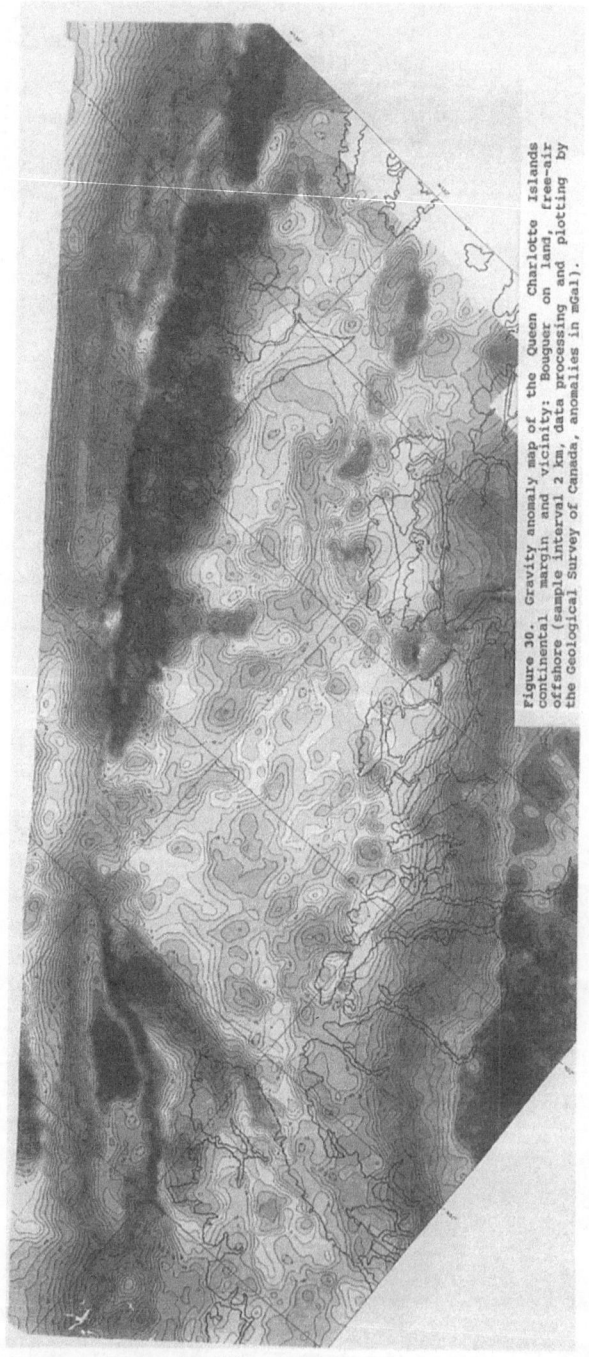

Figure 30. Gravity anomaly map of the Queen Charlotte Islands continental margin and vicinity: Bouguer on land, free-air offshore (sample interval 2 km, data processing and plotting by the Geological Survey of Canada, anomalies in mGal).

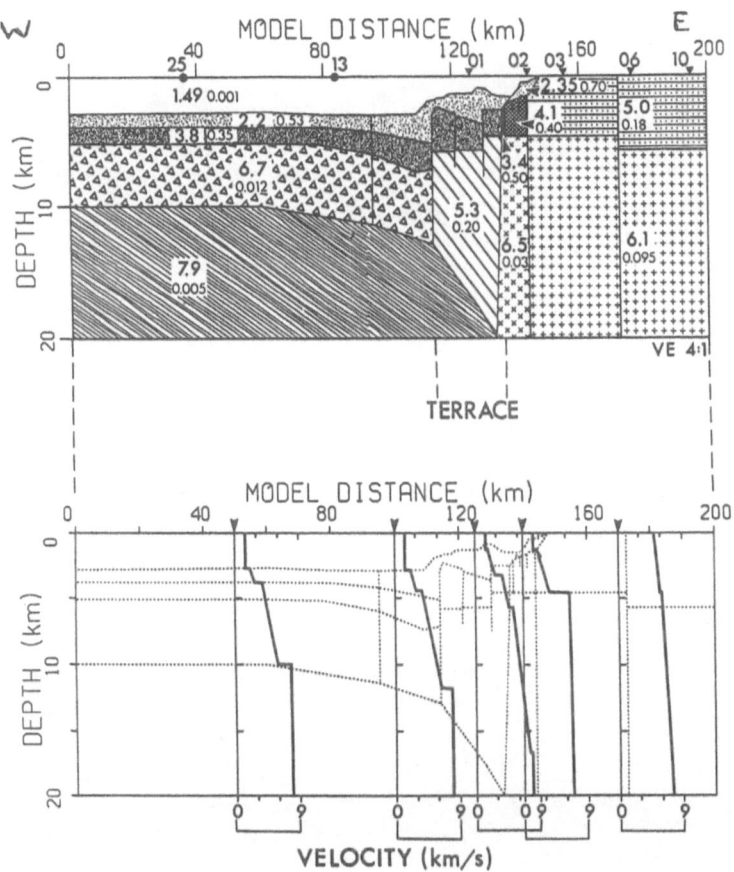

Figure 31. Crustal seismic refraction model across the Queen Charlotte Terrace (modified from Dehler and Clowes, 1988). The terrace is a distinct, fault-bounded crustal block. Bold numbers - velocities in km/s; small numbers - downward velocity gradients in km/s/km.

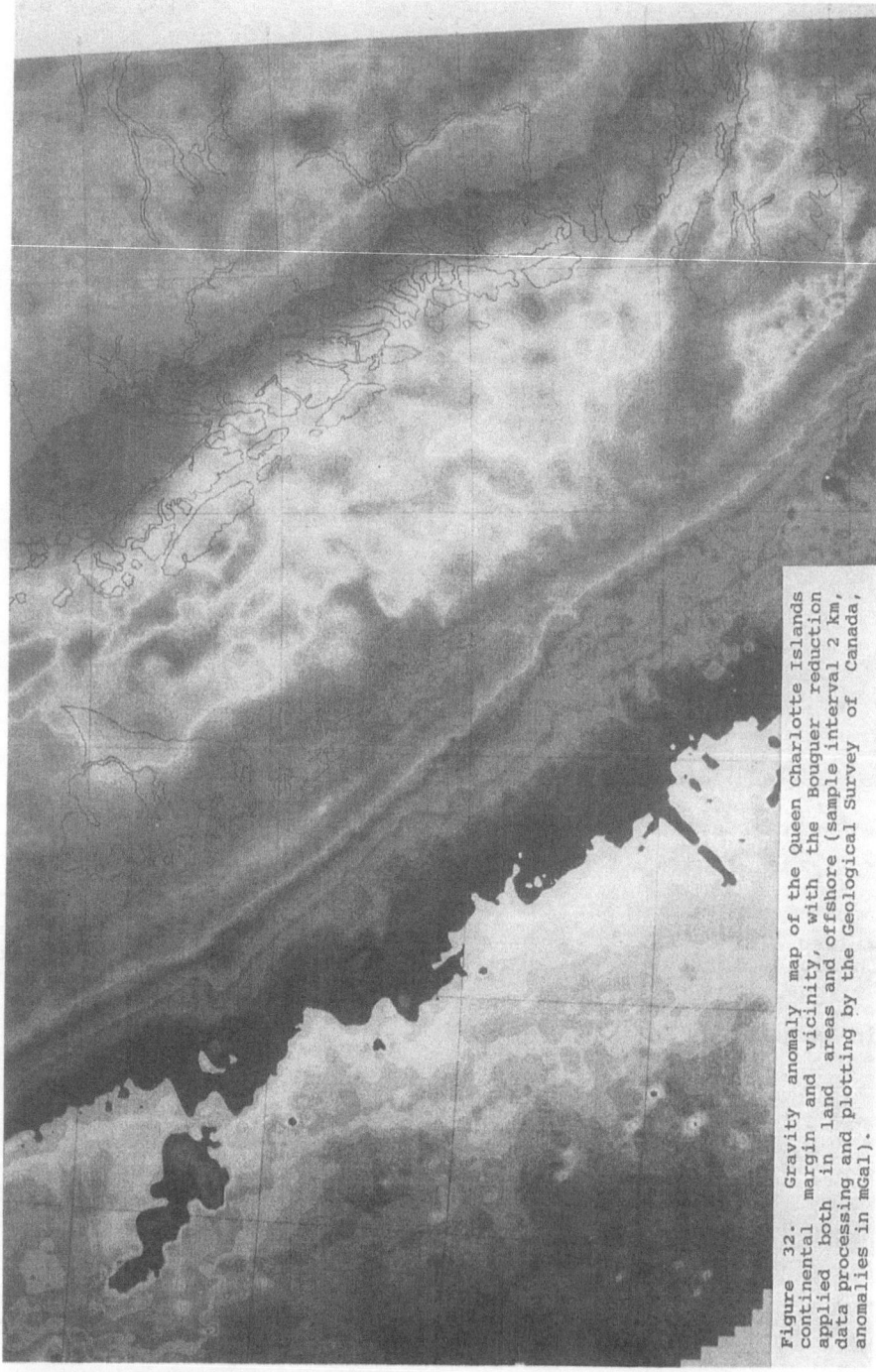

Figure 32. Gravity anomaly map of the Queen Charlotte Islands continental margin and vicinity, with the Bouguer reduction applied both in land areas and offshore (sample interval 2 km, data processing and plotting by the Geological Survey of Canada, anomalies in mGal).

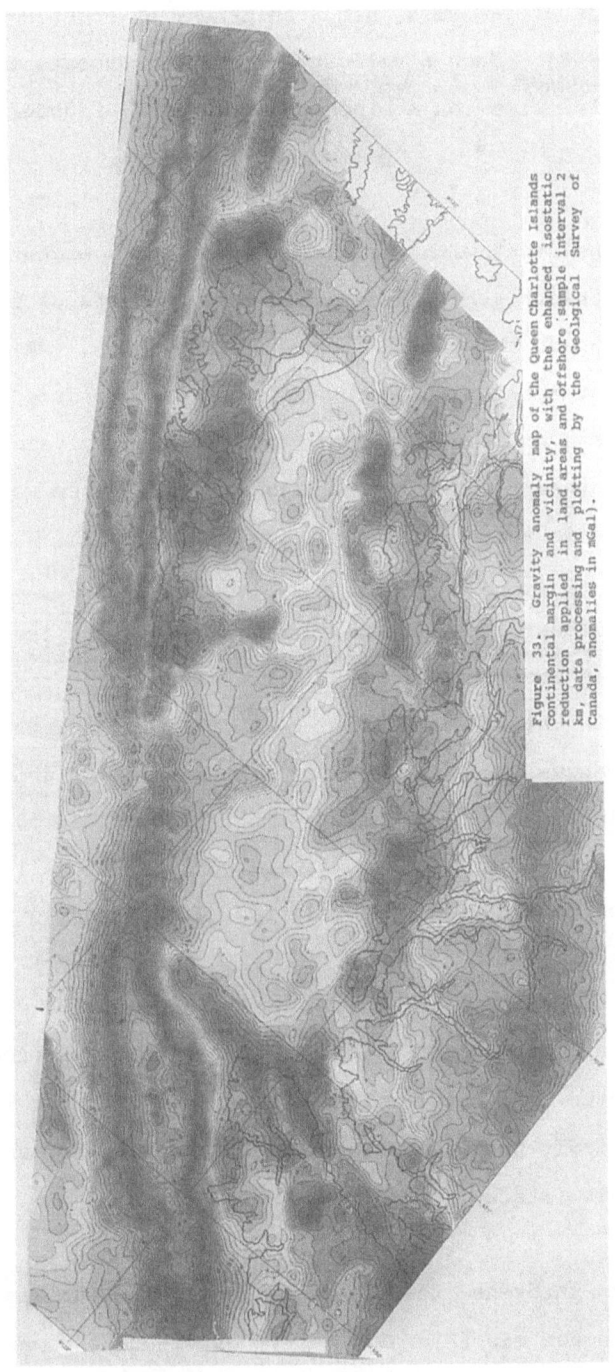

Figure 33. Gravity anomaly map of the Queen Charlotte Islands continental margin and vicinity, with the enhanced isostatic reduction applied in land areas and offshore sample interval 2 km, data processing and plotting by the Geological Survey of Canada, anomalies in mGal).

layer, with velocities of 4-5 km/s, has a thickness of about 3 km (Dehler and Clowes, 1988). Such a velocity structure suggests the terrace most probably lies on a block of attenuated, foundered continental crust.

This conclusion is consistent with the character of the magnetic field, as the broad blank zone includes the eastern parts of the Queen Charlotte Trough and the entire terrace (Fig. 27). Also consistent with the interpretation of the terrace as a downdropped block, both free-air and enhanced isostatic gravity anomaly values over it are lower than -80 mGal, and bounding faults are well expressed as gradients in the Bouguer map (Figs. 30, 32, 33).

Next to the terrace outboard lies the comparatively gentle free-air low over the Queen Charlotte Trough. Further outboard, free-air anomalies are generally near zero over the northeastern Pacific Ocean. The Oshawa Rise is marked by a broad high of about +15 mGal, with superimposed local highs of about +40 mGal over seamounts of the Kodiak-Bowie chain. Inboard, gravity anomalies in all reductions are positive over the uplifted Queen Charlotte Islands continental crustal block.

Off Moresby Island, at latutude 52°N, the Queen Charlotte Terrace gravity low is disrupted by a local relative high oriented NE. It lies on trend with a fault-bounded basement high, known as Moresby Ridge, in Hecate Strait (Stacey, 1975).

The continental crust thickens towards the continent stepwise (Fig. 31). In the Bouguer map (Fig. 32), steep gradient zones are related to abrupt changes in crustal properties across deep faults bounding the Queen Charlotte Terrace. A geothermal break has been reported at the outer bounding fault of the terrace: measured heat flow on the terrace has been found to be closer to that on the islands than in the trough (Hyndman et al., 1982). Under the Queen Charlotte Islands, Hecate Strait and Dixon Entrance, the Moho has been shown from seismic refraction and gravity data to undulate slightly but remain in the depth range of 25 to 32 km, much deeper than under the terrace (Johnson et al., 1972; von Huene et al., 1979; Hole et al., 1993; Spence and Asudeh, 1993). Crustal models based on interpretation of seismicity data confirm that eastward deepening of the Moho occurs across big faults: from 13 km in eastern Queen Charlotte Trough, to 16 km on the Terrace, to 27 km on the Islands (Hyndman and Ellis, 1981). A similar stepwise eastward increase in crustal thickness has been noted across the margin of southeastern Alaska (Johnson et al., 1972; Brew et al., 1991) and southern Vancouver Island (Mereu, 1990).

The landward, inner strand of the plate-boundary fault system off southeastern Alaska and the Queen Charlotte Islands is well expressed in patterns of seismicity (Rogers, 1986; Bérubé et al., 1989) and offsets of bathymetric features (Bruns and Carlson, 1987). However, the main changes in crustal properties occur at the outer strand. This suggests that historically this strand accommodated most of the displacement, though it is now inactive. The plate-boundary zone along the southeastern Alaska and Queen Charlotte Islands margins is deformed variably, with the amount of deformation increasing to the south, but the continental and oceanic crust are juxtaposed across the outer strand.

The Queen Charlotte Terrace must be cored by crystalline crust composed of rigid rocks, which enabled it to maintain structural integrity in such a complex zone. High-grade metamorphic rocks are exposed in the Westcoast belt on Vancouver Island, which runs along the western periphery of the island and beneath its northern exterior shelf. This belt and the terrace have a similar width, about 20-30 km, and they lie on trend with one another. This suggests their continuity. In this case, the terrace has a Mesozoic metamorphic basement, covered by compacted late Mesozoic or early Tertiary sedimentary rocks and younger sediments.

Southward extension of plate boundary off Queen Charlotte Sound

Similarity of Mesozoic formations on the Queen Charlotte and Vancouver islands suggests old geologic links along the Western Canada Archipelago. According to plate reconstructions, since 42 Ma dextral strike-slip movements have been occurring off the Queen Charlotte Islands, and the Cascadia subduction zone has existed to the south. Though in practice the exact extent of this subduction zone is unclear (Lister, 1991; Allan et al., 1993), in current models its northern end is placed off Queen Charlotte Sound (Riddihough, 1984; Riddihough and Hyndman, 1989).

This area is modeled as a point which separates two different tectonic regimes: strike-slip in the north, subduction in the south. A plate triple junction between North America, Pacific and Juan de Fuca (or its northern segment, Explorer) plates has been postulated in this area (Keen and Hyndman, 1979; Riddihough and Hyndman, 1989). The Explorer plate, detached from the larger Juan de Fuca plate around 3.5 Ma, is supposedly converging with North America at a rate and angle distinct from those of its parent plate farther south (Riddihough et al., 1980; Riddihough, 1984).

A left-lateral transform boundary - the Nootka fault zone - has been proposed between the Explorer and Juan de Fuca plates, and spreading centers between the Explorer and Pacific plates at Dellwood Knolls and Tuzo Wilson Seamounts (Figs. 3, 4).

Studies of volcanic-rock geochemistry of submarine basalts off Queen Charlotte Sound led to long-standing doubts about the validity of their interpretation as products of sea-floor spreading in this area (Chase, 1977; Cousens et al., 1985; Allan et al., 1993). Even for the Explorer Ridge to the west, simple spreading models do not provide adequate solutions (Cousens et al., 1984; Michael et al., 1989).

Dellwood Knolls and Tuzo Wilson Seamounts are not prominent features related to active sea-floor spreading. Bathymetrically, they are just a few of the many small mounds in a broad volcanic field on the ocean floor off Queen Charlotte Sound (Fig. 28). Local magnetic anomalies, on a subdued background, correspond to the many bathymetric highs scattered on the ocean floor (Fig. 27), and some volcanic mounds are draped by sediments.

Intercalated basalts and sediments seem to occur in Dellwood Knolls in equal proportion. Interbedding of Dellwood-sourced lava flows with sediments next to the knolls has been inferred from seismic reflection profiles (Davis, 1982). A low bulk density, just $2,500 \pm 100$ kg/m3, coupled with low magnetization, has been suggested for the knolls (Riddihough et al., 1980). A depression filled with hundreds of meters of sediments separates Dellwood Knolls from the foot of the continental slope. Geochemical differences in basalts have led to a suggestion that Dellwood Knolls have different mantle sources (Cousens et al., 1984).

Even smaller are Tuzo Wilson Seamounts to the north. These very small volcanic mounds have an edifice volume of only 12 km3. Their basalts, like those in the Dellwood Knolls, exhibit geochemical properties inconsistent with spreading-ridge origin (Cousens et al., 1985; Allan et al., 1993). Also contradicting sea-floor spreading is the morphology of these mounds. They are almost isometric and do not form coherent linear features. Detailed surveys show that sediment thickness increases from the mounds in all directions. An irregular pattern characterizes bathymetric and potential-field lineaments, which run NW, NE, N-S (Carbotte et al., 1989).

Absence of dramatic structural breaks off northern Queen Charlotte Sound is indicated by geophysical data. Whereas large negative free-air and enhanced isostatic gravity anomalies lie along the margin over the Queen Charlotte Terrace to the north and the Winona sedimentary basin to the south, only local anomalies with moderate amplitudes, usually from -40 to +10 mGal, are observed in this area. Enhanced isostatic anomaly values locally drop to -60 mGal, but generally remain around -40 mGal (Figs. 30, 32, 33).

The lack of regional isostatic anomalies on the Queen Charlotte Sound shelf has previously been cited as evidence that the crust in that area is isostatically compensated (Stacey, 1975). At the mouth of the sound, isostatic compensation has also been found to be "fairly good" (Srivastava, 1973, p. 1669). The absence of large free-air and enhanced isostatic anomalies in and off Queen Charlotte Sound suggests regional equilibrium, which would probably be disrupted if sea-floor spreading were occurring.

A prominent east-side-down gradient zone along the lower continental slope off Queen Charlotte Sound in the Bouguer map (Fig. 32) may reflect a sharp increase in Moho depth towards the continent. On the interior shelf under Queen Charlotte Sound, gravity and seismic refraction models show the Moho to lie at depths of 23 to 27 km (Yuan et al., 1992). This situation is similar to that at the Queen Charlotte Islands margin, where abrupt continentward deepening of the Moho also occurs. Indeed, the Bouguer gradient zone off Queen Charlotte Sound merges in the north with the similar gradient zone along the outer flank of the Queen Charlotte Terrace.

Off Queen Charlotte Sound, this gradient zone turns slightly towards the continental slope in the north, and then turns away from it in the south. The blank magnetic zone off the sound also seems narrower than off the Queen Charlotte Islands and northern Vancouver Island (Fig. 27). These patterns of gravity and magnetic anomalies are interpreted herein to be due to a small reentrant in the continental-oceanic plate-boundary zone. This reentrant may be occupied by a block of oceanic crust. Other examples of interlocking of oceanic and continental blocks along the plate boundary off central Vancouver Island are discussed in the next chapter.

Despite the local complexity related to interlocking of dissimilar crustal blocks, the plate-boundary fault zone recognized along the continental margin off southeastern Alaska and the Queen Charlotte Islands continues past Queen Charlotte Sound. The inner strand of this fault zone has been identified in this area from seismic reflection data (Line 88-03; Rohr and Dietrich, 1992), and detailed studies of the ocean floor show it continues at least as far as Dellwood Knolls. The outer strand continues towards Tuzo Wilson Seamounts, where it nearly meets the subparallel Revere-Dellwood fault (Carbotte et al., 1989).

In Line 88-03 (Fig. 34), the pelagic ocean floor shallows gently towards the foot of the slope. Sediments in the top 300 to 700 ms beneath the ocean floor are transparent to weakly stratified. The underlying basement is irregular, with many abrupt offsets up to 300-500 ms in amplitude. Inboard, between SP 900 and 1050, lies a series of small bathymetric ridges. Discontinuous reflections and numerous diffractions below these ridges suggest steep faults, which probably form the outer branch of the plate-boundary structural zone. Between SP 1050 and 1200, the ocean floor on the lower slope is smooth, but many short and dipping reflections are observed underneath. The inner strand coincides with the sharp bathymetric break at SP 1200-1300. Strata of the Queen Charlotte Basin to the east are truncated abruptly by this fault strand. If the short reflections under the lower slope are from downdropped Queen Charlotte Basin strata, the crustal sliver between strands of the plate-boundary zone is similar to that under Queen Charlotte Sound, and the contact with oceanic crust occurs at the outer strand. The total width of the plate-boundary zone in this area is only about 15 km.

Off northern Vancouver Island, continuity of the inner strand is indicated by the steepness of the upper continental slope (Figs. 4, 28) similar to that off the Queen Charlotte Islands (Tiffin et

161

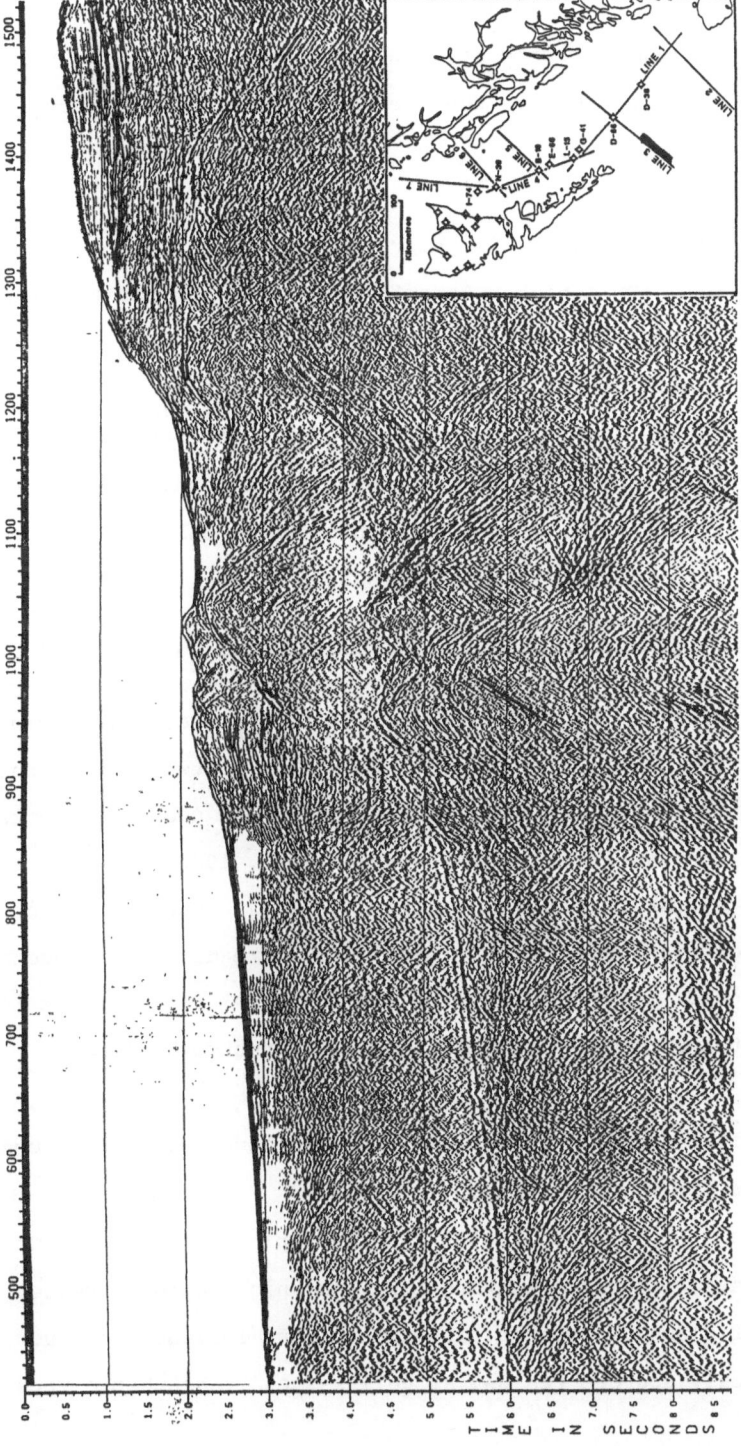

Figure 34. Structure of the submerged continental margin (from the abyssal plain to the outer shelf) off northern Queen Charlotte Sound, imaged in seismic reflection line 88-03 (the box gives the location; the data after Rohr and Dietrich, 1990). The interpretation, including two strands of the plate-boundary fault system, is discussed in text. The inner strand coincides with the abrupt steepening of the upper continental slope.

al., 1972; Chase et al., 1975). The outer strand, on the west side of the Queen Charlotte Terrace, ends only about 50 km north of Dellwood Knolls, where the Revere-Dellwood fault begins (Carbotte et al., 1989).

The Revere-Dellwood fault is represented prominently by a sea-floor ridge up to 500 m high. Seismic data show this fault has an oceanward-side-down, and sea-floor bathymetric features are known to be offset right-laterally across it by up to 6-8 km (Davis and Riddihough, 1982; Carbotte et al., 1989). Recent sea-floor mapping has shown that the Revere-Dellwood fault continues NW at least as far as Tuzo Wilson Seamounts, where only a small gap separates it from the bathymetric expression of the outer boundary fault the Queen Charlotte Terrace.

Various alternative tectonic models have been proposed in recent years, involving complex geological scenarios between for the oceanic plates and the continental margin off Queen Charlotte Sound (e.g., Cousens et al., 1984, 1985; Lister, 1989; Lyatsky et al., 1991; Lyatsky, 1993b; Allan et al., 1993). New data do not support the original idea about sea-floor spreading off Queen Charlotte Sound. Instead, the plate-boundary fault system is shown to continue through this area towards Vancouver Island.

Concept of plate rididity as applied to northern Juan de Fuca oceanic plate off western Canada
Two basic assumptions underlie the theory of plate tectonics. One is that magnetic anomaly patterns in oceanic regions reflect movements of slabs of oceanic lithosphere that include the crust, in which anomaly sources reside, as well as the lithospheric upper

mantle. At what crustal levels the magnetic anomalies are sourced is still unclear, but the sources must lie between the remanence-blocking isotherm somewhere near the crust-mantle interface, and the shallow sedimentary cover.

The oceanic crystalline crust has a normal thickness of about 7 km thick, so on a lithospheric scale the anomaly-causing layer is thin. Still, to the extent that crustal sheeted dikes and basalts are undisturbed records of production of the oceanic lithosphere at spreading ridges, and that the entire lithospheric plate moved as a single entity, magnetic anomalies are a powerful tool for reconstructing plate motions.

This tool relies on the second assumption: that lithospheric plates are stiff. They are assumed to interact with one another as rigid slabs, without internal deformation. They may become fragmented into smaller rigid plates or bend downward at subduction zones, but they are assumed to be strong enough to resist in-plate delamination, faulting or warping.

Because the available plate reconstructions generally omit in-plate deformation, they contain gaps and overlaps which in plate tectonics are unacceptable. To explain these discrepancies, attention is increasingly paid to deformation within plates and to diffuse plate boundaries (Stock and Molnar, 1988; DeMets et al., 1990). Evidence for such phenomena is now acknowledged worldwide.

Uncritical application of generalized assumptions to specific local circumstances may be misleading. With the benefit of modern data and methods of analysis, a more sophisticated approach is called for. The assumption that plates are rigid needs to be checked in each case.

Unusual characteristics of the Juan de Fuca plate have been noted for decades (e.g., McManus, 1971; Riddihough et al., 1983; Acharya, 1992). Particularly puzzling has been the absence of seismicity at the Juan de Fuca spreading ridge, or of Benioff-zone thrust seismicity along the Cascadia subduction zone. Accepting that this subduction zone is either aseismic or completely locked does not eliminate its specificity.

Stacey (1973) considered the western Canada continental margin to be different from typical active margins in three important respects: lack of a deep-sea bathymetric trench, absence of magmatic arc activity in southwestern British Columbia, and the lack of earthquakes on a megathrust between the continental and oceanic slabs. He speculated that subduction might have taken place earlier in the Tertiary, but has now stopped.

In the excitement of novelty that surrounded the general acceptance of plate tectonics, these local caveats seemed like undue conservatism. Riddihough and Hyndman (1976) argued that the western Canada margin possesses all the required attributes to postulate subduction, and this idea has guided subsequent geophysical interpretations.

New data, however, confirm that the Juan de Fuca plate and the Cascadia subduction zone are "unusual" (Acharya, 1992). The character of onshore magmatism, particularly in southwestern British Columbia and northwestern Washington, is atypical (Green,

1990; Sherrod and Smith, 1990). The Cascadia subduction zone is remarkably quiet seismically (McCrumb et al., 1989a,b). Onshore, seismicity occurs in clusters such as the one in Puget Sound (Crosson and Owens, 1987). Largely aseismic offshore are the Sovanco, Explorer and even Juan de Fuca ridges, which are generally regarded as active spreading or transform boundaries between oceanic plates (Riddihough et al., 1983; Davis and Currie, 1993). According to Wahlström and Rogers (1992, p. 953), "considerable seismicity occurs inside the Explorer plate, indicating internal deformation".

Allan et al. (1993) have proposed that tectonic movements are transferred off Queen Charlotte Sound from the inner fault of the Queen Charlotte Terrace to the Revere-Dellwood fault, without sea-floor spreading. R.P. Riddihough (in: Allan et al., 1993) has acknowledged that a transfer zone may connect the Queen Charlotte Terrace and northern Vancouver Island continental margins, and a diffuse boundary has been proposed between the Pacific and northern Juan de Fuca, or Explorer, plates (Furlong et al., 1994). More importantly, it is now accepted that the assumption of rigidity is inappropriate for the oceanic lithosphere off Queen Charlotte Sound (Carbotte et al., 1989; Davis and Currie, 1993).

A broad zone of deformation extends from the northern end of the Juan de Fuca Ridge to southern Moresby Island (Milne et al., 1978; Wahlström and Rogers, 1992), requiring alternative models for interaction of the Pacific and Juan de Fuca plates with each other (Furlong et al., 1994) and with North America (Lyatsky et al., 1991; Lyatsky, 1993). Evidence for sea-floor spreading off Queen Charlotte Sound is lacking, and the same continental-margin structures continue all along the Western Canada Archipelago.

Plate-boundary zone off northern Vancouver Island and the Winona Basin

The structural zone demarcating the western edge of the North American continental plate continues off northern Vancouver Island. The inner strand of the plate-boundary fault system has been recognized there as the Scott Islands fracture zone (Dehlinger et al., 1970). It controls the position of the upper continental slope, which is straight and steep. The outer strand, lying some 60 km outboard, is the Revere-Dellwood fault. Between these faults, and within the plate-boundary zone, lies the Winona sedimentary basin (Srivastava et al., 1971; Chase et al., 1975).

The Paul Revere Ridge and the Revere-Dellwood fault mark the contact of North American continental crust inboard with oceanic crust of northern Juan de Fuca (or Explorer) plate outboard.

The Paul Revere Ridge is asymmetric: its west side is steeper than its east side. The fault lies along the steep west side of the ridge, whereas the east side is covered by sediments. Sandstone and shale containing plant fragments have been dredged from the ridge crest, and basalt and gabbro from the west flank (Chase et al., 1975). In the northern part of the ridge, drilling at the DSDP site 177 encountered Lower Pliocene sediments underlain by basalt (Couch and Chase, 1973; Kulm et al., 1973). Total sediment thickness is 520 m. The lower 460 m contain a basal unit of fine-grained, consolidated, massive sandstone, overlain by rhythmic layers of sand, silt and clay likely of turbiditic origin. This Pliocene succession is covered by 60 m of hemipelagic silty clay of Pleistocene age, which contains deep-water fauna. Shallow-water fauna found in the Pliocene turbidite succession is probably displaced. Unlike typical oceanic crust, basalts that core the Paul Revere Ridge are not magnetic (Davis and Riddihough, 1982).

Conglomerate with pebbles of soft mud has been dredged from the Kwakiutl Ridge in the southeastern part of the Winona Basin close to the continental slope (Davis and Hyndman, 1989). Lower Cretaceous quartzite and conglomerate, similar to those on Vancouver Island, have been dredged from the upper slope to the east (Tiffin et al., 1972). A bathymetric break at mid-slope (1600-1800 m water depth) marks the eastern boundary of the basin.

An early gravity model showed the Winona Basin as a deep graben (Srivastava, 1973). Seismic data show its internal structure to be complex (Tiffin et al., 1972; Chase et al., 1975; Davis and Riddihough, 1982).

Lack of direct observations of the crystalline basement has left room for speculation. Plate reconstructions developed in the 1970s suggested this basin is underlain by oceanic crust. This block was initially considered a broken-off part of the Explorer plate (Davis and Riddihough, 1982). Recently it was raised to a status of a separate lithospheric plate (Davis and Currie, 1993). Models of plate restructuring (Davis and Riddihough, 1982) required the age of subsidence of the Winona block to be latest Pliocene to Pleistocene.

This interpretation encounters geological and geophysical difficulties. Seismic refraction and gravity surveys show that the crust thins oceanward under the continental margin gradually (Fig. 35). To the east, Moho depth is between about 40 km under Vancouver Island (McMechan and Spence, 1983; Drew and Clowes, 1990) and as little as 23 km under Queen Charlotte Sound (Yuan et al., 1992). The crystalline crust is 10-15 km thick under the eastern part of the Winona Basin and only 7 km thick under the basin's western part (Clowes et al., 1981; Davis and Riddihough, 1982). Such a gradual attenuation of crustal thickness indicates that the Winona Basin is probably underlain by continental crust. This conclusion is in agreement with the absence of magnetic stripes over the basin.

The Moho is mostly flat west of the Revere-Dellwood fault, which separates attenuated continental crust from oceanic crust outboard. This fault abruptly truncates the magnetic stripes, which are present over the oceanic crust throughout the northeastern Pacific region outboard (Fig. 27). Thickness of the oceanic crust is unusually large, varying between 7 and 11 km (Malecek and Clowes, 1978; Au and Clowes, 1982).

The Winona Basin is generally thought to contain about 8 km of sediments. A strong negative gravity anomaly, as low as -130 mGal, lies over it in free-air and enhanced isostatic maps (Figs. 30, 33). This anomaly delineates the shape of the Winona Basin and reveals its asymmetry: the anomaly widens to the south, and its minimum is offset slightly to the NE. The Bouguer map (Fig. 32) shows a major east-side-down gradient zone along the western flank of the basin. Seismicity is concentrated in the northern part of the Revere-Dellwood fault (Wahlström and Rogers, 1992), indicating ongoing activity in the plate-boundary zone.

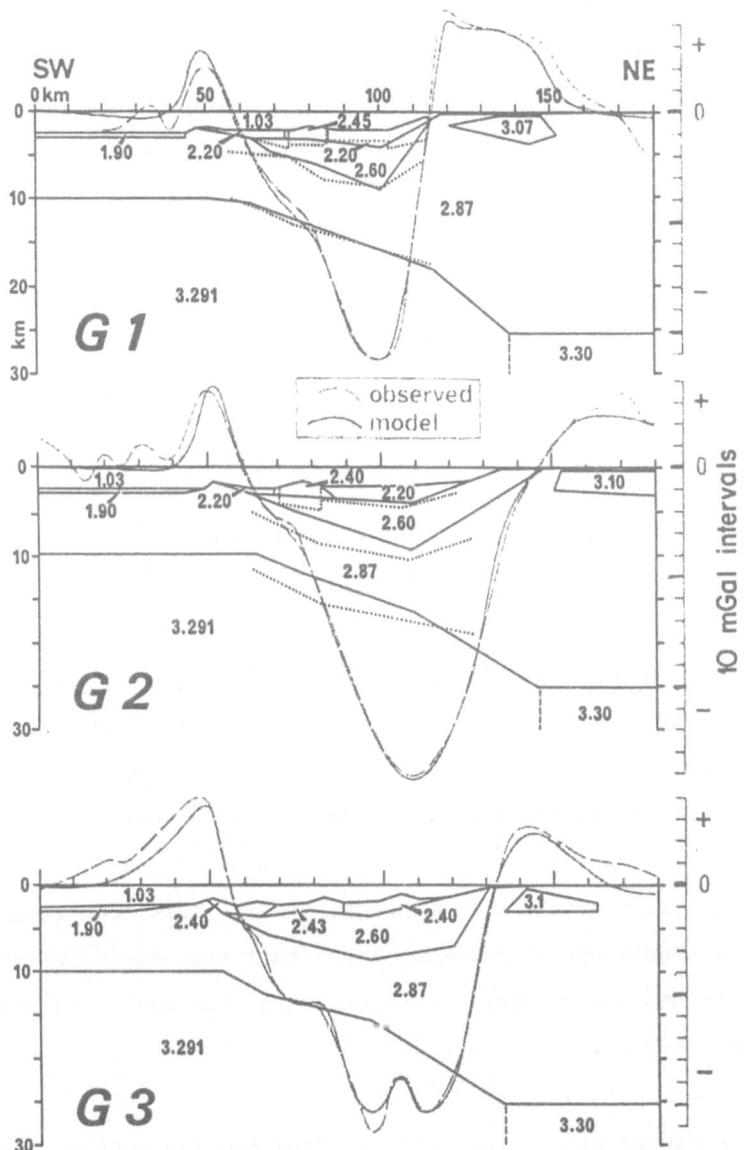

Figure 35. Structure of the northern, central and southern Winona Basin modeled from gravity data (modified from Davis and Riddihough, 1982). Profile locations are given in Fig. 38. Oceanward dip of the basement on the basin's east side is confirmed by seismic refraction data (dotted lines; Clowes et al., 1981). The cup shape of the Winona Basin is especially pronounced in the south.

In seismic reflection profiles across the basin, thickness of stratified sediments increases dramatically eastward. A new profile crossing the continental margin north of Vancouver Island (Line 88-02; Fig. 36) shows that stratified rocks dip towards the Scott Island fracture zone, which forms the inboard boundary of the Winona Basin. The dips increase downsection gradually, producing a fan-shaped pattern of seismic reflections. Such patterns, which typically result from syndepositional movements on listric normal faults, are expected in extensional regimes (e.g., Gretener, 1986).

The Winona Basin becomes symmetric and bowl-shaped to the south. A modern seismic reflection profile off Brooks Peninsula (Line 85-04; Fig. 37) reveals some new details. The acoustic basement lies at about 4 s traveltime on the flanks of the basin (at SP 450 and 1250) and deepens stepwise (steps at SP 550, 650, 800; and 1200, 1150, 1000) to about 6 s in the graben axis (around SP 900). A westward dip of the basement on the Winona Basin's east flank is apparent also from seismic refraction and gravity data (Fig. 35; Clowes et al., 1981; Davis and Riddihough, 1982). This contrasts with the idea (Davis and Riddihough, 1982) that the crustal block underlying this basin has been underthrusting Vancouver Island during the last 1-2 Ma.

Sedimentary rocks of Pleistocene and Pliocene age indeed form the upper part of this basin, but the age and nature of deeper rocks are untested and from some early plate reconstructions were also proposed to be no older than Late Pliocene (Davis and Riddihough, 1982). Extremely high rates of sedimentation would be required to

171

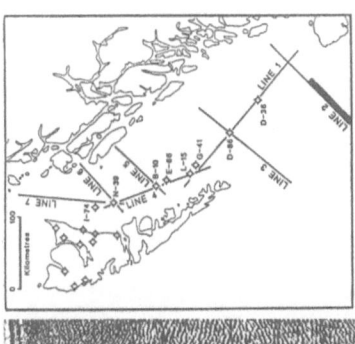

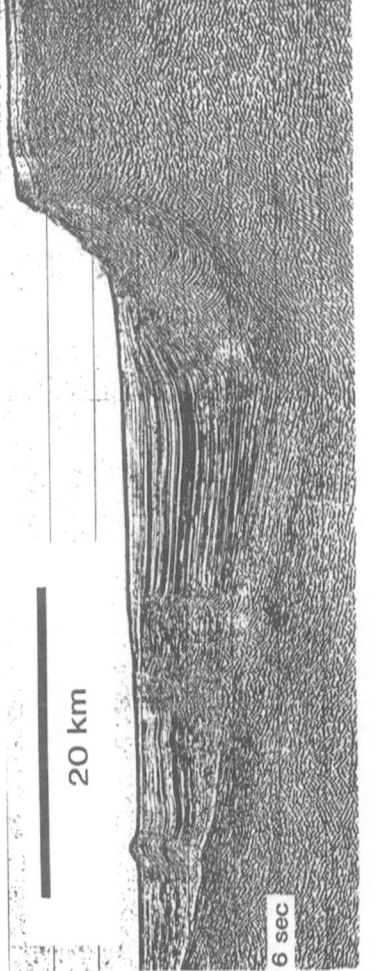

20 km

6 sec

Figure 36a. Large normal offsets on the steep, west-dipping Scott Islands fracture zone, imaged in seismic reflection line 88-02 (after Rohr and Dietrich, 1990), with location given in box. The fan-shaped pattern of reflections in the Winona Basin, with dips towards the normal fault increasing with depth, indicates syndepositional normal faulting. Such normal faults and fan-shaped reflection patterns in the hanging walls are typically associated with extensional tectonic regimes.

172

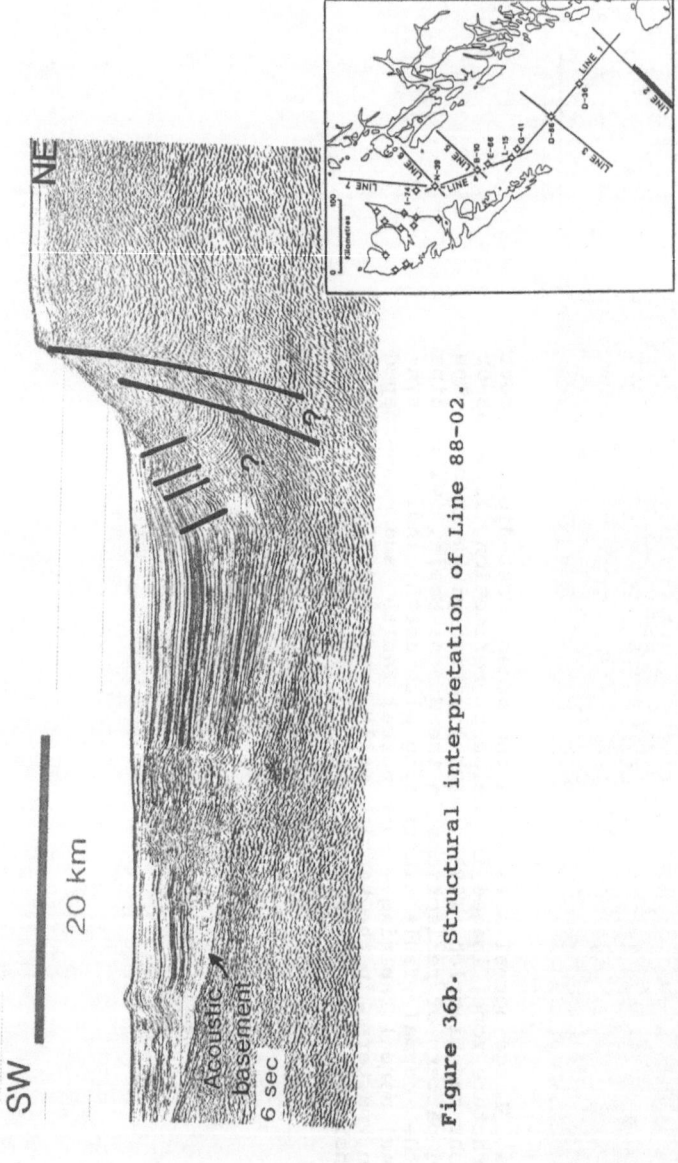

Figure 36b. Structural interpretation of Line 88-02.

fill an 8-km-deep basin in so short a time: to fill the Fuca Basin, of similar depth, took about 30 million years. The modern sedimentation rates at the base of the continental slope in this region are up to 1 m per thousand years, or 1 km per million years (Barnard, 1978; see also Hyndman et al., 1979), still much too low to fill the Winona Basin in just 1-2 Ma. Besides, an unusual kind of alteration of buried young sediments would be required to give them the high seismic velocities - in excess of 5 km/s - determined from reflection and refraction data.

Davis and Clowes (1986) have speculated that lithification of young turbidites proceeded unusually rapidly and reached an extremely high degree. Such a speculation invokes uncommon processes in unsampled rocks. In an alternative scenario proposed here, the Winona Basin is older. The velocities of >5 km/s suggested by seismic data for deep parts of the basin (Clowes et al., 1981; Davis and Clowes, 1986) may easily be correlated with Mesozoic and early Tertiary sedimentary rocks on Vancouver Island. Projecting such rocks into deep parts of the Winona Basin, which was formed by foundering of continental crust similar to that which created the Queen Charlotte Terrace, eliminates the need for abnormal geological assumptions.

Intensity of deformation in the Winona Basin increases downsection and from north to south. Broad folds are apparent in seismic profiles. Most faults are fairly steep, and reverse faults dip in various directions. Some breach the sea floor, producing mounds (Fig. 37, Line 85-04, SP 600) and ramps (SP 850). Many folds and faults have been inactive recently and are covered by undisturbed sediments. Synclines and anticlines (SP 1050-1150) affect all but

174

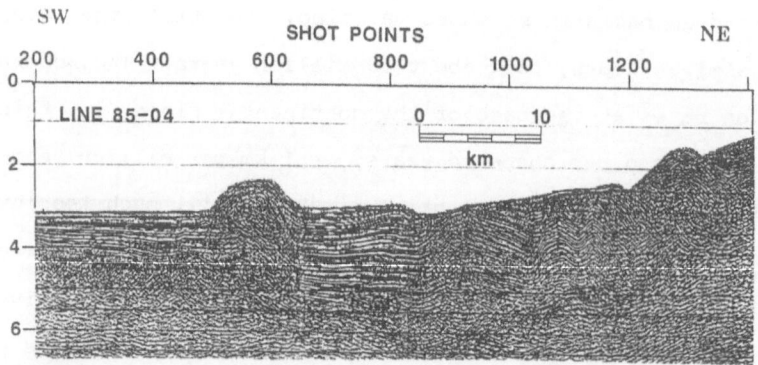

Figure 37. Faults and folds induced by sediment slumping and flowage in southern Winona Basin, imaged in seismic reflection line 85-04 (data after Yorath et al., 1987; Davis and Hyndman, 1989). Line location is given in Fig. 45.

the shallowest rocks. Some folds are detached at 4.5-5.0 s (SP 1050-1150). They were probably created by compressive stresses related to downslope gravity sliding of a sediment mass. The underlying acoustic basement in this area near the slope deepens stepwise to the west, towards the axis of the graben.

Sediment disturbances in the Winona Basin were interpreted by Davis and Riddihough (1982) and Davis and Hyndman (1989) to be results only of tectonic shortening caused by plate convergence. Later models, however, hold that convergence of the Winona block with North America has virtually stopped (Davis and Currie, 1993).

Presence of a deep graben beneath a stratified succession which despite the folding and faulting mostly remain coherent, suggests that the basin developed in two stages. Much of the deformation in the younger sedimentary unit overlying the graben might be a result of flowage of semi-consolidated, overpressured sediments. Though loading-related pressure is probably highest along the basin axis, where sediments are thickest, much of the flowage in Line 85-04 (Fig. 37) occurred near basin flanks. Some of it could have been triggered by basement irregularities and fault movements on the sides of the deep graben. The same phenomenon is reflected in the location of sedimentary anticlinal ridges on the sea floor - Haida, Kwakiutl - near the eastern flank of the basin (Fig. 38). Other ridges (SP 600) lie over the western flank of the deep graben. Thus, deformation in the Winona Basin had a variety of causes and was partially controlled by deep, steep faults.

Oblique to the predominant NW trend of the basin-bounding Revere-Dellwood fault and some of the sea-floor ridges, the Winona Ridge

176

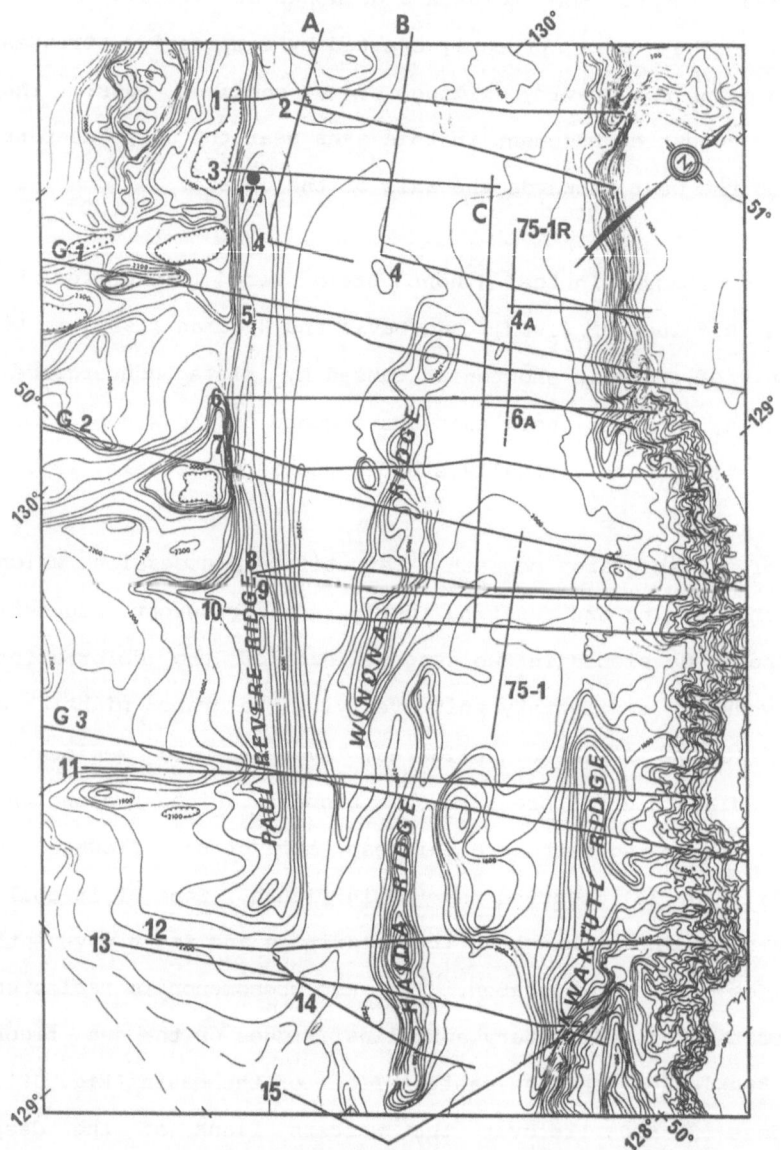

Figure 38. Major bathymetric features in the Winona Basin (in meters; modified from Davis and Riddihough, 1982), as well as the DSDP site 177 and locations of old seismic (Fig. 39) and gravity (Fig. 35) profiles.

trends NNW. Its trend is in line with strands of the plate-boundary fault system to the north. The Winona Basin deepens to the south. The axis of tilting of the continentward-dipping basement block is therefore not parallel to the Paul Revere Ridge, but rather trends NNW. Tilting in the Pliocene caused the upturn of this block's western flank, creating the Paul Revere ridge which was then onlapped from the east by Pleistocene sediments.

An interplay of NNW and NW structural trends is typical of the continental margin all along Vancouver Island.

Sediment deformation in the Winona Basin might have also been related to compressional shortening in the past. However, the amount of shortening was not large, and strata generally remained continuous along Line 85-04. No low-angle thrust faults or pervasive oceanward-vergent asymmetric structures are found in old seismic sections through this basin (Davis and Seemann, 1981). The reverse fault at SP 800 (Line 85-04) is steep and has and dips oceaward. In wide areas throughout the basin, sedimentary rocks between deformed zones are undisturbed altogether. At the Paul Revere Ridge and along the upper continental slope, the Winona Basin is bounded by two steep, west-dipping normal faults.

Localized magnetic anomalies as high as +500 nT on the basin's eastern flank are parallel to the continental slope (Fig. 27) and are probably caused by igneous rocks along a fault. Fault-controlled igneous activity might be the reason for the disturbance of buried sediments on the NE end of seismic line 7 (Fig. 39) in the area of these anomalies.

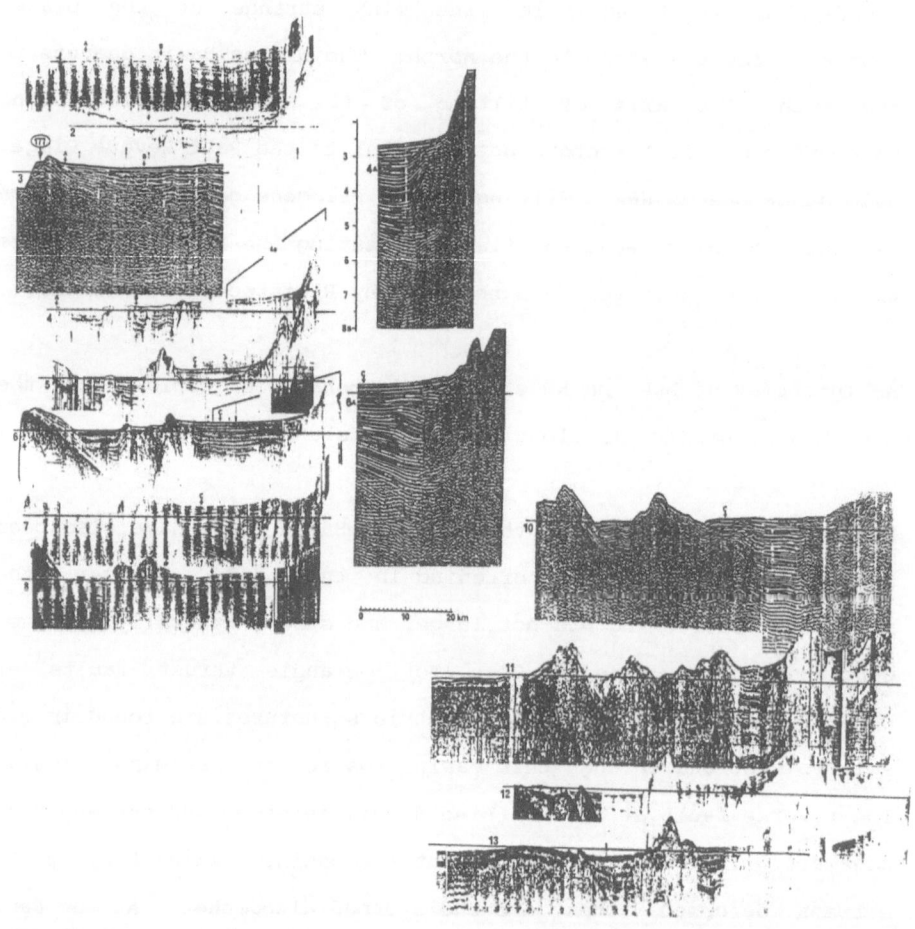

Figure 39. Seismic reflection profiles across the Winona Basin (data after Davis and Seemann, 1981; Davis and Riddihough, 1982). Line locations are given in Fig. 38. The DSDP site 177 is marked on line 3.

The southern part of the Winona Basin subsided deeper, and the Paul Revere Ridge loses bathymetric definition to the SE. Still, the Revere-Dellwood fault may continue there beneath a cover of sediments. Off Brooks Peninsula, seismic data show a buried west-side-down normal fault (Line 85-04, SP 420, traveltime 4.5 s), and magnetic stripes are terminated in that area.

Interlocking of continental and oceanic crustal blocks in the Brooks-Estevan Embayment

Bathymetric maps (Fig. 28) and sonar images of the ocean floor show that the continental slope south of Brooks Peninsula narrows between two NE-trending fault systems that continue into the submerged continental margin from Vancouver Island. One of them is the Brooks fracture zone (Tiffin et al., 1972; Muller et al., 1974); the other, named here Estevan, is the system of NE-trending faults beginning in an area between Hesquiat Peninsula and southern Nootka Island (Muller et al., 1981).

In the Brooks-Estevan embayment, as it is named herein, the continental slope narrows dramatically to just 30-35 km, from about 60 km to the north and south. Because its existence is hard to reconcile with simple tectonic models, this embayment has generally escaped notice in the modern literature. However, it offers several important observations that help understand the structure of the Vancouver Island continental margin.

The northern boundary of the Brooks-Estevan embayment is the broad Brooks fracture zone that runs across northern Vancouver Island (Muller et al., 1974) and the shelf (Tiffin et al., 1972; Chase et al., 1975). Strands of this fracture zone bound Brooks Peninsula.

The oceanward projection of the Brooks fracture zone lies at the south end of the principal Winona Basin depocenter, where the basin seems to dissipate into a series of small, elongated, fault-bounded depressions. The main Winona gravity low, with amplitudes <-100 mGal, also ends there.

The gravity domain to the southeast is characterized by several localized relative highs and lows whose amplitudes vary between -70 and -20 mGal (Fig. 40). Their trends are mostly NW-SE, but in the southern part of the embayment, at longitude 127°W to 127°40'W, anomalies run N-S. Magnetic anomalies in that area also have N-S trends (Fig. 41). Bathymetric trends in the area of the embayment form an orthogonal NW/NE pair; N-S trends are observed near the embayment's southern end (Fig. 28).

Inboard, beneath the shelf, the basement has been shown with seismic data to dip to the SW (Tiffin et al., 1972; see subsequent chapters for detailed review). Crystalline and sedimentary rocks of Mesozoic age, similar to those on Vancouver Island, are covered by Tertiary sedimentary rocks of the Tofino Basin. The crust on the shelf is continental, and the main structures continue into the submerged continental margin from Vancouver Island.

The Brooks-Estevan embayment contains a variety of structural trends that continue into it both continental regions inboard and from oceanic regions outboard. The NE trends are continental. Along the Brooks fracture zone in the interior of Vancouver Island, Muller et al. (1974) and Armstrong et al. (1985) have mapped basaltic flows and dikes of latest Miocene and Pliocene age. Outboard, faults of this fracture zone juxtapose Mesozoic

181

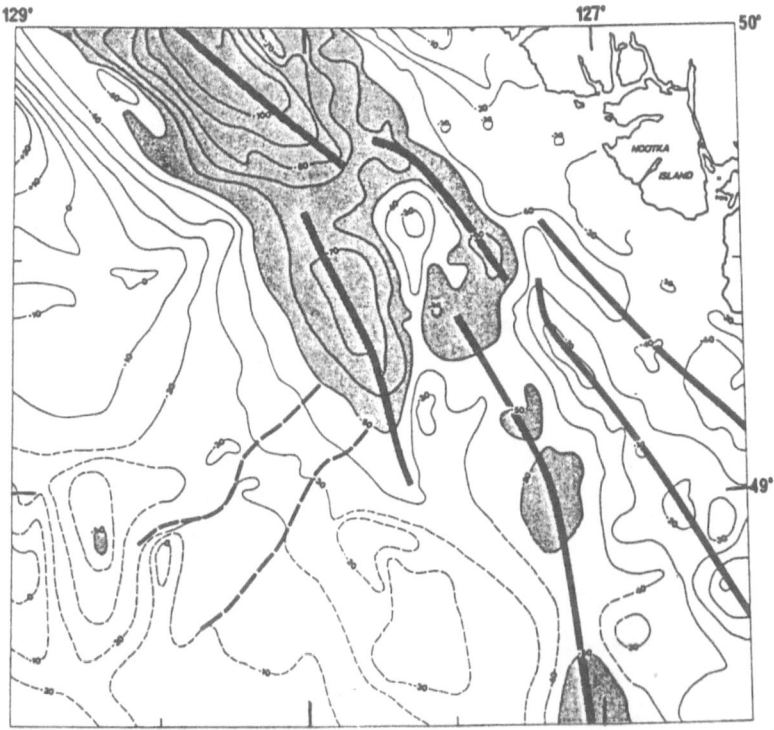

Figure 40. Free-air gravity anomaly map of the Brooks-Estevan embayment (in mGal; modified after Hyndman et al., 1979). Heavy lines indicate anomaly axes. Dashed lines indicate the location of the presumed "Nootka fault zone".

Figure 41. Magnetic anomaly map of the Brooks-Estevan embayment (in nT; modified from Hyndman et al., 1979; compare with regional magnetic anomaly maps in Figs. 3b, 27). Solid lines indicate some of the breaks in the anomaly pattern. Dashed lines indicate the presumed location of the broad "Nootka fault zone". In the southern part of the map area, note the partial coincidence of N-S-oriented magnetic anomalies with gravity anomalies in Fig. 40.

rocks exposed on Brooks Peninsula and Cenozoic rocks of the Tofino Basin on the submerged shelf (Tiffin et al., 1972).

On the south, the Brooks-Estevan embayment ends at the offshore projection of the NE-trending Estevan fault system, which on western Vancouver Island cuts both Mesozoic and Cenozoic rocks (Muller et al., 1981). Several mid- and lower-slope bathymetric breaks lie on trend with NE-oriented faults on Vancouver Island. South of the embayment, the continental slope widens again.

The NW-SE bathymetric, structural and potential-field trends in the Brooks-Estevan embayment are aligned with strands of the plate-boundary fault system to the northwest. However, the N-S trends continue into the embayment from oceanic regions.

A modern marine seismic reflection profile runs across the margin on trend with the Estevan fault system (Line 89-09; Fig. 42). It shows folds in stratified sediments on the lower slope (SP 500 to 700) with an amplitude of up to 0.5 s and a wavelength of about 4 km. They are similar to the folds in southern Winona Basin (Fig. 37), which are attributed to downslope slumping of sediments. Several reverse faults cut the sediments at the foot of the slope (Fig. 42). The biggest of them, a low-angle thrust with up to 1 km shortening at SP 400 (based on offset of reflections at traveltime of 4 s) dips towards the ocean.

Further outboard lie undisturbed, bedded sediments whose thickness corresponds to traveltime of 1.5 s. They overlie oceanic crystalline crust, which in this seismic line continues under the lower slope.

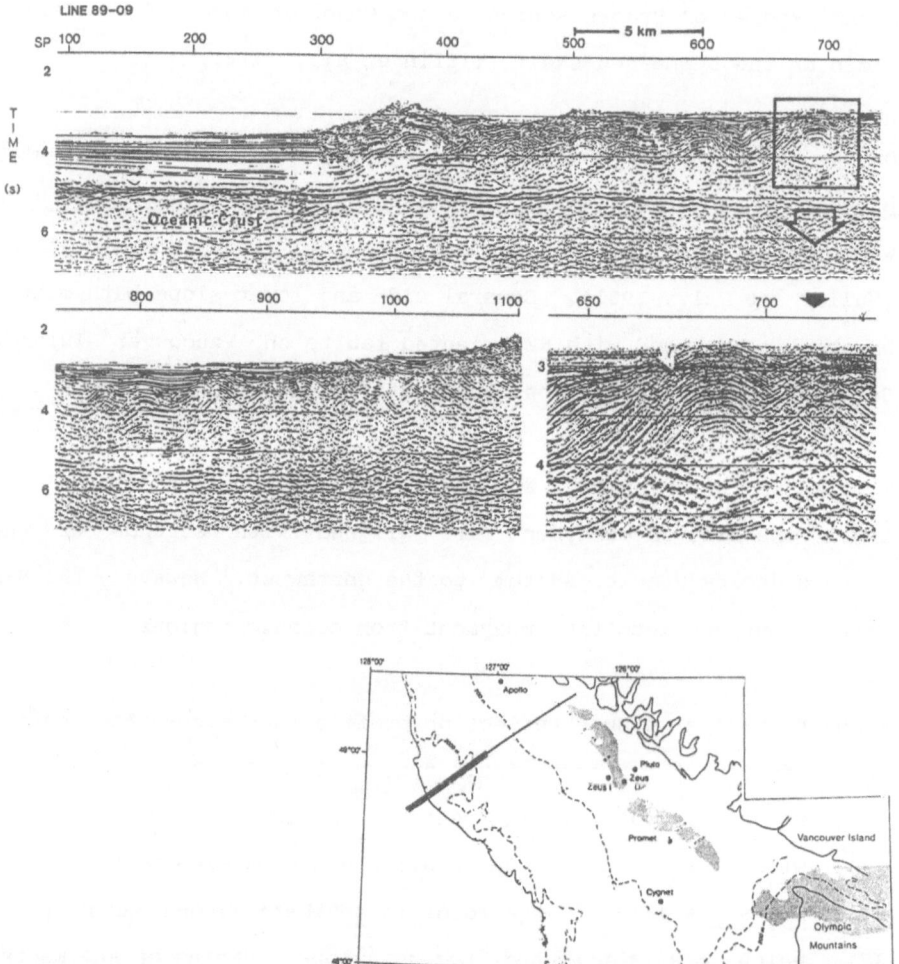

Figure 42. Structure of the submerged continental margin on the southeastern end of the Brooks-Estevan embayment imaged in seismic reflection line 89-09 (after Spence et al., 1991). The box gives line location; shaded areas represent magnetic highs; labeled dots represent drillholes.

Magnetic anomalies trending N-S continue into southern Brooks-Estevan embayment from the interior of the Juan de Fuca plate (Fig. 41). Near the continental slope, they coincide with elongated free-air gravity anomalies, whose orientation is also N-S (Fig. 40). These N-S potential-field anomalies also coincide, at longitude 127°-127°40'W, with N-S bathymetric features such as the foot of the continental slope and mid-slope ramps and ridges (Fig. 28). Thus, the configuration of southern Brooks-Estevan embayment is influenced by N-S-oriented structures.

Coupled with continuity under the lower continental slope of the oceanic-crust crystalline basement in Line 89-09 (Fig. 42), this continuity of potential-field anomalies from oceanic-crust areas suggests that the southern part of the Brooks-Estevan embayment is underlain by oceanic crust. In the northern part of the embayment, however, the crust is continental, with NW-trending faults and potential-field anomalies. This indicates that the Brooks-Estevan embayment is an area where small blocks of oceanic and continental crust are complexly juxtaposed and interlocked.

Some sort of previous convergence of the Juan de Fuca and North America plates, leading to local compression of the oceanic plate against the continent, might account for the invasion of oceanic-crustal blocks into the Brooks-Estevan embayment. To assess whether such processes are occurring on a large scale, the next chapters will examine the character of block movements on Vancouver Island and adajacent submerged margin, using outcrop and drillhole information as well as geophysical data to develop a model of geologic evolution of this region.

CHAPTER 7 - CRUSTAL BLOCKS UNDER VANCOUVER ISLAND
AND THE EXTERIOR SHELF

Variations in crustal thickness along the Insular Belt

A regional seismic refraction profile along Vancouver Island confirms that the island is underlain by continental crust 36-42 km thick (McMechan and Spence, 1983; Drew and Clowes, 1990). It is 10 to 15 km thicker than the crust under the Olympic Mountains to the south and the Queen Charlotte Islands to the north.

Refraction and gravity data show that under Queen Charlotte Sound the Moho may be as shallow as 23 km (Yuan et al., 1992). Farther north, it deepens again, to 29 km under the Queen Charlotte Islands and 27-29 km under Hecate Strait (Mackie et al., 1989; Sweeney and Seemann, 1991; Spence and Asudeh, 1993; Hole et al., 1993). Under Dixon Entrance, the Moho lies at 25-30 km depth (Johnson et al., 1972), and similar values have been reported from southeastern Alaska (Brew et al., 1991).

Thus, along the Insular Belt, thickness of the continental crust varies greatly. The crust is attenuated from the mainland towards the ocean stepwise, and the Moho becomes shallower, step by step, under the continental shelf and slope (Johnson et al., 1972; Mereu, 1990; Brew et al., 1991).

The dramatic Moho relief along the Insular Belt in Canada has been interpreted as a recent phenomenon, a product of tectonism in the Neogene. Large extension (factor 1.5 to 3.3) was proposed to be the cause of the Moho rise under Queen Charlotte Sound (Yorath and Hyndman, 1983; Hyndman and Hamilton, 1993). This idea arose from applying to the overlying Queen Charlotte Basin in this area the theoretical basin-formation model of McKenzie (1978), which suffers from unrealistic assumptions about properties and dynamics of the lithosphere leading to exaggerated estimates of extension (Lyatsky and Haggart, 1993; Lyatsky, 1994).

No tectonic sutures (Yorath and Chase, 1981) or regional tilt (Yorath and Hyndman, 1983), predicted for the Queen Charlotte Islands in models of Mesozoic terrane accretion and late Cenozoic extension and subduction, are expressed in outcrops (Hickson, 1991; Thompson et al., 1991; Haggart, 1991). Differential subsidence of crustal blocks bounded by pre-existing faults, with extension no greater than 10%, was responsible for the formation of the Queen Charlotte Basin (Lyatsky, 1993a). Besides, it has been shown on other continental margins that the Moho is not a passive marker that simply rises in proportion to crustal extension (Rosendahl et al., 1992).

Large Tertiary extension and associated strike-slip faulting were proposed for the Queen Charlotte Basin (Yorath and Chase, 1981). However, subsequent detailed mapping has revealed no major strike-slip faults or faults accommodating large extension (Thompson et al., 1991; Lewis et al., 1991b; Tribe, 1993). Only ≤10% extension is indicated by modern seismic reflection profiles across the basin on the interior shelf. Cross-cutting relationships of major faults and dike swarms on the Queen Charlotte Islands and the interior shelf also preclude large strike-slip movements (Lyatsky, 1993a). In a modified model, Rohr and Dietrich (1992) proposed that right-lateral movements were distributed into the basin from the Queen Charlotte fault. However, strands of the plate-boundary fault system lie west of the Queen Charlotte Islands and are distinct from the fault network on the interior shelf (see also Lyatsky, 1993a,b).

Unlike in British Columbia, the continental crust thins oceanward gradually in western Washington and Oregon, where Moho-depth contours run parallel to the margin without major disruptions (Mooney and Weaver, 1989). In Washington (Taber and Lewis, 1986; Crosson and Owens, 1987; Owens et al., 1988; Lapp et al., 1990; Finn, 1990), the crust is about 40 km thick in the North Cascades, about 30 km thick under the Olympic Mountains, and about 20 km thick at the coast. On the shelf and upper slope, under about 2-3 km of sedimentary cover, the crystalline crust thins oceanward gradually, to 10 or even 7 km, and the leading edge of the continental plate is thus fairly blunt.

The continental-crust slab over the undergoing oceanic slab in western Washington and Oregon was smoothed from below by Cenozoic subduction. The lack of smoothness along the base of the crust in western British Columbia suggests that in that area, such a process did not operate.

Geological shortcomings of existing seismic models of Vancouver Island crust

The crust-mantle interface under Vancouver Island was once modeled as a flat-lying, 5-km-thick zone with velocities around 7.5 km/s (McMechan and Spence, 1983). It was later remodeled as a zone 2 km thick, with velocities from 7.4 to 7.9 km/s (Spence et al., 1985; Drew and Clowes, 1990). This zone separates the lower crust with velocities 6.4 to 6.95 km/s from the continental upper mantle whose velocities are ≥7.9 km/s. Thickness of the upper crust was modeled by McMechan and Spence (1983) to vary along the island from 17 to 22 km, its velocity from 5.5 to 6.75 km/s.

In the models of Spence et al. (1985) and Drew and Clowes (1990), at the top of the crust lies a 3-km-thick surface layer with velocities 5.3 to 6.4 km/s (Fig. 43). It extends from the Coast Belt on the mainland, over the entire Vancouver Island, across the Tofino Basin on the exterior shelf. Thus, statics corrections applied to the refraction data unfortunately undermined the possibility of correlating the modeled velocity structure with the observable, variable surface geology.

In the Insular Belt in general and on Vancouver Island, the Phanerozoic volcano-sedimentary succession at the top of the crust is about 15 km thick (see Chapters 2-4; Fig. 16). Geophysically the most distinctive are Upper Triassic mafic volcanic rocks of the Karmutsen Formation, which are up to 6 km thick. Their density on Vancouver Island is around 2,950 kg/m3 (Currie and Muller, 1976), so their seismic velocity must also be high. These basalts are underlain by Paleozoic sedimentary and volcanic rocks whose total thickness is more than 5.5 km (Massey and Friday, 1989), and overlain by Upper Triassic to Lower Cretaceous rocks more than 4 km thick. Strong contrasts in acoustic impedance probably mark the top and base of the Karmutsen Formation, and at its base a velocity inversion is expected.

The post-Karmutsen Mesozoic succession is lithologically diverse and thus contains many acoustic-impedance contrasts. It includes carbonate and clastic units, as well as volcanic rocks of the

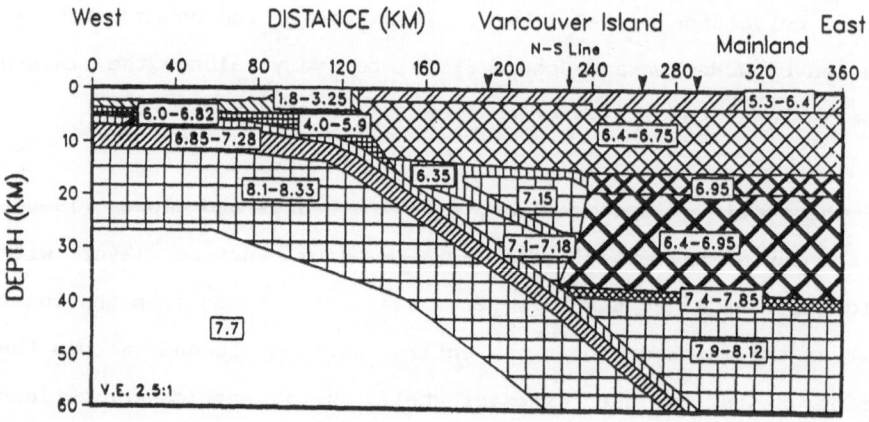

Figure 43. Seismic refraction model of Drew and Clowes (1990) for the LITHOPROBE profile across Vancouver Island and the adjacent submerged continental margin. Profile location is given in Fig. 45; numbers give P-wave velocities in km/s. The steep crustal-scale boundary in the middle of Vancouver Island coincides with the Alberni-Cowichan Lake structural zone (see geologic maps in Figs. 14, 16). To illustrate the non-uniqueness of this model, some of the existing alternative refraction models along this profile are presented in Figs. 47, 48.

Early Jurassic Bonanza Group. Comagmatic with these volcanics are widespread granitoids of the Island Intrusions suite. Field mapping (Jeletzky, 1976) and analysis of aeromagnetic data (Arkani-Hamed and Strangway, 1988) suggest these plutons might merge in the subsurface into bigger bodies.

The Georgia Basin, which lies between Vancouver Island and the mainland and on parts of eastern Vancouver Island, contains low-velocity clastic sedimentary rocks intercalated with coals. This succession, of Late Cretaceous and Tertiary age, is many kilometers thick (Pacht, 1984; Mustard, 1991; England and Hiscott, 1992). In the Tofino Basin on the exterior shelf, semi-consolidated Tertiary mudstone and sandstone have been drilled to a depth of almost 4 km (Shouldice, 1971); their seismic velocities are usually less than 2.5 km/s (Yorath et al., 1987; Calvert and Clowes, 1991). To represent such diverse rock types as a uniform crustal layer covering the entire region is misleading.

A diverse volcano-sedimentary package 15 km thick may cause many seismic reflections. Drilling in the Sevier Desert in Utah has shown that a stratigraphic contact in that area accounts for reflections previously interpreted as a large low-angle fault (Anders and Christie-Blick, 1994). On Vancouver Island, stratigraphic contacts in seismic sections may similarly be confused with low-angle faults.

Analysis of gravity anomalies on Vancouver Island
Comparison of gravity signatures in the Insular Belt with those elsewhere in the Canadian Cordillera suggested to Stacey (1973) that, overall, the Vancouver Island crust may be abnormally dense.

Indeed, seismic refraction and gravity models suggest in many parts of the Insular Belt mid-crustal layers of high seismic velocity and density (Spence et al., 1985; Yuan et al., 1992). These layers may be oceanic-lithosphere material underplated by tectonic processes related to subduction. On the other hand, the shape of many of these lenses suggests that magmas from the upper mantle might have produced sill-like bodies in the crust.

Consistent with variations in the geologic structure of the upper crust, three principal domains are seen in Bouguer gravity maps of Vancouver Island (Fig. 44). From north to south, they are:
(1) positive (0 to +30 mGal), roughly north of latitude 50°N;
(2) negative (0 to -30 mGal), roughly between 50° and 49°N; and
(3) positive (0 to +30 mGal), south of 49°N.

In the northern domain, positive anomalies are extensions of a strong (>+60 mGal) gravity high over southern Queen Charlotte Sound (Stacey and Stephens, 1969; Lyatsky, 1991a). A raised basement block, probably containing thick Karmutsen basalts, underlies this part of the interior shelf, and Tertiary strata in marine seismic profiles are thin (Rohr and Dietrich, 1992). On northern Vancouver Island, the gravity high splits up in accordance with the Karmutsen outcrop pattern (Fig. 16).

The central domain, where Bouguer anomaly values are negative, is dominated by low-density rocks. Granitoid Island Intrusions are widespread in outcrop in its western part. Incongruously, the dense Karmutsen basalts are common at the surface on the east side of this domain, but their thickness is reduced by erosion. Occurrence in that area of outcrops of Paleozoic rocks prompted

193

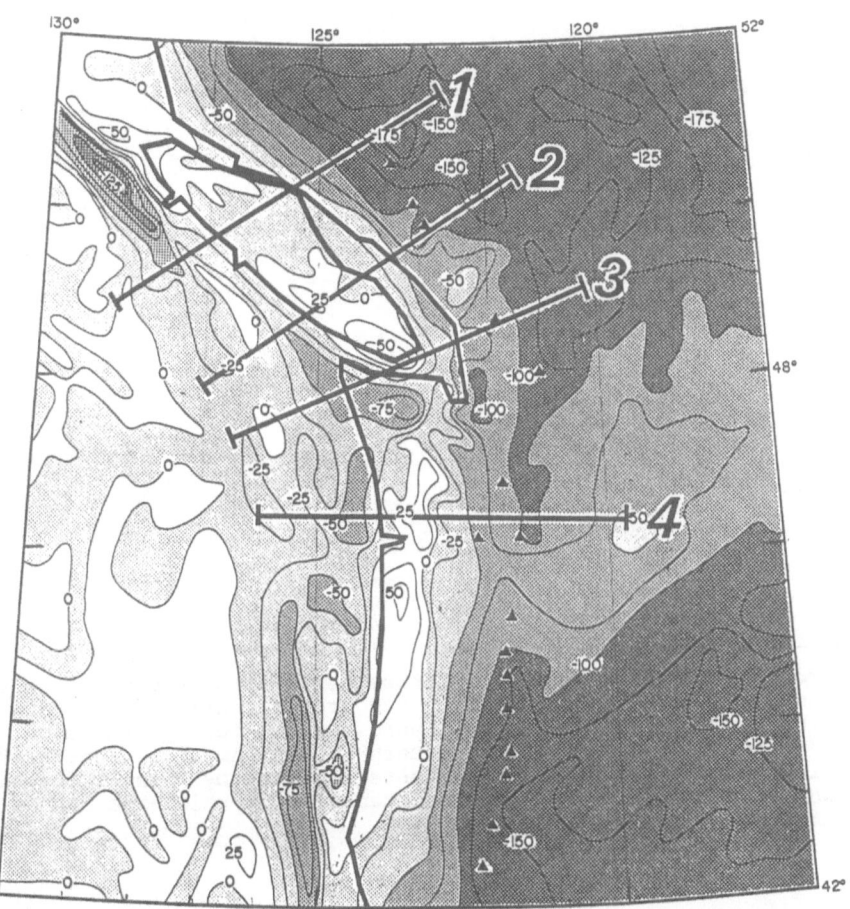

Figure 44a. Gravity anomaly map and models of lithospheric structure of Vancouver Island and western Washington and Oregon (in mGal; Bouguer on land, free-air offshore; after Riddihough, 1979). Dark triangles represent Quaternary volcanoes. Dark lines mark locations of the 2-D gravity models in Fig. 44b.

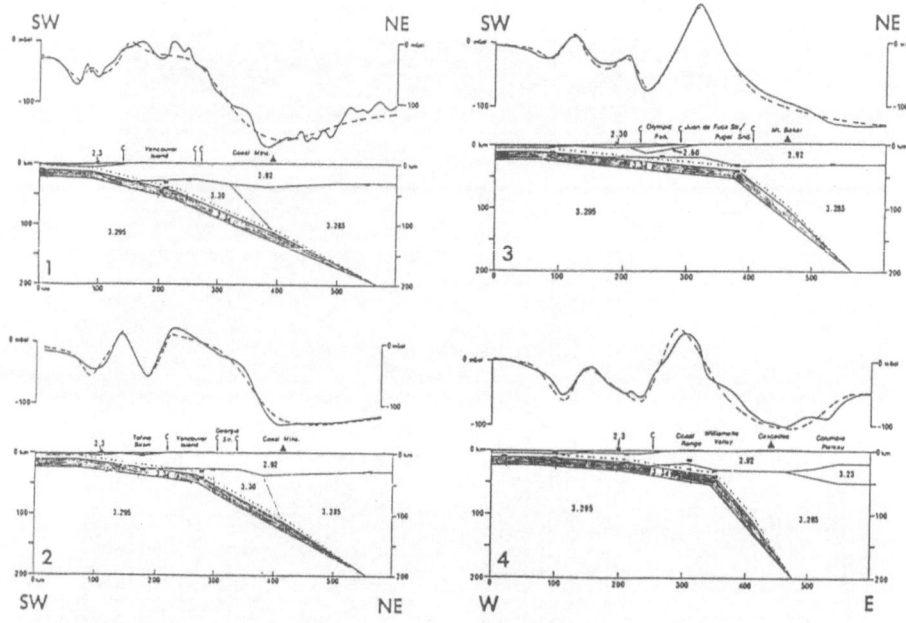

Figure 44b. Gravity models of lithospheric structure of Vancouver Island and western Washington and Oregon continental margin by Riddihough (1979). These models were seminal for postulation of ongoing subduction at the western North America continental margin, but subsequent modeling tests reported by Finn (1990) have shown that presence of a subducted slab in this area cannot be established from gravity data.

Stacey and Stephens (1969, their Fig. 5) to propose that the Karmutsen Formation is thinned in an uplifted axial zone of an anticlinorium. Existence of such an anticlinorium is confirmed by field mapping (Massey and Friday, 1989).

The boundary with the southern positive Bouguer domain runs WNW, and gravity anomalies reflect the existence on southern Vancouver Island of two distinct crustal blocks. Positive anomalies are observed over the Western block, whereas over the Eastern block anomalies remain negative. The positive anomalies correspond to areas where Karmutsen basalts are exposed in outcrops or lie close to the surface under a cover of only Lower Jurassic Bonanza Group. On the east, Karmutsen basalts are largely eroded and Paleozoic rocks are exposed in the core of an anticlinorium. Island Intrusions are also present; if their volume increases at depth, density of the crust is reduced further. Along the island's east coast, all rocks are covered by light Upper Cretaceous sediments of the Nanaimo Group.

The gravity high of >+60 mGal at the southern tip of the island is related to the Metchosin igneous massif.

Seismic refraction constraints on deep crustal structure

At lower-crustal levels, where constraints from surface geology are lacking, geophysical models are more speculative. The quality of seismic data also decreases with depth, and Morgan and Warner (1990, p. 41) noted that "the refraction data alone do not resolve the deep structure". But still, if interpreted with due caution and in conjunction with other information, the seismic transect across southern Vancouver Island and the adjacent submerged continental margin (Figs. 14, 45) offers many useful insights into the structure of this region.

The Moho under Vancouver Island and the western mainland is placed in most refraction models (Green, ed., 1990) at depths of about 40 km. Iwasaki and Shimamura (1990) modeled the entire continental crystalline crust for these areas to be horizontally layered, with velocities between 6.4 and 6.85 km/s. Similar parameters, albeit with some intracrustal velocity inversions between horizontal layers, were used by Fowler and Pandit (1990). According to Mereu (1990), the upper crust has velocities of 6.5-6.6 km/s, and the lower crust of 6.8 km/s.

The two main crustal blocks on southern Vancouver Island, Eastern and Western, found an expression in the refraction model of Drew and Clowes (1990). The steep boundary between them is located along the Alberni-Cowichan Lake structural zone which runs NW-SE through the middle of the island (Figs. 14, 43). Unlike other workers, Drew and Clowes (1990) modeled the Vancouver Island lower crust as different to the east and west of this boundary. To the east, the modeled lower crust is a single block with velocities of 6.4 to 6.95 km/s. To the west, it contains a wedge-shaped set of four thin prisms with a combined thickness of up to 20 km. Two of these prisms were modeled with velocities as low as 6.35 km/s, two others with 7.1-7.2 km/s. This model is different from an earlier interpretation, which assumed that a single body with an extremely high velocity of 7.7 km/s exists in this area in the middle part of the crust (Spence et al., 1985).

The composite lower-crustal wedge of Drew and Clowes (1990) is

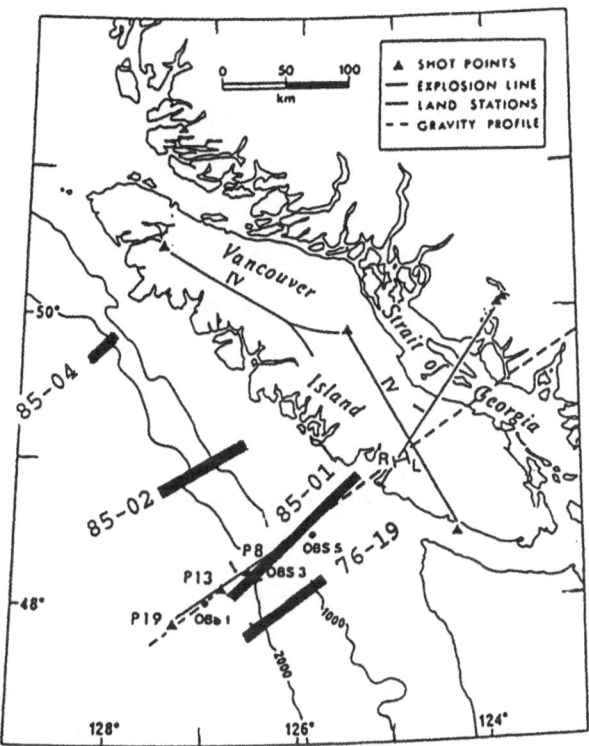

Figure 45. Location of LITHOPROBE and U.S. Geological Survey seismic reflection and refraction profiles on and off southern Vancouver Island (modified from Green, ed., 1990). Heavy lines indicate reflection profiles or their segments reproduced in other figures. Some of the alternative models for the refraction profile I are shown in Figs. 43, 47, 48, 55. Refraction profile IV along Vancouver Island was used by McMechan and Spence (1983) and Drew and Clowes (1990) to model the main characteristics of the Vancouver Island crust.

about 90 km wide and thins to zero under the shelf. In their along-island refraction model, four layers of this wedge, some just 2 km thick, were extended all along Vancouver Island for a distance of 350 km. These layers were interpreted as underplated material. However, velocities of 7.1-7.2 km/s are compatible with normal averages for lower continental crust (Pakiser and Mooney, eds., 1989). To expect the fine structure of a wedge produced by a dynamic tectonic process like underplating to continue along the continental margin for hundreds of kilometers is unrealistic.

Initial interpretations of Vancouver Island structure from seismic reflection data

Two deep reflection profiles are available on Vancouver Island near the refraction profile (Figs. 14, 45). Line 84-01 runs NE-SW from coast to coast, roughly parallel to the refraction transect. Line 84-03 is short and runs N-S but does not intersect the long reflection line.

Line 84-01 is shown in Fig. 46. Under the mostly transparent upper crust, two broad low-angle bands of seismic events have been detected, separated by another transparent zone in the middle crust (e.g., Clowes et al., 1984, 1987).

The mid-crustal transparent zone was interpreted as "underplated", thrust-imbricated basalts and sediments scraped off the undergoing Juan de Fuca plate (Yorath et al., 1985b; Yorath, 1987). The upper reflection band was interpreted as a thrust zone at the base of Wrangellia, which was assumed to be just 15 km thick. The lower band was interpreted as the subduction megathrust zone. In keeping with the subduction-complex model, east-dipping seismic

199

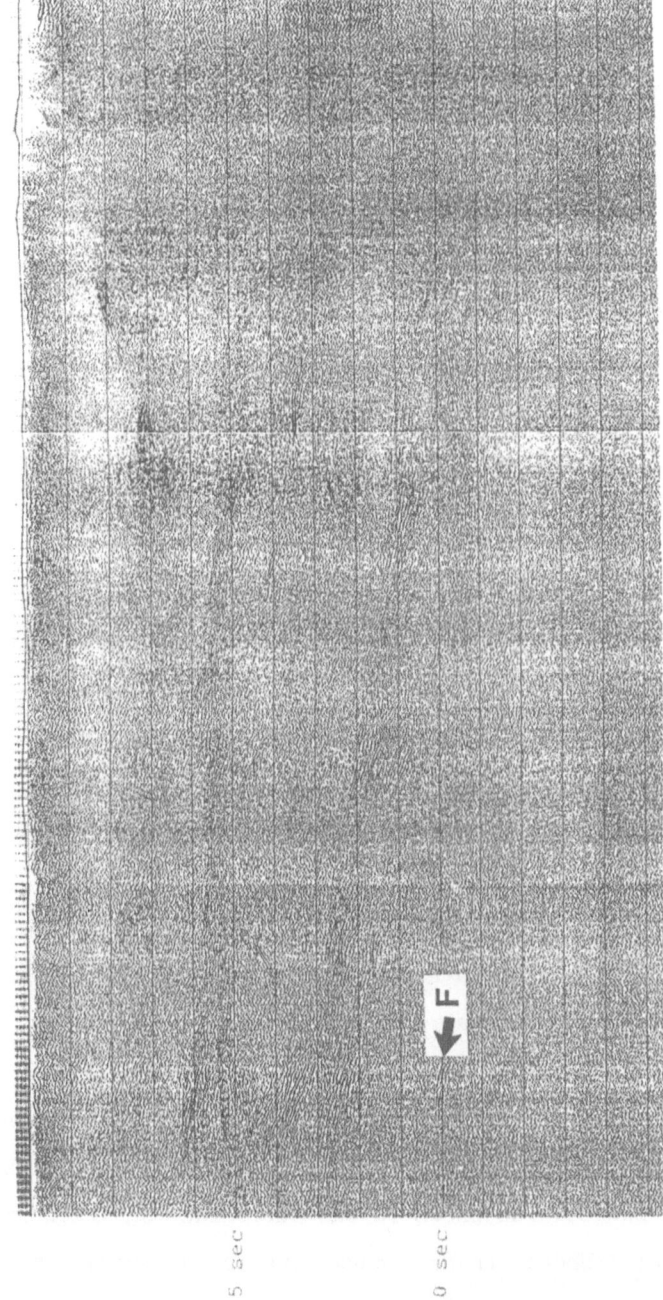

5 sec

10 sec

two-way traveltime

Figure 46. Deep structure of Vancouver Island imaged in the seismic reflection profile 84-01 across the island (after Green, ed., 1990). Profile location in relation to geology is given in Fig. 14. Note the two broad bands of seismic events in the middle and lower crust. The deep event "F" had been interpreted variously as the top (Hyndman et al., 1990) or base (Clowes et al., 1987) of the subducting Juan de Fuca slab, but processing tests showed it to probably be off-line noise (Hawthorne, 1990; Levato et al., 1990). Its correlation with the event labeled "F" in the unconnected Line 84-04 on southern Vancouver Island (Fig. 23), suggested by Clowes et al. (1987), is highly questionable.

events were emphasized and interpreted as thrusts, but west-dipping events received less attention.

In a later interpretation (Clowes et al., 1987), the top of the subducting slab was moved down, to the base of the lower reflection band. The subducted oceanic Moho was placed at the short event at 10 s on the west end of Line 84-01 (event "F"). Rocks in the transparent zone between two main reflection bands were presumed to be slices of mafic and ultramafic rocks and oceanic sediments, metamorphosed to blueschist grade. However, no blueschist-grade or ophiolitic rocks are exposed on Vancouver Island, and event "F" was shown by subsequent processing tests to probably be off-line noise (Hawthorne, 1990; Levato et al., 1990).

Interpretation of this reflection profile soon changed again. All sediments were assumed to be scraped off the incoming Juan de Fuca plate into an accretionary prism on the continental slope, and the idea of underplating was rejected (Davis and Hyndman, 1989). The two main reflection bands were interpreted as non-structural, representing metamorphic fronts and porous intracrustal zones containing water (Hyndman, 1988; Hyndman et al., 1990). The top of the subducting slab was moved still further down and placed at event "F", which is probably noise.

Structure of the upper crust was interpreted from the reflection data in geological terms by Yorath (1987, his Fig. 13). A thick wedge of Metchosin basalts was shown in the subsurface on the west end of Line 84-01 (where these rocks are not exposed). The Westcoast metamorphic complex, mapped in outcrop as a belt just 15-30 km wide (Muller et al., 1974, 1981), was interpreted as a broad, east-dipping zone continuing beneath central Vancouver Island. In the shallow (<4 s) transparent seismic zone above the upper reflection band, many low-angle, west-vergent thrust faults were shown slicing through an upper crust made up chiefly of Jurassic Island Intrusions. The entire succession of Paleozoic and Mesozoic volcanic and sedimentary rocks, including Karmutsen basalts, was shown as small pendants, only about 2 km thick, on top of these granitoids.

In the refraction models, the Vancouver Island crust has normal continental thicknesses (about 40 km) and velocities (about 7 km/s even in the lower crust; Green, ed., 1990). Such parameters suggest that the original continental lower crust is still in place, and a more-or-less full thickness of Wrangellian crust lies under Vancouver Island.

Steep discontinuities are not always imaged well in seismic reflection profiles. On Vancouver Island two major, high-angle, NW-trending fault systems are associated with diffractions and changes in reflection pattern at various depths. One set of such faults lies in the Westcoast belt, where its strands bound blocks of high-grade metamorphic rocks brought to the surface from mid-crustal levels. The other set, the Alberni-Cowichan Lake fault system, is located in the interior of the island (Figs. 16, 45).

Between these steep fault systems, the two main reflection bands in Line 84-01 diverge (Fig. 46). The upper band, at about 15 km depth (4-5 s traveltime), is almost horizontal. Events in it are discontinuous. The lower band, between 22 and >30 km (7 to 10 s), dips towards the continent.

Amphibolite- and granulite-grade metamorphism is common in the middle and lower crust, and these seismic bands might represent metamorphic fronts. Presence of bright spots and discontinuity of events in these bands (Milkereit et al., 1990) may be consistent with local presence of metamorphic water.

Geology-based interpretation of Vancouver Island seismic data
The boundary between the Coast and Insular belts lies under the Strait of Georgia, where refraction data have revealed a steep crustal-scale discontinuity (White and Clowes, 1984). On Vancouver Island, the Alberni-Cowichan Lake fault system is of lesser significance, as it lies entirely within the Insular Belt and separates the Eastern and Western Vancouver Island blocks.

Two phases of thrusting that might have affected the Eastern block are revealed by geologic mapping. In the mid- to Late Cretaceous, the Northwest Cascade thrust system, exposed on the San Juan Islands, was created (Whetten et al., 1980; Brandon et al., 1988). The early Tertiary saw the creation of the Cowichan thrust system, exposed on the Gulf Islands and eastern Vancouver Island (England and Calon, 1991; England, 1991). Thrusts related to these systems may disrupt seismic reflection patterns in the crust of eastern Vancouver Island (Fig. 46).

East of the Alberni-Cowichan Lake fault system, in the core of a geanticlinal uplift, lie exposures of Paleozoic rocks flanked by Upper Triassic basalts of the Karmutsen Formation (Muller, 1977a,b). This anticlinorium developed in Late Jurassic to mid-Cretaceous time (Massey and Friday, 1989). Its eastern limb was affected by block movements in the Late Cretaceous (Pacht, 1984), where thick Nanaimo Group sediments of the Georgia Basin were deposited in a series of grabens.

Blocks on the west side of the anticlinorium are also dowdropped. West of the Alberni-Cowichan Lake fault system, a synclinorium disrupted by a narrow raised zone in the middle is apparent from geologic maps (Fig. 16): two belts of Upper Jurassic Bonanza Group rocks are separated by a central belt of Karmutsen basalts.

In Line 84-01 (Fig. 46), this synclinorium is indicated above 4 s by dips of seismic events towards its center from both sides. East-dipping events on the west end of the profile are weak but fairly coherent. West-dipping events on the east flank of the synclinorium are brighter but shorter, perhaps due to disruption by thrusts. The inferred metamorphic fronts represented by the subhorizontal band at 4-5 s (about 15 km) ignore these shallower structures. In the short seismic line 84-03, which crosses the eastern flank of the synclinorium and the Alberni-Cowichan Lake structural zone, similar events lie at 5-6 s, but their nature is unclear because the southern half of this profile runs along the NE-oriented Nitinat fault system.

Continentward deepening of the lower reflection band in Line 84-01, from 7 s (22 km) on its SW end to 10 s (>30 km) on the NE, may suggest regional tilting of Vancouver Island towards the interior of the continent. This band, about 1 s wide, despite the laterally variable amplitudes (Milkereit et al., 1990), persists along the profile. Tilting of Vancouver Island, especially of the Eastern block, may be related to tectono-magmatic loading of the crust in the Coast Belt during the Cretaceous (Crawford et al., 1987; Rusmore and Woodsworth, 1991), and/or to west-vergent thrusting on the island's east side in the Late Cretaceous and early Tertiary (Brandon et al., 1988; England and Calon, 1991).

If the subhorizontal reflection band at 4-5 s has a recent metamorphic origin, that it ignores this tilt is not surprising. The deep seismic events, which might represent older metamorphic fronts, shear zones or primary compositional variations in the lower crust, are tilted with Vancouver Island.

At least some refraction models suggest that the Alberni-Cowichan Lake structural zone continues, steeply, to the base of the crust (Fig. 43; Drew and Clowes, 1990). In the reflection Line 84-01 (Fig. 46), many short events, dipping in opposite directions, lie on the east flank of the synclinorium between 2 and 10 s. Shallow events at 2-3 s dip mostly towards the synclinorium.

On the north end of Line 84-03, short events dip towards the synclinorium to a traveltime of 5 s. However, these profiles do not intersect, and correlations between them are complicated by presence of both NW- and N-S-oriented faults (Figs. 14, 16).

On the SW end of Line 84-01, in the Westcoast belt area, short seismic events are observed between traveltimes of 3.5 and 7.5 s. Mostly they dip towards the synclinorium. Thus, both the Alberni-Cowichan Lake and Westcoast structural zones, as well as the synclinorium and the anticlinorium which they bound, are rooted in the lower crust.

This conclusion is consistent with the suggestion from refraction models that the continental crust of Wrangellia is about 40 km thick. It is diffucult to reconcile with those interpretations which assume a thin Wrangellia - modeled to be just 12 km thick on the west side of the island and 18 km on the east side (Dehler and Clowes, 1992).

Presence on Vancouver Island of Phanerozoic rocks of various ages rules out large thinning of the crust by massive surface erosion. Though the island was uplifted during most of the Tertiary, the rate of uplift in most areas was slow. The ideas that Wrangellia was anomalously thin to begin with (Drew and Clowes, 1990) or abraded from below by subducting oceanic slabs (Yorath et al., 1985b; Clowes et al., 1987) - later to be thickened again by underplating of oceanic-derived material - are inconsistent with refraction models showing the Vancouver island lower crust to have continental-type properties.

These and other inconsistencies of Vancouver Island geophysical models with each other and with observable geology illustrate a cardinal rule of interpretation: geophysical solutions are not unique, and they need to be tested by joining various types of geophysical data with direct geological constraints (e.g., Pakiser and Mooney, eds., 1989).

Inconsistencies in current tectonic models of evolution of Vancouver Island and adjacent submerged margin
Models of the structure of the upper crust can be tested directly by geological observations at the surface and by drilling. For deeper crust and mantle, verification of geophysical models is

more difficult. This proved to be the case with the high-velocity, high-density body sometimes modeled at mid-crustal depths under western Vancouver Island and adjacent submerged margin. Existence of a body was suspected originally from gravity data (Stacey, 1973), but to this day its presence has not been confirmed unequivocally.

Early gravity models of Riddihough (1979) showed a slab of oceanic lithosphere underthrusted beneath the largely undifferentiated continental lithosphere of Vancouver Island and western Washington. This general idea became the basis for many subsequent interpretations of regional geophysical data (Keen and Hyndman, 1979; Clowes et al., 1984, 1987; Yorath et al., 1985a,b). Though it has been demonstrated in western Washington that the gravity data by themselves cannot resolve a subducted slab (Finn, 1990), in British Columbia its presence is still considered evident (Hyndman et al., 1990; Dehler and Clowes, 1992).

Plate reconstructions indeed require convergence of the Juan de Fuca and North America plates. The Cascadia subduction zone, though "unusual" in many respects (Acharya, 1992), is well developed in western Oregon and most of western Washington, where internal disruptions of the Juan de Fuca plate are small. By contrast, on both the southern and northern ends of this oceanic plate, magnetic stripes are broken or curved and earthquakes occur in the plate interior (Wilson, 1986; Stoddard, 1987, 1991; Atwater and Severinghaus, 1989; Wahlström and Rogers, 1990, 1992). These parts of the Juan de Fuca plate may no longer be subducting (Couch and Riddihough, 1989; Davis and Currie, 1993).

In a seismic refraction model across the Vancouver Island margin, Spence et al. (1985) included, at a depth of 17 to 29 km within the crust under western Vancouver Island and the exterior shelf, a body of material with a very high velocity of 7.7 km/s (Fig. 47). This body, placed beneath a thin continental crust of Wrangellia and corresponding roughly to the transparent zone between the two reflection bands in Line 84-01, was interpreted as an underplated sliver of oceanic crust detached from a subducted slab (see also Clowes et al., 1984, 1987). This idea was judged premature (Sobczak, 1988), and from the same seismic data many alternative models were developed later (see papers in Green, ed., 1990).

Mereu (1990) modeled a single Vancouver Island crustal block continuing as far as the mid-shelf, where a steep fault separates it from an outboard block of transitional crust beneath the outer shelf and upper slope (Fig. 48). This model contains no distinct high-velocity body in the middle crust. No such body was included in many other models, which showed the Vancouver Island crust to have a layered structure with velocities no higher than 6.95 km/s (Fowler and Pandit, 1990; Iwasaki and Shimamura, 1990). This high-velocity body was reinterpreted by Drew and Clowes (1990) as a wedge of four layers with seismic velocities from 6.35 to 7.1-7.2 km/s (Fig. 43). It is terminated under central Vancouver Island by the Alberni-Cowichan Lake fault system, but to the west it thins gradually. Such a wedge might be explained by presence in the crust of intrusive igneous rocks with parameters similar to, say, the Karmutsen Formation.

On a cautionary note, even the postulated subducted slab has not yet been reliably imaged with refraction seismic data. Though the

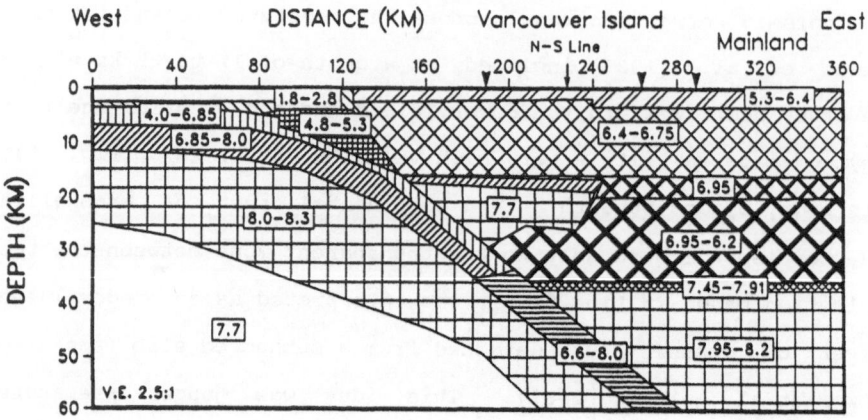

Figure 47. The original seismic refraction model of Spence et al. (1985) for the LITHOPROBE profile across Vancouver Island and the adjacent submerged continental margin (profile location is given in Fig. 45). Numbers give P-wave velocities in km/s. Note the mid-crustal body with velocity of 7.7 km/s, and the steep crustal-scale boundary in the middle of Vancouver Island coinciding with the Alberni-Cowichan Lake structural zone (geologic maps of Vancouver Island are presented in Figs. 14, 16).

209

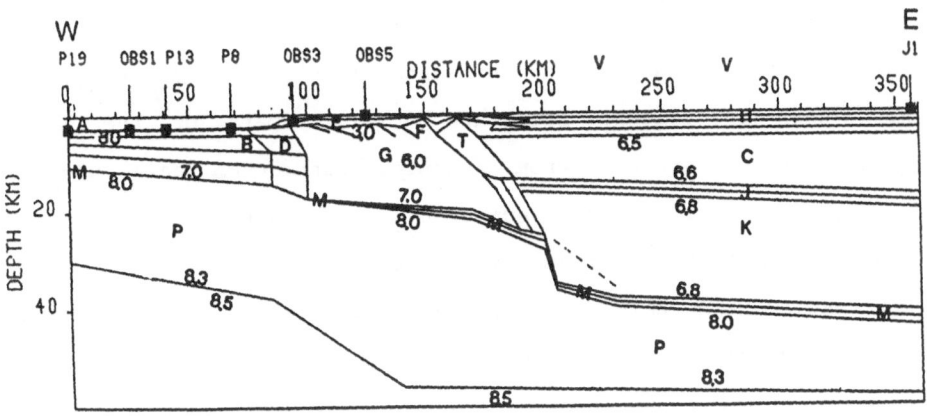

A. — Pacific Ocean
C. — Intermediate continental crust
E. — Young low velocity sediments
G. — Basement rocks
J. — Intermediate discontinuity
M. — Moho
V. — Vancouver Island.

B. — Oceanic sediments and crust
D. — Deformation front
F. — Folded/faulted sediments
H. — Upper continental crust
K. — Lower crust with zero gradient
P. — Juan de Fuca plate
T. — Tofino fault

Figure 48. An alternative seismic refraction model of Mereu (1990) for the LITHOPROBE profile across Vancouver Island and the adjacent submerged continental margin (profile location is given in Fig. 45). Numbers give P-wave velocities in km/s. The outer continental shelf and upper slope in this model are underlain by a block of transitional crust. The crust thickens continentward stepwise, across steep faults.

reason for this may be poor data quality at depth (Morgan and Warner, 1990; Iwasaki and Shimamura, 1990), Thybo (1990, p. 69) noted that "the velocity structure of this part of the model cannot be determined from existing refraction data and is constrained mainly by the assumption [sic] that the subducting slab is continuous between different parts of the model". It is interesting, in this regard, that the top of the presumed subducting slab was modeled by Mereu (1990) to be deepening towards the continent stepwise and that a block of transitional crust was modeled under the outer shelf and upper slope (Fig. 48).

Fundamental disagreements remain also in the interpretation of seismic reflection data, whether the broad reflection bands beneath Vancouver Island have a tectonic origin (Yorath et al., 1985b; Clowes et al., 1987; Calvert and Clowes, 1990) or represent metamorphic fronts (Hyndman, 1988; Hyndman et al., 1990). Thus, constraints on the structure of Vancouver Island and the continental margin available from seismic data are only loose.

Geologic sketch of the Vancouver Island exterior shelf
Geologically, the crust under the shelf was initially considered an extension of the Vancouver Island crust (Shouldice, 1971), as was suggested also in the early gravity models (Stacey and Stephens, 1969; Riddihough, 1979). In current refraction models, continental (Figs. 43, 47) or transitional (Fig. 48) crust is shown under the shelf and even the upper slope.

In other interpretations, however, continental crust is shown ending at the Westcoast fault, i.e. at the outer boundary of the Westcoast metamorphic belt (Davis and Hyndman, 1989; Hyndman et al., 1990). To the west, running along Vancouver Island roughly parallel to the present-day shoreline, three tectonic zones have been postulated: from east to west, the Pacific Rim "terrane", the Crescent "terrane", and the still-growing sedimentary accretionary wedge (Davis and Hyndman, 1989; Dehler and Clowes, 1992). The Tertiary Tofino Basin, overlying the presumably accreted rocks, was interpreted as a fore-arc basin.

Many NW-oriented faults are indeed present on the coast and on the shelf. The Tofino fault on the inner shelf (e.g., Brandon, 1989a) is probably a local branch of the Westcoast fault. A larger fault (Fig. 15) runs from the Olympic Peninsula offshore, towards the Prometheus magnetic anomaly and farther WNW (MacLeod et al., 1977; Snavely, 1987).

But presence of faults does not imply presence of exotic terranes. Pacific Rim rocks are exposed along the Westcoast fault in a narrow tectonic slice (Brandon, 1989a,b), and the Pandora Peak unit on southern Vancouver Island is confined largely to slices in the San Juan fault zone (Rusmore and Cowan, 1985). Emplacement of these rocks probably occurred by strike-slip faulting in the Late Cretaceous or early Tertiary. Basaltic massifs of the Crescent Formation are distinct bodies produced by eruptions from separate centers in a rift setting (Brandon and Vance, 1992; Babcock et al., 1992, 1994). Seismic refraction models show the Vancouver Island shelf to be underlain by uniform crust of continental type. Various lines of geological and geophysical evidence suggest that like on Vancouver Island, this crust is broken into blocks.

That a large raised block lies in southern Queen Charlotte Sound is suggested by exposures of Jurassic Island Intrusions and Bonanza Group rocks on the small Scott Islands (Muller et al., 1974), by free-air gravity anomaly values in excess of +60 mGal (Stacey and Stephens, 1969; Lyatsky, 1991a), and by thinness of Tertiary sediments in marine seismic lines. This block, named by Rohr and Dietrich (1992) Cape Scott High, extends westward as far as the edge of the shelf. There it is truncated by the Scott Islands fracture zone (Dehlinger et al., 1970), which the seismic line in Fig. 36 shows to be a west-dipping normal fault.

West of northern Vancouver Island, on the exterior shelf north and south of Brooks Peninsula, lie two blocks tilted towards the ocean (Tiffin et al., 1972). The Northern block (new name) is covered by 100 to 400 m of Neogene sedimentary rocks. The rectangular Broooks Peninsula, bounded by strands of the Brooks fracture zone, is composed of Jurassic metamorphic rocks of the Westcoast complex (Muller, 1977a,b; Nixon et al., 1995). The submerged Kyuquot block to the south, extending as far as Nootka Island, is covered by folded Paleogene rocks of unknown thickness and more than 600 m of west-dipping Neogene strata (see also Chase et al., 1975; Yorath, 1987). Magnetic maps and geologic correlations suggest the basement in both these shelf blocks contains rocks of the Westcoast complex.

The outer boundary of these blocks is a NW-trending fault parallel to the Scott Islands fracture zone, which also controls the edge of the shelf. All along the shelf edge, seismic data show that sediments have slumped down the steep upper continental slope (Tiffin et al., 1972).

In the Northern block, the basement is very shallow. In seismic sections, it lies at 100 m below seabed on the inner shelf and at 400 m on the outer shelf. High free-air gravity anomaly values (+15 to +30 mGal, similar to Brooks Peninsula; Fig. 49) confirm that the low-density cover is thin. In seismic sections, the slightly irregular basement and the overlying sedimentary strata dip towards the shelf edge. Quartzite and conglomerate dredged on the slope are similar to Lower Cretaceous rocks on Vancouver Island (Tiffin et al., 1972; Chase et al., 1975). Lower Miocene to Pliocene sedimentary-rock ages have been confirmed by dredging on the upper slope.

Free-air gravity anomalies over the Kyuquot block are lower, -10 to -20 mGal (Fig. 49). The sedimentary cover in marine seismic lines is thicker and consists primarily of folded Upper Eocene and Oligocene strata (Tiffin et al., 1972). Oligocene rocks dipping 25° to the SW are exposed onshore locally north of Esperanza Inlet (Muller et al., 1981). Near the shelf edge, Eocene-Oligocene rocks are overlain, with an angular unconformity, by Upper Miocene and Pliocene strata more than 600 m thick. The unconformity and the Neogene beds dip >5° towards the shelf edge.

The southern end of the Kyuquot block, at the NE-trending fault which runs onto the shelf from Esperanza Inlet (Fig. 50), is reflected in potential-field maps. South of this fault, free-air gravity anomaly values drop to -30 mGal (Fig. 49), and the Apollo J-14 well penetrated 3095 m of Neogene sedimentary rocks but failed to reach the basement (Shouldice, 1971; Yorath, 1980).

Jurassic metamorphic rocks, exposed on Brooks Peninsula, cause

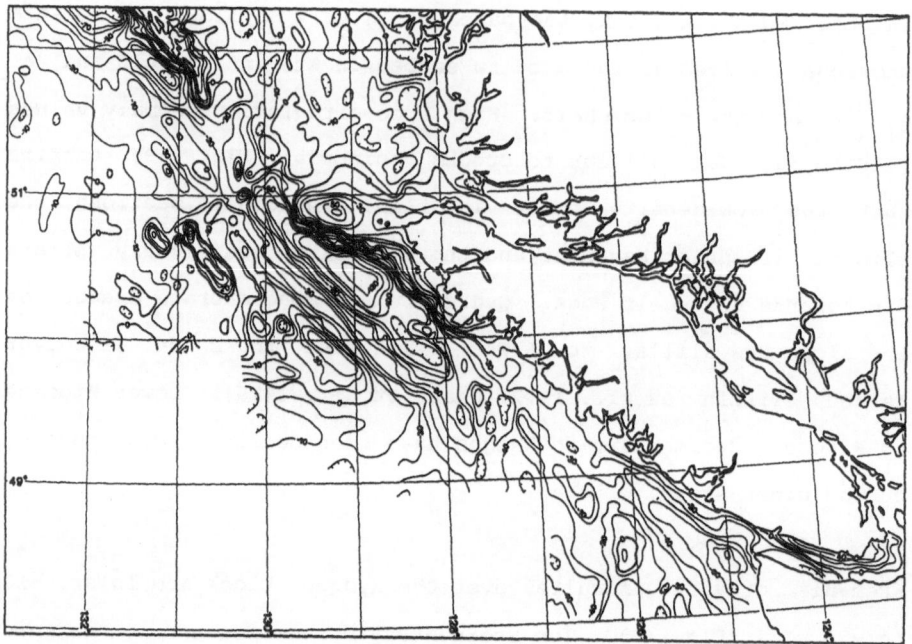

Figure 49. Free-air gravity anomaly map of the submerged continental margin off Vancouver Island and Queen Charlotte Sound (in mGal; after Currie et al., 1983b).

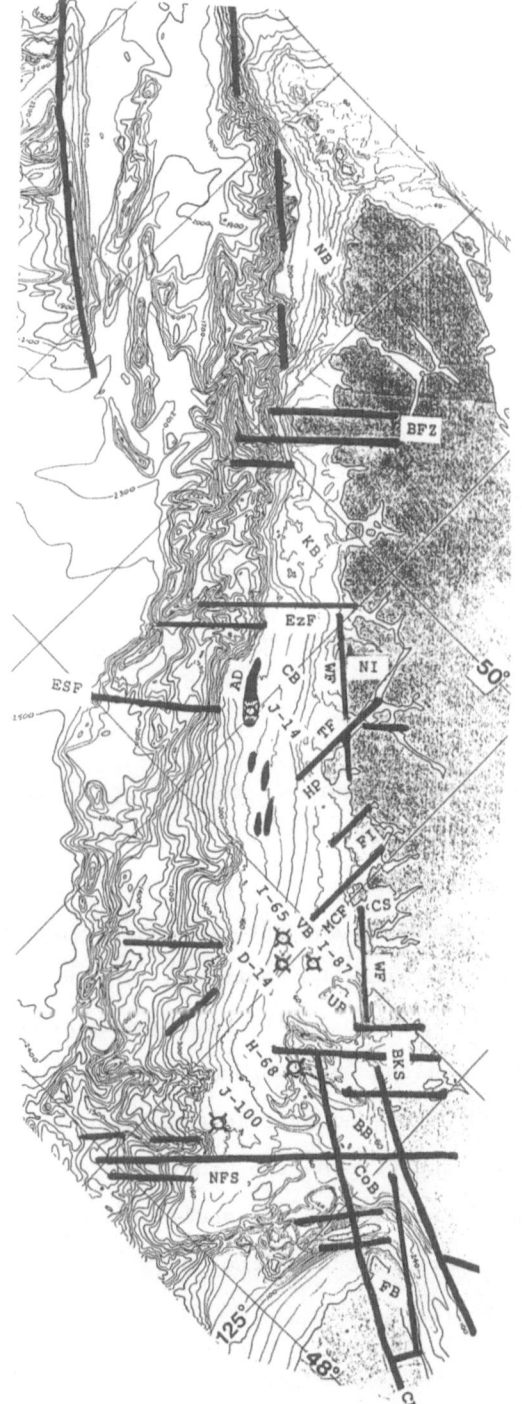

Figure 50. Main blocks and their bounding faults on the Vancouver Island continental shelf. Bathymetry is given in meters. Dark ovals indicate elongated diapirs as mapped by Tiffin et al. (1972). AD – Apollo diapir; BFZ – Brooks fracture zone; BKS – Barkley Sound; CF – Calawah fault; CS – Clayoquot Sound; EsF – Estevan fault; EzF – Esperanza fault; FI – Flores Island; HP – Hesquiat Peninsula; NFS – Nitinat fault system; NI – Nootka Island; MCF – Millar Channel fault; TF – Tahsis fault; WF – Westcoast fault; BB – Bamfield block; CB – Cove block; CoB – Clo-oose block; FB – Flattery block; KB – Kyuquot block; NB – Northern block; UB – Ucluth block; VB – Vargas Block.

magnetic anomalies elsewhere on Vancouver Island (Arkani-Hamed and
Strangway, 1988). For this reason, continuity of NW-trending
magnetic anomalies across Brooks Peninsula onto the Northern and
Kyuquot blocks (Fig. 27; see also Geological Survey of Canada Map
NM-9-10-M) probably indicates continuity of the Westcoast complex.
These anomalies stop abruptly near Esperanza Inlet and reappear
onshore, along with outcrops of Westcoast complex metamorphic
rocks. This suggests a 15-km left-lateral tectonic displacement
of the Westcoast belt. Farther inland, however, the Jurassic
Island Intrusions and Bonanza volcanics are not known to be
similarly offset by NE-trending faults (Muller et al., 1981). It
thus seems that movements on the Esperanza fault took place in the
Early Jurassic, when the Westcoast complex was still at deep
crustal levels, but prior to the Jurassic magmatism. Reactivation
of this fault in the Tertiary accounts for dissimilar subsidence
of shelf blocks to the north and south.

The Westcoast belt thus consists of two parts. The northern part
has as its outer boundary the Scott Islands fracture zone, which
is also the inner strand of the plate-boundary fault system. In
the southern part of the Westcoast belt, the outer boundary is the
Westcoast fault, which is connected with the northern strand of
the OWSZ. Parallel to it, the Calawah fault continues from the
Olympic Peninsula onto the shelf along the Prometheus magnetic
anomaly, which runs along southern Vancouver Island (MacLeod et
al., 1977; Snavely, 1987). This fault forms the southern strand
of the OWSZ. Thus, the central Vancouver Island shelf is the area
where these two very large structural zones, the OWSZ and the
plate-boundary fault system, meet.

Tectonic information from deep drilling in the Tofino Basin

Six deep hydrocarbon-exploration wells drilled in the 1960s by Shell Canada Ltd. on the Vancouver Island shelf, as well as coastal outcrops, provide direct information about the Tertiary sedimentary package in the Tofino Basin (Shouldice, 1971). In the wells and outcrops, the lithology is broadly similar - semi-consolidated siltstone and mudstone, with subordinate sandstone. A major difference occurs only in the Neogene rocks: mainly mudstone in the wells, as opposed to sandstone and conglomerate in coastal outcrops. This confirms that the shoreline during at least the Neogene lay near its present location. In the wells, age determinations are only approximate because of poor fossil control. Sandstone lenses, usually between 50 and 200 m thick, are not correlative between wells and cannot serve as markers. Large stratigraphic differences between the six wells reflect the complexity of block movements in the Tofino Basin.

Three wells off Clayoquot Sound, Pluto I-87, Zeus D-14 and Zeus I-65, though just 10-15 km apart, have encountered very different stratigraphic sections. The Zeus D-14 well penetrated about 2200 m of Early Miocene to Plio-Pleistocene rocks, underlain by basalts similar to those in the Eocene Crescent Formation. Three unconformities separate Lower and Middle Miocene (about 1800 m depth), Middle and Upper Miocene (1700 m), and Upper Miocene and Pliocene (1200 m) strata; sandstone bodies occur at the base of the sedimentary section (in the Lower Miocene) and at the Miocene-Pliocene boundary. Well Zeus I-65 penetrated about 2500 m of Oligocene and about 500 m of Lower Miocene rocks. This section is conformable and contains five sandstone intervals: one at the base of the section, two higher up in the Oligocene, one in the undifferentiated Upper Oligocene-Lower Miocene, and one in the Lower Miocene. The Pluto I-87 well contains almost 2000 m of Eocene and Oligocene sedimentary rocks, overlain by a similar thickness of Lower and Middle Miocene rocks. Unconformities occur between Eocene and Oligocene (depth 3200 m) and Lower and Middle Miocene (600 m) rocks. Three thin sandstone bodies are found in the Eocene, Oligocene, and at the Lower-Middle Miocene boundary.

The Prometheus H-68 well has a section similar to that in the Zeus D-14 drillhole, except that the intra-Miocene unconformities merge and the Middle Miocene is absent. The unconformity between the Lower and Upper Miocene lies at about 1600 m. The Upper Miocene section ends at an unconformity at about 1200 m, above which lies a conformable Plio-Pleistocene section. No Paleogene rocks occur, and at about 1800 m depth Crescent volcanics lie directly beneath Neogene sedimentary rocks. Four sandstone intervals have been encountered: at the base of the sedimentary section in the Lower Miocene, in the Upper Miocene, in the Pliocene, and in the undifferentiated Upper Pliocene-Pleistocene.

On the outer shelf off southern Nootka Island, the Apollo J-14 well penetrated a Pleistocene mud diapir. It encountered nearly 3100 m of Early Miocene to Pleistocene rocks, with unconformities between the Lower and Middle Miocene (about 2600 m), Middle Miocene and Lower Pliocene (1500 m), and Lower Pliocene and undifferentlated Plio-Pleistocene (900 m). Two sandstone intervals lie in the Lower Miocene, one in the Middle Miocene, and one at the Middle Miocene-Lower Pliocene boundary. The Cygnet J-100 well, on the outer shelf off Cape Flattery, penetrated 2460 m of Upper Miocene to Early Pliocene rocks, with an unconformity between the Upper Miocene and Lower Pliocene (1900 m) and two sand bodies in the Upper Miocene and at the Miocene-Pliocene boundary.

Such a stratigraphic variability suggests the Tofino Basin did not develop in the fore-arc setting. Fore-arc basins are usually broad and shallow, whereas the Tofino Basin is relatively narrow and deep. The Pluto I-87 well penetrated almost 4 km of sedimentary rocks and did not reach the basement. Several seismic refraction models (Iwasaki and Shimamura, 1990; Egger and Ansorge, 1990) suggest a sediment thickness of about 5 km. A similar thickness is indicated by presence of seismic reflections to 3-3.5 s two-way traveltime in Line 89-06 (SP 1700 and 2050; Fig. 51). Reflections under Crescent volcanics (1900 ms at SP 1840) suggest these basalts may be underlain by sedimentary rocks.

The Tofino Basin is graben-like, flanked by raised crustal blocks. Seismic data show under the shelf edge a buried basement ridge (Tiffin et al., 1972), which in Line 85-01 is apparent around SP 1400 (Fig. 52). Inboard of the basin lies the uplifted Vancouver Island. Marking the graben axis, the gravity low off central and southern Vancouver Island lies at mid-shelf. The bounding gradient zones are associated with major faults some of which are related to the OWSZ.

Well data show the Tofino Basin is overpressured (Shouldice, 1971), and elongated mud anticlines in it trend NNW (Tiffin et al., 1972). In defense of the subduction-complex model, Yorath (1980, 1987) postulated that shallow, east-dipping thrust faults which presumably cut the Tofino Basin were responsible for creating these anticlines. Neither wells nor seismic profiles (Figs. 51, 52) indicate a thrust structure of this basin.

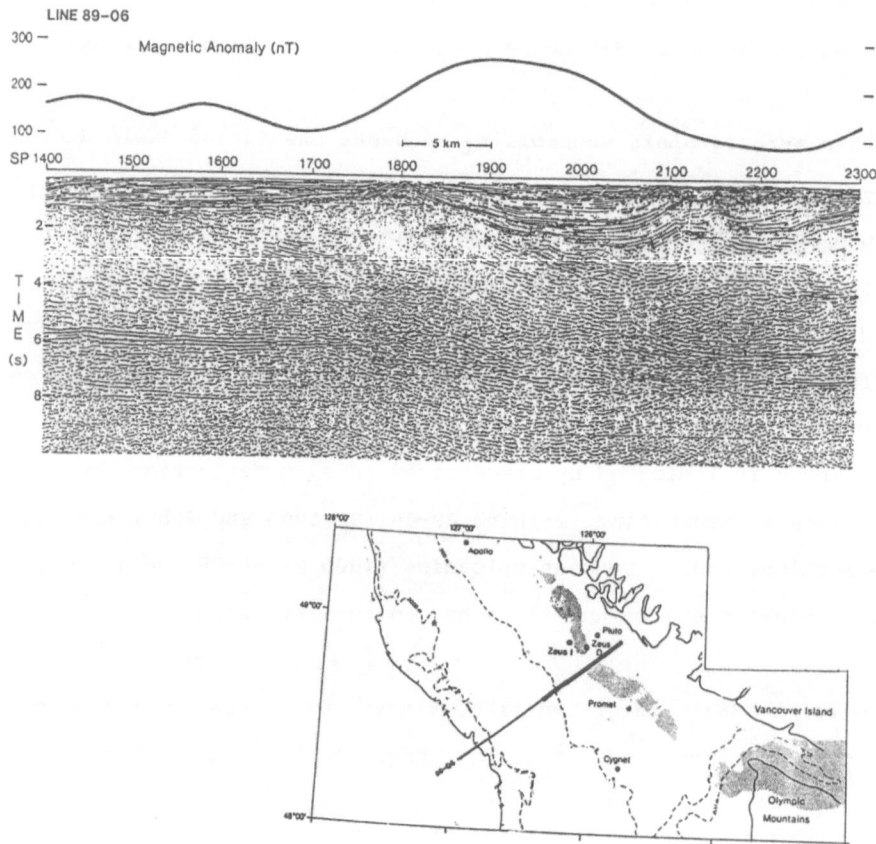

Figure 51. A buried igneous body in the Tofino Basin off central Vancouver Island, imaged in marine seismic reflection line 89-06 (modified from Spence et al., 1991). The box gives line location; shading represents magnetic highs; labeled dots represent wells. Reflections at 1900 ms beneath a body of Eocene Crescent Formation basalt (SP 1800-1840) suggest the basalt may be underlain by stratified sedimentary rocks. Similar stratigraphic relationship of Crescent volcanics with sedimentary rocks has been mapped on the Olympic Peninsula onshore (Babcock et al., 1994).

221

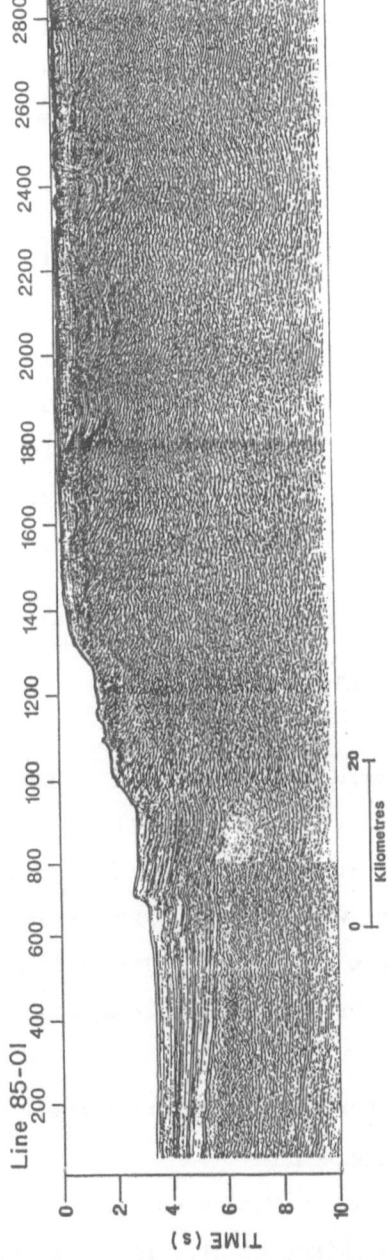

Figure 52. Submerged continental margin off southern Vancouver Island, from the shelf to the abyssal plain, imaged in seismic reflection line 85-01 (data after Yorath et al., 1987; Hyndman et al., 1990). Line location is given in Fig. 45; interpretation is discussed in text. Note the absence of thrust faults in the Tofino Basin, and a steep reverse fault on the outboard side of the Juno depression.

The coherent NNW orientation of mud anticlines (Fig. 50) indicates that diapiric movements were probably triggered by deep, steep faults. The Apollo J-14 well, for which detailed information is available, was drilled into such an anticline on the outer shelf (Shouldice, 1971; Yorath, 1980). It penetrated 3095 m of Neogene sedimentary rocks divided by unconformities into four sedimentary units: Lower Miocene (below 2561 m), Middle Miocene (2561 to 1540 m), Lower Pliocene (1540 to 896 m) and Plio-Pleistocene (above 896 m). This might indicate a complicated history of jerky, rapid subsidence and uplift. But despite the stratigraphic gaps, no major breaks are observed in wireline logs even at unconformities. The main lithology throughout the section is semi-consolidated grey mudstone and minor siltstone, with abundant pyrite and locally glauconite. This suggests that movements occurred rapidly, allowing shelfal conditions (probably below the storm wave base) to be re-established quickly.

Tectonic instability is also indicated by the many faults that cut this 3-4-km-wide, 17-km-long, NNW-trending anticline (Yorath, 1980). From old seismic data, it has been suggested that the Apollo structure is controlled by west-dipping, steep reverse faults (Shouldice, 1971). On trend with this anticline to the NNW lies the straight shelf edge of the Brooks-Estevan embayment.

Transverse faults and crustal structure of the exterior shelf
<u>Identification of blocks and bounding faults</u>
Transverse faults, commonly recognized on Vancouver Island (Muller et al., 1974, 1981; Jeletzky, 1976; Nixon et al., 1995), continue

on the shelf, where they bound a series of continental-crustal blocks characterized by dissimilar histories of subsidence and uplift. Large stratigraphic variations between even adjacent wells make it impossible to explain the development of the Tofino Basin in terms of simple two-dimensional models.

Transverse faults on western Vancouver Island control the NE and N-S orientation of many coastline features. Brooks Peninsula, which lies within a broad, NE-trending fracture zone extending all across Vancouver Island, separates the Northern and Kyuquot crustal blocks on the shelf. Farther outboard, this zone controls the north end of the Brooks-Estevan embayment, where the continental slope is abnormally narrow.

Other faults are indicated by bays with straight coastlines or by long, narrow fiords - Esperanza or Alberni inlets, Millar Channel, Nitinat Lake, Barkley or Clayoquot sounds. Topographic lineaments were accentuated in the Quaternary by glaciers. Geologically, transverse faults on Vancouver Island are manifested as breaks in outcrop patterns, shear zones, or linear belts of volcanic rocks and plutons.

Many large transverse faults continue on the submerged continental margin. Some of them are expressed in the bathymetry. Across the Esperanza fault, for example, the shelf deepens abruptly to the SE by about 40 m, as depth contours from 60 to 160 m jump 10-15 km inland. This shelf embayment, which bears no relation to the Brooks-Estevan embayment ouboard, continues as far as Barkley Sound (Fig. 28).

Two SW-tilted blocks on the shelf off northern Vancouver Island, Northern and Kyuquot, separated by the Brooks fracture zone (Tiffin et al., 1972), are cored by high-grade Jurassic metamorphic rocks of the Westcoast complex. However, a dissimilar Cenozoic history of their subsidence and uplift is reflected in their different Tertiary stratigraphy.

Across the Esperanza fault, to the south, the Westcoast complex is offset inland, and a different basement inderlies the central and southern parts of the shelf. Six continental-crustal blocks are distinguished there.

These blocks are most apparent on the inner shelf. Off southern Vancouver Island, their outboard boundary is the southern strand of the OWSZ. It is represented by the WNW-trending Calawah fault, expressed as a west-side-down gravity gradient zone and a west-side-up magnetic gradient zone (Dehler and Clowes, 1992, their Figs. 3-5). It separates inner-shelf blocks from the deeper part of the Tofino and northern Hoh basins at mid-shelf. Nevertheless, some of the transverse faults bounding can be traced far into these basins, across the shelf.

Unlike on the Northern and Kyuquot blocks, magnetic anomalies off central and southern Vancouver Island are mostly negative, commonly less than -100 nT. Only narrow, variously oriented positive anomalies up to +150 nT in amplitude lie between Flores Island and Nitinat Lake. Wells Zeus D-14 and Prometheus H-68 confirmed that they are caused by Eocene volcanic rocks. Unlike the Metchosin massif on southern Vancouver Island, these igneous bodies have no strong gravity signature and may lack deep roots.

Reflections under the basalt in seismic line 89-06 suggest they may be underlain by stratified sedimentary rocks.

The crystalline basement rocks on central and southern shelf cause no magnetic highs. On Vancouver Island, only the Leech River metamorphic complex is consistently associated with negative magnetic anomalies (Arkani-Hamed and Strangway, 1988). Material with seismic velocities virtually identical to those in the Leech River complex was included in some refraction models (Waldron et al., 1990; Drew and Clowes, 1990; cp. Mayrand et al., 1987) across the submerged continental margin off Barkley Sound.

Admittedly, K/Ar dates from the Leech River Complex of only about 40 Ma (Fairchild and Cowan, 1982) make it difficult to extend these rocks onto the shelf, because the unmetamorphosed, Early to Middle Eocene volcanics in the Tofino Basin are older. On the other hand, these K/Ar dates may reflect the age of cooling and uplift rather than metamorphism: in the Fuca Basin to the south, clasts from the Leech River complex are found in the Lyre Formation whose age is 42-40 Ma (Babcock et al., 1994). At any rate, presence under the Vancouver Island upper slope and shelf of similar but older metamorphic rocks would be consistent with all the geophysical evidence.

The existence of several basement blocks with dissimilar Tertiary history of relative movements is suggested by the drillhole information. Based on wells, coastal outcrops, and geophysical data, six crustal blocks have been newly recognized in the Tofino Basin. They are reviewed below, from north to south.

Cove block

The Cove block is separated from the Kyuqout block by the NE-trending Esperanza fault, and retains similar geophysical characteristics as far SE as Flores Island.

Still, this block is crosses by several other transverse faults. On trend with the Estevan fault, the shelf widens as its edge steps about 10 km outboard off southern Nootka Island. Farther outboard, the Estevan fault bounds from the south the Brooks-Estevan embayment. Another big fault, oriented N-S, runs into the Cove block without causing significant offsets. On Vancouver Island, it follows Tahsis Inlet, cutting Karmutsen basalts, Island Intrusions and the Westcoast complex. The latest movements on it occurred in the Neogene, because this N-S fault forms the western boundary of Hesquiat Peninsula, truncating Oligocene rocks of the Carmanah Group (Muller et al., 1974, 1981).

In coastal outcrops, metamorphic rocks of the Westcoast complex predominate, intruded by Jurassic and Eocene-Oligocene granitoids. Tertiary sedimentary rocks in places overlap the older rocks or are separated from them by the Westcoast fault.

Up to 1300 m of conformable Upper Eocene and Oligocene sedimentary rocks of the Escalante and Hesquiat formations (belonging to the Carmanah Group) have been mapped on Hesquiat Peninsula as well as on Nootka and Flores islands. This "unsorted slump-conglomerate", sandstone and shale was probably deposited in shelf to upper-slope settings. Only locally, on western Nootka Island, are they overlain unconformably by the Neogene Sooke Formation, found on a small seaward promontory named Bajo Point. These well-sorted conglomerate and sandstone, in this area just 20 m thick, were deposited in very shallow water. The entire Carmanah Group dips and thickens seaward (Muller et al., 1981).

An old seismic line across the shelf off Esperanza Inlet (Tiffin et al., 1972, their Fig. 7) showed about 200-250 m of Pliocene sedimentary rocks unconformably overlying older rocks. On the outer shelf, well Apollo J-14 penetrated 1540 m of Plio-Pleistocene rocks and 1555 m of Middle and Lower Miocene rocks (Shouldice, 1971; Yorath, 1980). No Paleogene rocks were reached in this well, and Upper Miocene rocks are missing. Presence of unconformities between the Lower and Middle Miocene, Middle Miocene and Pliocene, and Pliocene and undifferentiated Plio-Pleistocene indicates that Neogene subsidence was interrupted by several pulses of uplift. The Apollo diapir formed in the Pleistocene, creating an anticline that involved all older strata.

Deepening of the Tofino Basin is consistent with the development in this area of mud diapirs (Tiffin et al., 1972). They have been mapped mostly with seismic data, but some of them (e.g., Apollo) breach the sea floor. Concentration of the diapirs off central Vancouver Island may be caused by great basin depth, overpressuring due to impermeability of sedimentary rocks, and availability of basement faults to trigger mud flowage. That faults played a role is suggested by a common NNW elongation and alignment of the diapirs (Fig. 50).

Free-air gravity anomaly values over this part of the Vancouver Island shelf are negative, -20 to -30 mGal near the coast and -30 to -40 mGal at mid-shelf (Fig. 49). This also indicates that the

Cove block is downdropped relative to the Kyuquot block, where anomaly values are only -10 to -20 mGal.

Magnetic anomalies over the Cove block are mostly negative, from 0 to <-100 nT (Fig. 27). This suggests that rocks causing magnetic highs, such as Jurassic Island Intrusions or Westcoast complex on Vancouver Island, are probably absent beneath the sediments on the shelf. A small WNW-trending positive anomaly (0 to +50 nT) off Flores Island is probably caused by igneous rocks similar to the Eocene basalts that lie within the thick Tertiary sedimentary succession and are known to cause the Prometheus magnetic high farther south.

A WNW-trending linear magnetic low begins south of Hesquiat Peninsula and runs past Flores Island to Clayoquot Sound. This narrow low and the high off Flores Island end on the southward projection of the Tahsis fault. The magnetic low might be caused by metamorphic basement rocks, which on the flank of the Tofino graben lie close to the sea bottom.

Vargas block

North-south faults probably separate the Cove and Vargas blocks. On both sides of Flores Island lie N-S inlets, and on trend with them in the interior of Vancouver Island outcrops of Karmutsen volcanics and Island Intrusions trend N-S along longitudes 126°05'W to 126°25'W (Fig. 16; Muller et al., 1981). The Carmanah Group crops out on southern Flores Island, but the Pacific Rim complex is exposed on Vargas Island and Ucluth Peninsula to the southeast.

The N-S Millar Channel fault zone on the east side of Flores Island even seems to offset the WNW-oriented Westcoast fault left-laterally about 5 km (cp. Brandon, 1989a, his Figs. 1, 2). To the SE, the Westcoast fault separates the Westcoast and Pacific Rim complexes. The latter was believed by Brandon (1989a) to be a tectonic slice 15 km wide between the Westcoast and Tofino faults.

The Pacific Rim complex, exposed in coastal outcrops, has no strong signature in magnetic maps (Arkani-Hamed and Strangway, 1988; Dehler and Clowes, 1992, their Fig. 12). Over the Vargas block, magnetic anomaly values remain around -50 nT on the inner shelf, and become positive 10-15 km from the coast (Fig. 27). Well Zeus I-65 at mid-shelf penetrated 3042 m of sedimentary rocks, which are mostly Oligocene: only about 500 m of Lower Miocene strata lie at the top of the section (Shouldice, 1971). This indicates that the Vargas block was downdropped in the Paleogene and early Neogene but is now uplifted.

Relative uplift of this block is consistent with gravity data, as free-air anomaly values are higher than over the adjacent blocks to the north and south (Fig. 49). A broad relative high of 0 to -10 mGal lies off Clayoquot Sound, and a narrow and segmented low, <-30 mGal in amplitude, lies at mid-shelf. To the south, over the Ucluth block, gravity values are lower, declining rapidly from -10 mGal at the coast to -40 and -50 mGal at mid-shelf.

Ucluth block

The boundary between the Vargas and Ucluth blocks is unclear. Large stratigraphic differences between three closely spaced wells - Pluto I-87, Zeus D-14 and Zeus I-65 (Shouldice, 1971) - confirm that this area was tectonically active during the Tertiary.

Zeus D-14 encountered, above Eocene volcanics, some 2200 m of Neogene sedimentary rocks with unconformities between the Lower and Middle Miocene, Middle and Upper Miocene, and Upper Miocene and Pliocene. Paleogene sedimentary rocks are absent. Pluto I-87, in contrast, bottomed out in Eocene sedimentary rocks, separated by an unconformity from the Oligocene to Lower Miocene, in turn overlain unconformably by the Middle Miocene. Paleogene sedimentary rocks in this well are almost 2 km thick, covered by a similar thickness of Neogene rocks. Seismic line 89-06 in this area contains reflections suggesting stratified rocks to a depth of about 5 km (traveltimes of 3-3.5 s, SP 1700 and 2050, Fig. 51).

Such a great thickness of sedimentary rocks confirms that the Ucluth block is relatively downdropped. The relatively shallow Eocene basalts at the bottom of the Zeus D-14 well do not form the basement. Magnetic highs related to these volcanics are narrow and change their orientation: NNW in the Vargas block, mostly WNW in the Ucluth block. Presence of stratified rocks beneath the volcanics is suggested by reflections at 1900 ms at SP 1840 (Line 89-06). The boundary between these magnetic highs, and between the Eocene volcanic bodies they represent, coincides with a fault junction in the are of the three wells.

This junction is also indicated in gravity maps (Fig. 53). To the southeast, the Tofino free-air low broadens and increases in amplitude, from about -30 mGal off Clayoquot Sound to <-50 mGal off Barkley Sound. Its western boundary is a gentle, NNW-trending gradient zone which continues to the fault junction. Outboard of

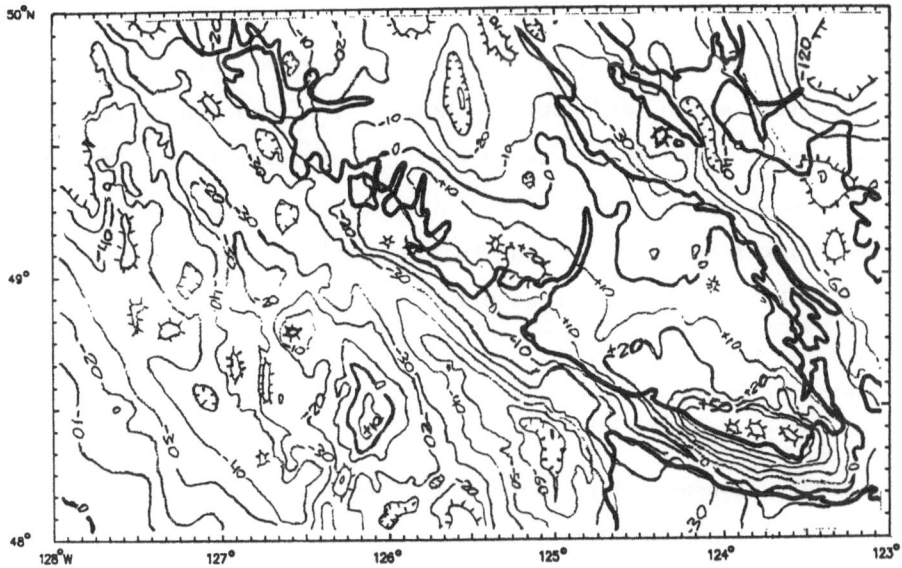

Figure 53. Gravity anomaly map of southern and central Vancouver Island and adjacent submerged continental margin; Bouguer on land, free-air offshore (in mGal; simplified from Fig. 3 of Dehler and Clowes, 1992).

the junction lie local gravity anomalies oriented NE and N-S,
suggesting that faults with these trends may be present on the
continental slope and outer shelf. The junction of all these
faults was an area of increased tectonic activity in the Tertiary,
perhaps accounting for the large stratigraphic differences between
three neigboring wells. In this junction lies the boundary
between the Ucluth and Vargas blocks.

Magnetic anomaly values over the Ucluth block drop below -100 nT
some 12 km from the coast (Fig. 27). This confirms that
relatively non-magnetic rocks of the Pacific Rim complex are
confined to a narrow slice. Between the Pacific Rim complex and
the Eocene volcanic bodies indicated by magnetic highs, strong
negative anomalies suggest presence of metamorphic rocks.

Bamfield block

The uplifted Bamfield block lies between central Barkley Sound and
Nitinat Lake. Inboard, Jurassic metamorphic rocks of the
Westcoast complex dominate coastal outcrops, and the Pacific Rim
complex is absent. The Bamfield block, likes the blocks to the
SE, lies on the inner shelf between strands of the OWSZ.

A steep magnetic gradient near the coastline separates the
magnetic low over this block from the high over the Westcoast
complex onshore (Arkani-Hamed and Strangway, 1988). This gradient
follows the Westcoast fault under shallow water on the inner
shelf. The Westcoast fault in this area is in line with the
northern strand of the OWSZ.

Outboard of the Bamfield block lies the mid-shelf Prometheus magnetic high of +150 nT. It is caused by a buried body of Eocene volcanics whose northern boundary is the Calawah fault (MacLeod et al., 1977; Snavely, 1987). A straight gradient zone trending WNW separates this magnetic high from the inner-shelf linear low of <-100 nT. This gradient zone, marking the outer boundary of the Bamfield block, coincides with the Calawah fault, which is part of the southern strand of the OWSZ (see Chapter 5).

The Bamfield block is relatively uplifted, and free-air gravity anomalies on the inner shelf are as high as 0 mGal. At mid-shelf, anomalies drop to -50 mGal over the Tofino graben. The boundary between the Bamfield block and the graben is marked by a gravity gradient zone associated with the Calawah fault.

At mid-shelf, on the outer flank of the Tofino gravity low and in the area of the magnetic high, the Prometheus H-68 well penetrated Eocene basalts at a depth of only about 1800 m, and remained in them to its total depth of 2335 m. No Paleogene rocks were encountered in this well, and Neogene strata overlie the basalts directly. Unconformities separate Lower and Upper Miocene, and Upper Miocene and Pliocene rocks (Shouldice, 1971). This suggests that subsidence occurred in the Neogene and was interrupted by occasional pulses of uplift.

On the outer shelf, the Cygnet J-100 well penetrated about 1900 m of Plio-Pleistocene sedimentary rocks, separated by an unconformity from Upper Miocene sedimentary rocks in which the well reached its total depth of 2460 m. The conformable Plio-Pleistocene sequence suggests that subsidence during that time was continuous. It had been preceded by alternative pulses of subsidence and uplift in the Miocene. Significantly, the Cygnet J-100 well penetrated bedded sedimentary rocks but did not encounter any mélange.

The submerged continental margin in this area is imaged in the seismic reflection profile 85-01 (Fig. 52), which forms an offshore continuation of Line 84-01 across Vancouver Island. It runs NE-SW from Barkley Sound to pelagic oceanic areas outboard of the continental slope. On the shelf, it shows strata of the Tofino Basin to be fairly continuous, though folded over deeper structures. The floor of the Tofino Basin is mostly undefined due to poor signal penetration. Locally, between SP 2100 and 2320, in the area of the Prometheus magnetic high, the strong SW-dipping event at 1500 to 2200 ms is associated with diffractions of the kind commonly observed along tops of igneous horizons. This event marks the top of the basaltic body: the Prometheus H-68 well, where volcanics were penetrated, lies at SP 2280.

Inboard, at SP 2320 to 2450, lies a fault-bounded half-graben more than 3 s deep, probably associated with the Calawah fault zone. The gravity minimum occurs over this half-graben. Inboard from it, the resolved stratified package thins dramatically over the raised Bamfield block, to about 500 ms on the NE end of this seismic profile.

The outer parts of the Tofino Basin, containing up to 2500 ms of resolved stratified sedimentary rocks, lie outboard of the main gravity minimum. Increase in gravity anomaly values on the west side of the basin may be caused by reduced sediment thickness (compared to that in the half-graben) or presence in places of dense Eocene volcanics.

Besides the OWSZ-related faults trending WNW, the Bamfield block is also bounded by NE-oriented faults. On the northwestern end of this block, presence of such faults is suggested by the NE orientation of the Barkley Sound coastlines. On the inner shelf, a local bathymetric depression as low as 200 m is also oriented NE (Fig. 28). The middle shelf shallows to less than 80 m, and the shelfal embayment which begins at Esperanza Inlet ends off Barkley Sound. Outboard, several NE-trending channels cut across the continental slope.

In the middle of Barkley Sound, positive magnetic anomalies related to the Westcoast metamorphic complex jump about 15 km inland to the north (Arkani-Hamed and Strangway, 1988; Dehler and Clowes, 1992, their Fig. 5). This suggests that the Westcoast belt is disrupted by transverse faults.

The southeastern boundary of the Bamfield block is the NE-trending Nitinat fault system, which begins on Vancouver Island and crosses the entire continental margin. On the continental slope and outer shelf, just south of the Cygnet J-100 well, it is represented by an elongated relative free-air gravity low, some 20 km wide and <-30 mGal in amplitude, between highs to the north and south (Fig. 53). On western Vancouver Island, in the area of the narrow, NE-trending Nitinat Lake, this fault system is associated with a parallel, narrow relative Bouguer gravity low. On the middle shelf, its projection passes between the elongated, NW-trending Tofino free-air gravity low of about -55 mGal off Vancouver Island and the more isometric low of <-80 mGal off the Olympic Peninsula.

This low is associated with the northern depocenter of the Hoh Basin (see Chapter 5). Thus, the Nitinat fault system is considered to be a boundary between the Tofino and Hoh basins.

On the inner shelf, the Nitinat fault system is imaged in seismic reflection Line 85-05, which runs parallel to the coast (Yorath et al., 1987). Though signal penetration and resolution are extremely poor (for that reason, the line is not reproduced here), at SP 950-1000 the resolved stratified section deepens from about 900 ms on the NW to about 1200 ms on the SE. Dip of some of the reflections in the direction of deepening suggests that sedimentary rocks may be draped over some step-like structure in the underlying basement. Consistent with this interpretation, free-air gravity values along the profile drop from about -15 mGal on the NW of the step to -30 mGal on the SE.

Clo-oose block

The Clo-oose block, downdropped and rectangular, lies SE of the Nitinat fault system. It is outlined by a box-shaped gravity low 30 km long and 15 km wide. This low, -30 to -40 mGal in amplitude, is flanked by steep gradient zones trending WNW and NE.

Coastal outcrops in this area contain Jurassic rocks of the Westcoast complex and Island Intrusions. Positive magnetic anomalies associated with these rocks end at a steep gradient zone along the coast, outboard of which lies a WNW-oriented magnetic low up to -150 nT in amplitude. The Nitinat fault system on the NW side of the Clo-oose block terminates the Prometheus magnetic anomaly, which is found only to the north.

The outboard boundary of the Clo-oose block is the Calawah fault. A steep WNW-trending magnetic gradient zone along it separates the strong inner-shelf low from moderate anomalies at mid-shelf, whose amplitude is only 0 to -50 nT. A gravity gradient zone along the Calawah fault separates the rectangular free-air anomaly on the inner shelf from the strong, negative anomaly at mid-shelf, related to the Hoh Basin.

Flattery block

This uplifted, rectangular block, 35 km long and 15 km wide, lies partly on the exterior shelf and partly in the Strait of Juan de Fuca. Gravity anomalies over it are as high as -10 mGal. A sharp WNW-trending gradient zone in the Strait of Juan de Fuca, which forms the northern boundary of the Flattery block, lies within the OWSZ. The southern boundary of this block is the gradient zone along the Calawah fault.

The western boundary of this block, near Cape Flattery, is marked by NE-trending gravity and magnetic gradient zones. The gravity gradient zone separates the rectangular anomalies over the Clo-oose and Flattery blocks. The NE-trending magnetic gradient zone marks the end of the negative magnetic domain on the inner shelf, and anomalies over the Strait of Juan de Fuca are strongly positive (>+500 nT) due to presence of Metchosin volcanics. A NE-trending fault is inferred to run across the continental shelf in this area, separating the Flattery and Clo-oose blocks.

An isolated magnetic high of >+100 nT is found west of Cape Flattery, on trend but unconnected with the Prometheus high. It may represent a small body of volcanics in the Hoh Basin. The western boundary of this magnetic anomaly is oriented NE and lies on trend with the inferred fault on the northwestern flank of the Flattery block. This fault may thus continue SW into the northern depocenter of the Hoh Basin. It may control the large sea-floor canyon which originates at the mouth of the Strait of Juan de Fuca and runs SW across the entire submerged continental margin.

The Flattery block is probably a fault-bounded structural high. Its basement has a complex composition, as indicated by exposures to the north and south. The Metchosin massif to the north contains at least 7 km of volcanic rocks (Muller, 1977c) and is probably underlain by a large intrusive body. To the south, outcrops at Point of the Arches on the west coast of the Olympic Peninsula contain Jurassic crystalline rocks of the continental-crust basement (see Chapter 5).

Crucial role of geological information in validation of geophysical models

On the scale of lithospheric plates, whose thickness may reach hundreds of kilometers, surface outcrops or drillholes a few kilometers deep may seem insignificant. In many recent papers on the Vancouver Island continental margin, information from the Tofino Basin wells is rarely mentioned. A notable exception is an old paper by Yorath (1980), which attempted to interpret the data from one well (Apollo J-14) in terms of ongoing subduction and related thrusting in the Tofino Basin. Minor distuptions in sedimentary rocks in this well, which was drilled into a mud diapir, were interpreted as low-angle thrusts.

Such shallow thrusts were shown by later surveys (mostly seismic;

Figs. 51, 52) to be absent. Perhaps it was this initial lack of success that led to disuse of the well data. Subsequent papers offered tectonic models for this area where the well data were barely mentioned (Hyndman et al., 1990; Spence et al., 1991; Dehler and Clowes, 1992). Still, well control is indispensable because it provides the only direct geologic information about the structure of the subsurface.

In some cases, even drilling in remote regions may provide information relevant to local models. In the lower crust under Vancouver Island, a zone of high electrical conductivity has been revealed by magnetotelluric surveys (Kurtz et al., 1986, 1990) along the deeper band of seismic reflections in Line 84-01 (Fig. 46). To explain this coincidence, Hyndman (1988) hypothesized that the high reflectivity and the high conductivity have a common cause: intracrustal saline fluids, presumably derived from the Pacific Ocean, carried down the subduction zone with the Juan de Fuca slab, and then expelled into porous zones along metamorphic fronts in the overriding continental crust.

Hyndman (1988) estimated the porosity in these lower-crustal reservoirs to be very high, in the range of 1 to 4 percent. This estimate is probably exaggerated, but small amounts of metamorphic water are now thought to be present in the lower crust of intracontinental regions that have nothing to do with subduction processes (e.g., Bailey, 1993). But due to lack of constraints, speculations about the structure of the lower crust are notoriously loose. A common cause of electrical conductivity is graphite (Jones et al., 1994), and intracrustal reflectivity may have a variety of geologic causes (Smithson and Johnson, 1989; Meissner, 1989; Meissner and Wever, 1992).

In the upper crust, however, geophysical models are in some areas constrained by drilling. The Kola superdeep well in the Fennoscandian Shield, far away from any continental margins, encountered several zones in the continental crystalline crust where fluids migrate freely. Intervals of crystal dissolution and "hydraulic disaggregation" were found at depths of 7-10 km. These zones are assiciated with seismic reflections and refractions, some of which had initially been misinterpreted as fundamental compositional boundaries in the continental crust, such as between the hypothetical "granitic" and "basaltic" layers (Kozlovsky, ed., 1984; Pavlenkova, 1989). On the other hand, drilling into a Cordilleran metamorphic core complex in Arizona showed that reflectivity in that area is caused by thin compositional layering in gneisses (Thompson et al., 1989).

Several continental drilling projects showed that porosity in fracture systems in the upper crust may occasionally be as high as 2-3% at shallow depths, but is usually 1% or less (Kozlovsky, ed., 1984; Zimmermann et al., 1992). Unexpectedly, it was found that circulation of fluids, along with variations in thermal properties of rocks, in places causes sharp increases in geothermal gradients which prior geophysical models failed to predict (Clauser and Huenges, 1993).

The Vancouver Island continental margin was modeled to contain three belts of oceanic-crustal origin outboard of the Vancouver Island "backstop": the Pacific Rim "terrane", the Crescent "terrane", and the modern sedimentary accretionary prism (Davis and Hyndman, 1989; Hyndman et al., 1990; Dehler and Clowes, 1992). These belts were thought to continue along the island for hundreds of kilometers. The Westcoast, Tofino and Hurricane Ridge faults were postulated to be low-angle thrusts dipping towards the continent. The small half-graben associated with the Calawah fault in the area of the Bamfield block (Line 85-01, SP 2320-2450; Fig. 52) was interpreted as a "fossil trench" over a former subduction megathrust. The Tofino Basin was regarded as a fore-arc basin lying on top of these deeper structural belts.

In defense of this scenario, several closely spaced gravity and magnetic models across the southern Vancouver Island margin were presented by Dehler and Clowes (1992), but parameters assumed in these models for the postulated structural belts are questionable. The Pacific Rim "terrane", for example, was assigned an extremely high density of 2,900 kg/m3. This is incompatible with results of outcrop mapping, which show this complex to contain a mélange of sedimentary and volcanic rocks cut by a granitoid pluton, with ultramafic rocks found only locally in displaced fragments (Brandon, 1989a,b). A high density like 2,900 kg/m3 implies that ultramafic rocks are present in abundance at shallow depth, but this is inconsistent with the subdued magnetic field over the Pacific Rim slice. By contrast, a presumed sediment-volcanic mélange off Washington has been modeled with a density of only 2,700-2,750 kg/m3 (Finn, 1990).

Also unrealistic in the models of Dehler and Clowes (1992) is the presentation of the Crescent "terrane" as a continuous, uniform basaltic panel 4 km thick. The Crescent Formation along the margin onshore is known to consist of many discontinuous igneous bodies with great lateral variations in composition and thickness (Babcock et al., 1992, 1994), whose distribution is reflected in gravity and magnetic maps (Finn, 1990). On the Vancouver Island shelf, isolated magnetic anomalies probably reflect the presence of only localized igneous bodies. Absence of correlative gravity anomalies confirmes these bodies are local.

The elongated Prometheus body sits on the Calawah fault (MacLeod et al., 1977), which is part of the southern strand of the OWSZ. The regional extent of this strand is apparent from the horizontal-gradient gravity map of Dehler and Clowes (1992, their Fig. 3), which highlighted a gravity lineament running from northwestern Olympic Peninsula along the exterior shelf as far as Barkley Sound. Continuity of the Calawah fault on the shelf was also noted by Snavely (1987), as illustrated in Fig. 15.

Distinct magnetic highs on the shelf (Fig. 27) indicate the presence of several volcanic bodies. Their trends vary between WNW and NNW, reflecting a complex interplay of faults along which these basalts were probably erupted.

Well data show that the Prometheus body is local and non-uniform. Volcanics were penetrated at a depth of about 1800 m in the Prometheus H-68 well, but at 2200 in the Zeus D-14 well. Just 10-15 km away from the Prometheus location, wells Zeus I-65 and Pluto I-87 encountered no igneous rocks at all, even though the Pluto well is nearly 4 km deep. Though basalts in the Prometheus well are up to 500 m thick, they have only a local distribution.

In seismic line 89-06 (Fig. 51), reflections under the basalts

(1900 ms, SP 1840) suggest that volcanic flows may be underlain by sedimentary rocks. Field mapping on the Olympic Peninsula has confirmed that sedimentary rocks underlie Crescent volcanics in that area (Babcock et al., 1994).

Existence on the shelf of eight crustal blocks whose bounding faults continue from Vancouver Island confirms that the crust in this area is probably continental. The same is indicated by exposure of Jurassic metamorphic rocks on Brooks Peninsula, and by refraction models across the margin off southern Vancouver Island. Large relative block movements, recorded in the stratigraphic differences between the offshore wells, are inconsistent with the development of the Tofino Basin in a fore-arc setting. Also contradicting the subduction-complex model, the Cygnet J-100 well drilled on the outer shelf to a depth of 2460 m did not encounter any mélange.

Distribution of crustal blocks on the continental slope is discussed in the next chapter.

CHAPTER 8 - STRUCTURE OF CONTINENTAL SLOPE OFF VANCOUVER ISLAND

Regional overview of Vancouver Island continental slope

The outer strand of the continental-oceanic plate-boundary structural zone, the Revere-Dellwood fault, continues under a cover of undisturbed sediments into northwestern Brooks-Estevan embayment. The continental slope in the embayment narrows, but off central and southern Vancouver Island the lower slope steps out and again lies on trend with the Revere-Dellwood fault.

The southern Vancouver Island continental slope, oriented NW-SE, is about 60 km wide. Seismic and sonar data show numerous sediment ridges, channels, slumps, ponds. Broad plateaus lie at water depths of about 1400 m and 2000-2200 m. Many anticlinal ridges hundreds of meters high, surrounded by areas of smooth seabed, are found on the lower slope (Fig. 28). The upper slope off southern Vancouver Island is steeper than the lower slope. A sharp break separates it from the shelf, which is smooth but cut locally by Pleistocene channels and Holocene glacial scours (Carter, 1974). The shelf widens to the south, to about 80 km.

The Tertiary sedimentary succession, drilled on the shelf to a depth of almost 4 km, continues on the upper continental slope. The Tofino Basin was initially thought to extend to the foot of the slope (Shouldice, 1971). One seismic profile showed late Tertiary strata continuing from the outer shelf onto the upper slope with only minor disruption (Snavely and Wagner, 1981). Later, however, the slope was reinterpreted as a thrust-faulted accretionary prism of sediments continually scraped off the subducting Juan de Fuca plate by the Wrangellian backstop (Davis and Hyndman, 1989; Hyndman et al., 1990).

Gravity and magnetic anomalies along the continental slope off northern Washington and southern British Columbia

Near-zero free-air gravity anomaly values characterize the upper continental slope off northern Washington and southern British Columbia. On the middle shelf lie negative anomalies: -30 to -55 mGal over the Tofino Basin, less than -80 mGal over the northern Hoh Basin. A low of -40 to -50 mGal marks the lower slope. Between these lows, near-zero anomalies on the upper slope and outer shelf form a relative high (Fig. 53).

Over the pelagic northeastern Pacific off Washington and British Columbia, free-air anomaly values decrease towards the continental margin gradually, in response to downbending of the oceanic crust and thickening of the overlying sediments. The sedimentary layer over the Juan de Fuca plate is 500 m thick on average but as much as 2500-3000 m thick in the Astoria and Nitinat fans at the foot of the slope (Carlson and Nelson, 1987). This sedimentary layer is conventionally denoted Cascadia Basin.

A pronounced gravity minimum of about -50 mGal lies along the foot of the slope off southern Vancouver Island. It is associated with the deepest part of the Cascadia Basin, and with the sediment-filled Juno structural depression (new name) in the oceanic crust under the lower slope. On the north, the Juno depression ends at the system of N-S-oriented faults, expressed as coincident gravity and magnetic anomalies; this structural zone also marks the southern end of the Brooks-Estevan embayment. On the south, the Juno gravity low is terminated by a NE-trending lineament related to the Nitinat fault system, which crosses the continental margin.

In a gravity model across the central Washington margin, Finn (1990) included under the shelf and upper slope about 1 to 2 km of sediments with a low density of 2,530 kg/m3 (Fig. 54). These sediments overlie material which had previously been modeled with a density of 2,800 kg/m3 (McClain, 1981). Finn considered this value exaggerated and assigned to this body a density of 2,700 to 2,750 kg/m3. Below this body she placed the undergoing oceanic slab. To explain negative gravity anomalies on the Washington and Oregon shelf, Couch and Riddihough (1989) had previously proposed a series of sediment-filled depressions. Finn (1990) thought the cause of such a low off central Washington is the material with the density of 2,700-2,750 kg/m3, which she modeled as a wedge thickening to as much as 20 km under the inner shelf. She interpreted it as a mix of sediments and basalts scraped off the subducting Juan de Fuca plate. Confusingly, though, such densities are also consistent with crystalline continental crust.

In the gravity models of Dehler and Clowes (1992), a wedge with densities of 2,410 to 2,610 kg/m3, some 10 km thick, was included under the outer shelf and slope off southern Vancouver Island. Such densities are similar to those measured in the Tofino and Hoh basins, but Dehler and Clowes (1992) interpreted this modeled material as an accretionary sedimentary prism. To compensate for this assumed low-density pile, they assigned a high density of 2,880 kg/m3 to the underlying body, which they presumed to be the oceanic Juan de Fuca slab. By comparison, Finn (1990) modeled this slab under the Washington margin to have a lower density of 2,850 kg/m3.

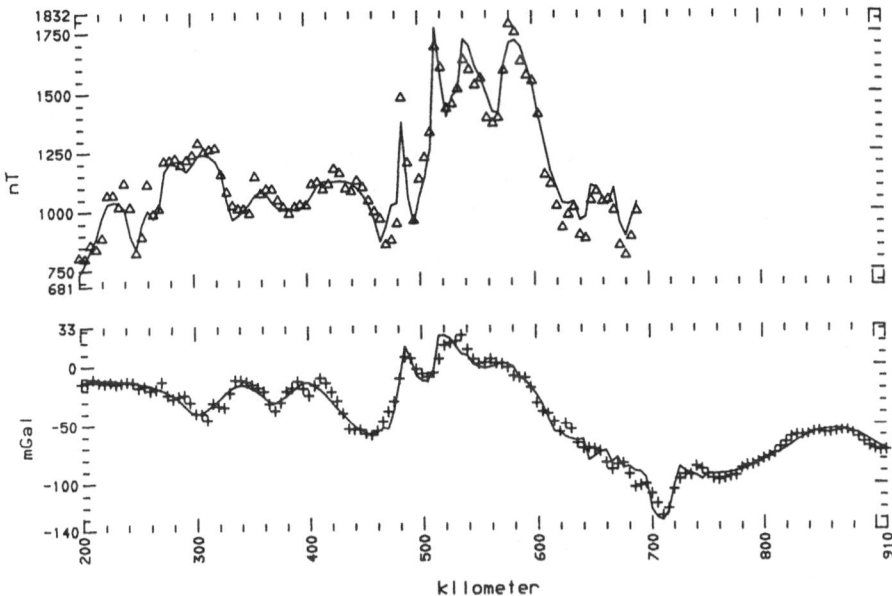

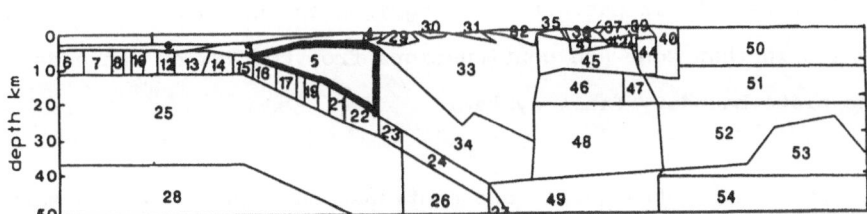

Figure 54. Crustal structure of central Washington continental margin, in an E-W gravity and magnetic model of Finn (1990). Under the shelf and slope, body 5 (in heavy outline) with density of 2,700-2,750 kg/m3 was interpreted by Finn as a mélange consisting of sheared sedimentary and volcanic rocks. However, as discussed in text, its modeled parameters are similar to those typical for attenuated continental crust.

Previous chapters showed that placement under the Vancouver Island shelf of underthrusted oceanic-crustal bodies is inconsistent with geological observations from field mapping and drilling, and with seismic refraction models. Rather, the shelf is underlain by variously uplifted and depressed, fault-bounded blocks with continental-crustal affinity. Steep faults accommodated thousands of meters of differential uplift and subsidence of crustal blocks on Vancouver Island and in the inboard parts of the submerged continental margin. Such a tectonic style has characterized the Cenozoic evolution of most of the Insular Belt, where long-lived fault networks were inherited from at least the Mesozoic (Lyatsky, 1991a, 1993a). A similar tectonic style indicates that continental crust extends beyond Vancouver Island into the submerged margin.

A different fault pattern is recognized on the lower slope. Many faults in that zone are continuations from the NNW of strands of the plate-boundary fault system.

Another major structural zone continues onto the sumberged margin off Vancouver Island from the ESE. The OWSZ, an intracontinental zone of crustal weakness stretching from the Cordilleran interior in Idaho and Oregon, crosses the Washington Cascades and persists along the Strait of Juan de Fuca onto the shelf (Fig. 7). New and rejuvenated old faults were incorporated into the tract of this zone. The North Olympic fault system continues on the shelf offshore, where the Calawah fault bounds several crustal blocks, controls the Prometheus body of Eocene volcanic rocks, and even disrupts Pleistocene deposts. The South Vancouver Island fault system continues WNW as the Westcoast fault.

On the eastern flank of the elongated Tofino gravity low on the
shelf lies the WNW-oriented gradient zone coinciding with the
OWSZ-related Calawah fault. On the western flank of this gravity
low lies a gentler, slightly irregular gradient zone that trends
NNW. Such an interplay of WNW and NNW trends occurs all over the
shelf (Fig. 50). Positive narrow magnetic anomalies caused by
Eocene basaltic bodies have both these orientations, suggesting
that faults along which these basalts erupted have similar trends.
The magnetic low on the inner shelf, thought to be associated with
metamorphic rocks, also trends WNW. The interplay of NNW and WNW
orientations idicates that branches of both the plate-boundary
fault system and the OWSZ have propagated on the submerged margin
off southern Vancouver Island.

On the lower continental slope and on the abyssal plain to the
west, the main magnetic anomalies are stripes oriented N-S. Their
amplitudes range from -300 nT to +400 nT. On the upper slope, the
only N-S magnetic anomaly is a faint low of about -100 nT at
longitude 126°W. Geophysical evidence presented in this chapter
suggests that whereas oceanic crust underlies the lower slope off
southern Vancouver Island, the crust beneath the upper slope is
most probably continental.

Deep structure of the southern Vancouver Island continental slope
The submerged continental margin off Vancouver Island and the
Olympic Peninsula has been imaged in seismic reflection profiles
acquired between 1976 and 1989. Signal penetration on the upper
slope is extremely poor, with a notable exception of the U.S.

Geological Survey Line 76-19 which is discussed below. In other profiles, the reflection character on the upper slope is chaotic.

Without good seismic reflection images, the most reliable information about the deep structure of the upper slope comes from a refraction profile across the continental margin off Barkley Sound (Fig. 45). However, interpretations of this profile are contradictory, as very different models have been produced from these data by different workers (Green, ed., 1990).

In the model of Mereu (1990), under the outer shelf and upper slope lies a block of crust with velocities increasing downward from 6 to 7 km/s, to the Moho at about 18 km (Fig. 48). On the upper slope, this block is covered by 1-2 km of low-velocity sediments. The faults bounding it are steep, and the Moho deepens across them towards the continent stepwise. In the model of Weber (1990), the crust under the upper slope has velocities between 4.8 and 5.7 km/s and is covered by up to 3.5 km of sediments with velocities from <3.3 km/s to 3.9 km/s. Thybo (1990) modeled a similar upper-slope sediment section but gave the underlying crystalline crust velocities of 6 km/s.

The detailed model of Waldron et al. (1990) showed the sedimentary cover as being 3 to 4 km thick, with velocities increasing with depth from 1.9 to 3.2 km/s (Fig. 55). Below lies material whose velocities increase downward from 4.8 km/s to about 5.5 km/s. It is underlain by an oceanic slab, which deepens from 7 km under the middle slope to 12 km under the shelf edge.

Such a diversity of models indicates that the refraction profile

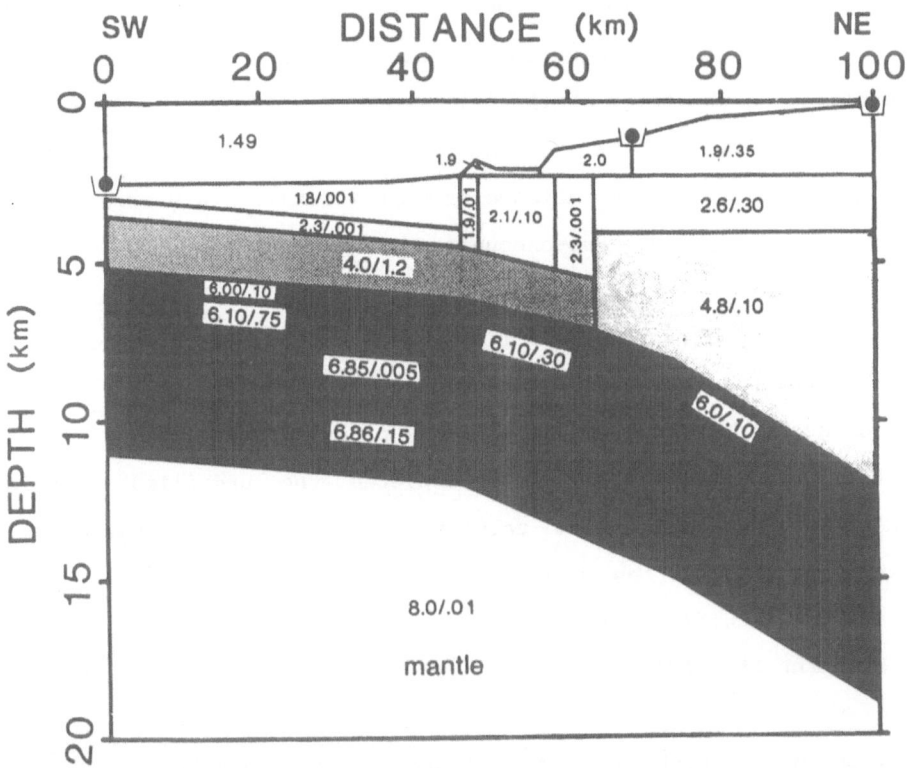

Figure 55. Velocity structure of the southern Vancouver Island continental slope and adjacent areas, as modeled from LITHOPROBE seismic refraction data by Waldron et al. (1990). The first number in each block is the P-wave velocity along the top boundary in km/s, the second is the linear velocity gradient in km/s/km. On the lower slope, the Juno depression between vertical faults is well expressed. Beneath the upper slope, the velocity of 4.8 km/s (gradient 0.10 km/s/km) is almost identical to that in the Leech River metamorphic complex on southern Vancouver Island. The Juno depression lies between 45 and 63 km.

offers no unique solution, and a comprehensive approach should involve other considerations. Just 18 km SE of this seismic profile, on the outer shelf, well Cygnet J-100 penetrated 2460 m of stratified Late Miocene and Pliocene sedimentary rocks (Shouldice, 1971). No mélange or thrust faults, expected in an accretionary prism, were encountered.

Waldron et al. (1990, p. 112) cautioned that "existing data and their interpretations are inadequate to clarify what tectonic processes are taking, or have taken, place". Still, these workers (their velocity profile C) interpreted the undrilled material below 4 km depth, whose velocity they modeled to be between 4.8 km/s and 5.5 km/s, as an accretionary wedge. However, a total sediment thickness of about 10 km required by such an interpretation (Fig. 55) is inconsistent with other evidence.

By proposing thick sediments off Oregon and Washington, whether in depocenters or in accretionary wedges, Couch and Riddihough (1989) and Finn (1990) sought to explain the very low gravity values (commonly -60 mGal or less) on the continental shelf. In contrast, on the southern Vancouver Island upper slope and outer shelf free-air anomaly values are generally near zero; in the area of the refraction profile, they exceed +10 mGal (Fig. 53). The relative gravity high along the outer shelf and upper slope contains several local maxima, of which the one crossed by the refraction transect is the highest. These maxima form a NW-trending band between longitudes 126°W and 127°W. They are separated by transverse gradient zones trending NE and N-S, whose causative faults seem to continue to the mid-shelf and contribute to the long-lived structural complexity indicated by large stratigraphic differences in the three neighboring wells in the Vargas and Ucluet blocks (see previous chapter). Presence of old transverse faults on the upper slope is hard to explain unless this area is underlain by old continental crust.

On southern Vancouver Island, seismic velocities almost identical to those modeled by Waldron et al. (1990) under the upper slope have been found to characterize the Leech River complex (4.6 to 5.5 km/s; Mayrand et al., 1987), which consists of high-grade metamorphic rocks. Surface measured densities of the Leech River complex are 2,670 to 2,750 kg/m3 (Walcott, 1967; see also Dehler and Clowes, 1992), almost identical to the values used by Finn (1990) for the wedge she modeled under the submerged margin off Washington (Fig. 54). Seismic refraction data show that velocities in the deep parts of this wedge may reach 6 km/s (Taber and Lewis, 1986). Thus, in the geophysical sense, presence of crystalline rocks - such as those in the Leech River complex - underneath parts of the submerged margin off Washington and Vancouver Island is consistent with all the evidence.

Above these deep rocks, the model of Waldron et al. (1990) shows up to 4 km of sediments with velocities between just 1.9 and 3.2 km/s. Stratified fine-grained sedimentary rocks in the Cygnet J-100 well near the shelf edge, drilled to a depth of 2460 m, suggest that fairly stable marine conditions existed there during the Late Miocene and Plio-Pleistocene. There is only one disconformity (between the Upper Miocene and Lower Pliocene), and no indication of thrust faults or mélange.

The conformable Plio-Pleistocene succession in this well is about

1900 m thick. Seismic line 76-19 (Fig. 56), which crosses the continental margin near the Cygnet location, shows that Plio-Pleistocene strata continue for about 30 km on the upper slope and thin out to zero only at mid-slope (Snavely and Wagner, 1981; Snavely, 1987). There, a bathymetric break at about 1000 m water depth coincides with a series of continentward-dipping thrust faults. Inboard from this break, late Tertiary strata are draped over structures in older rocks. Outboard, all sediments except for the very youngest (Quaternary?) are involved in penetrative deformation, and at the foot of the slope even these sediments are cut by a thrust fault. This fault separates the lower-slope succession from stratified sediments of the Cascadia Basin further west. Due to poor signal penetration, the details of deformation on the lower slope are hard to interpret.

A mélange is present, however, on the Washington shelf to the south. The Shell Oil Co. well P-0155, along with several other wells and seismic reflection profiles, indicates that a thrust-faulted mélange lies beneath 1-1.5 km of undisturbed Upper Miocene and younger strata (Snavely, 1987, his Fig. 9). Because the deepest of the wells reached only 4017 m, the depth extent of this mélange is unclear. It was probably thrusted over any underlying crystalline rocks, and its Miocene age is consistent with the timing of contraction of the Hoh Basin on the Olympic Peninsula. The northward extent of the Washington mélange is unclear, but there is no evidence that it continues beyond the Nitinat fault system. In the area of the refraction profile off southern Vancouver Island, presence under the upper slope of attenuated continental crust is consistent with all the geophysical evidence.

255

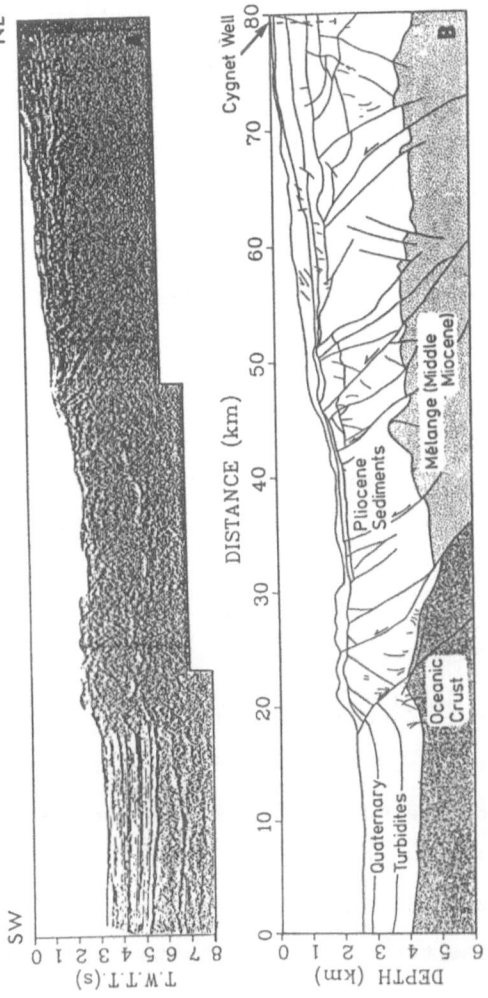

Figure 56. Continental slope off southern Vancouver Island and northern Olympic Peninsula, imaged in the U.S. Geological Survey seismic reflection line 76-19: (a) data; (b) interpretation (both after Snavely and Wagner, 1981). Profile location is given in Figs. 2b, 45. Stratified Plio-Pleistocene sedimentary rocks, penetrated on the outer shelf by the well Cygnet J-100, continue on the upper slope. This well, drilled to a depth of 2460 m, encountered no mélange.

Off northern Vancouver Island, the narrow shelf is underlain by Jurassic metamorphic rocks, and its outer edge is defined by a crustal-scale normal fault which belongs to the plate-boundary fault system. Off southern Vancouver Island, the shelf is broader, but the underlying crust is continental. Attenuated, foundered continental crust continues to the mid-slope, where a fault separates it from a downdropped block of oceanic crust under the Juno depression.

Sediment-filled Juno depression on lower continental slope

The structural boundary between the continental and oceanic crust is expressed in the bathymetry as the mid-slope break. To the west, under the lower continental slope, lies the Juno structural depression in the oceanic crust.

The existence of a structural depression filled with thick stratified sediments is indicated by seismic profiles across the lower continental slope south of the Brooks-Estevan embayment. In the detailed refraction model of Waldron et al. (1990), the Juno depression lies between about 45 and 63 km profile distance (Fig. 55). In this model, from base to top, four units are included: the lower unit, with P-wave velocities ≥6.0 km/s; the next unit, 1.8 km thick, with velocities of 6.0 km/s at the base and 4.0 km/s at the top; the 2.5- to 3-km-thick unit which has velocities of 2.4 km/s at the bottom and to 2.1 km/s at the top; and the uppermost veneer 300 m thick whose velocity is only 1.9 km/s. The vertical faults bounding the Juno depression were modeled to lie 18 km apart and continue to a depth as much as 6 km below seafloor. The two lower units were assigned by Waldron et al. (1990) to the oceanic crystalline crust of the Juan de Fuca plate, and the two upper units to a bipartite sedimentary package.

Detailed velocity analysis of modern seismic reflection data
(lines 89-04 and 89-07; Yuan et al., 1994) showed that the Juno
depression contains a sedimentary package up to 5 km thick, with
velocities increasing from 1600 m/s at the seafloor to more than
4000 m/s at the base. In the reflection line 85-01 (Fig. 52)
along the refraction profile, the Juno depression lies at 2000 m
water depth between SP 700 and 1000. Sediments are about 4 km
thick, though the floor of the sedimentary package is unclear.
The depression is bowl-shaped. Strata dip towards its axis from
the east and west, but continentward dips are predominant.

The western boundary of the Juno depression in this seismic line
is a steep, east-dipping fault at SP 700 which cuts the entire
sedimentary section and produces on the ocean floor a ramp 400-500
m high. This fault is reverse, but because the sediments thicken
across it to the east, it must have intially been normal but was
inverted. Other reverse faults lie just outboard at SP 600, or
inside the depression at SP 810. The same faults are observed in
the nearby, parallel Line 89-04 (see Yuan et al., 1994). A small
west-dipping reverse fault just outboard of the Juno depression in
observed in the adjacent Line 89-03 (see Spence et al., 1991).

Shallow sediments in the Juno depression are stratified, whereas
deeper ones are semi-transparent (Fig. 52). A similar
distribution of seismic facies is observed in the adjacent eastern
parts of the Cascadia Basin, but traveltime through each facies in
the Juno depression is about 50% longer.

On the inboard side of the fault-bounded Juno depression, seismic reflections become chaotic between 3300 and 3900 ms at SP 900. Stratified seismic facies, however, continue eastward both above and below this chaotic zone. Sandwiched between bedded sediments, this zone probably represents buried sediment slumps. The sediments slumped down the paleo-slope from the east: westward dips of seismic events at >4 s traveltime are observed as far as SP 1000, but further east the seismic image is chaotic. The Juno depression's inboard boundary in Line 85-01 is thus indistinct. If the depression is 18 km wide, as the refraction model of Waldron et al. (1990) suggests, its inboard boundary should lie near SP 1100.

About 55 km to the north, the Juno depression has also been imaged in seismic reflection lines 85-02 and 89-07. In Line 85-02 (Fig. 57), it begins at SP 1870. It can be traced inboard reliably to SP 1700, but may continue as far as the mid-slope bathymetric break around SP 1300. The ocean floor between SP 1700 and 1300 lies at about 1400 m water depth. In the narrowest definition (also SP 500 to 680 in Line 89-07; Fig. 58), the northern Juno depression is only about 8 km wide, less that half its width in the south.

A strong basement seismic event under the depression is apparent at 5.5 s in Line 89-07. Despite the disruption under the bathymetric ridge at the outer boundary of the depression (SP 500), this event seems similar to the oceanic-crust basement of the Cascadia Basin outboard. It can be traced inboard to SP 680.

The ridge at SP 500 (Line 89-07) is an anticlinal structure in

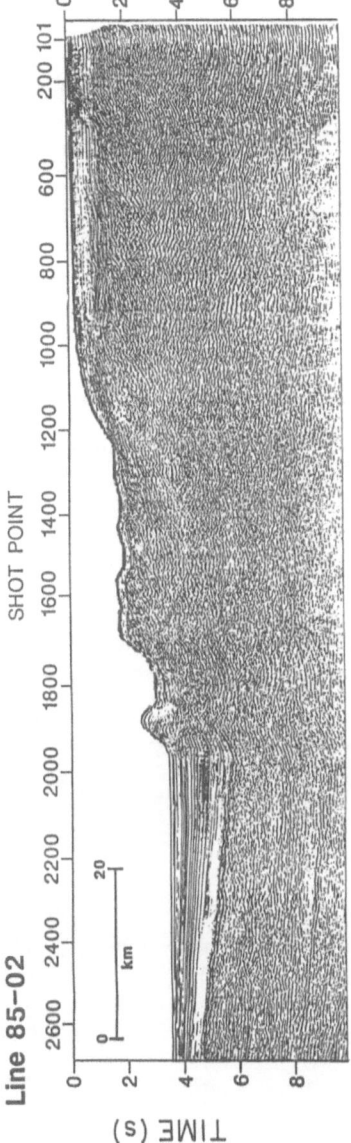

Figure 57. Structure of the submerged continental margin off Vancouver Island, from the shelf to the abyssal plain, imaged in seismic reflection line 85-02 (data after Yorath et al., 1987; Hyndman et al., 1990). Line location is given in Fig. 45.

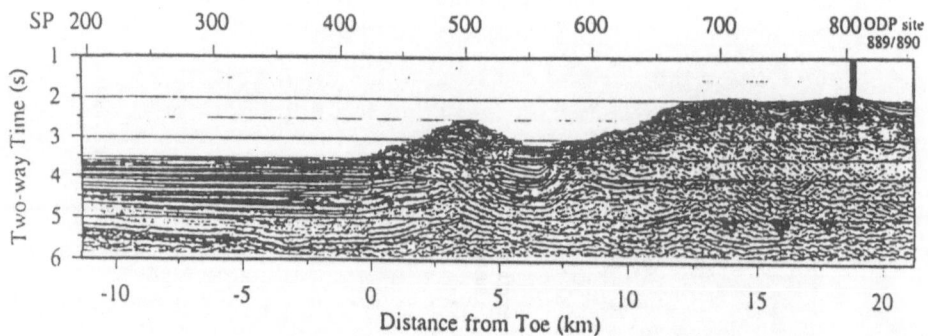

Figure 58. Structure of the northern part of the Juno depression on the lower continental slope off southern Vancouver Island, imaged in seismic reflection line 89-07 (data after Yuan et al., 1994). This line runs parallel to the reflection line 85-02 (Fig. 57), some 5 km to the south.

semi-consolidated but thinly stratified sedimentary rocks. Nearby, at SP 650, inside the Juno depression, lies a syncline up to 0.5 s in amplitude. A narrow anticline at SP 620 is a small disturbance on the inboard side of this system. East of it begins a subhorizontal seismic event at 4 s which, unlike the basement event, can be traced inboard as far as SP 720. Compared with the southern part of the Juno depression (Fig. 52), the northern part is more deformed.

A drillhole near SP 810 in Line 89-07 (SP 1550 in Line 85-02), at ODP site 889/890, penetrated 345 m of Pliocene to Quaternary silt, clay and fine sand (Carson et al., 1993; MacKay et al., 1994). Bedding is subhorizontal to a depth of 104 m, but in sediments below the dips are as much as 70°. Sedimentary rocks are fractured pervasively below 150 m. Significantly, however, no faults have been found in this well off southern Vancouver Island. This deformation probably degrades the quality of seismic images: between SP 1700 and 1300 (Line 85-02), no deep interpretation can be made. No refraction data are available in this area, but some constraints are provided by gravity maps (Fig. 53).

The free-air gravity low over the Juno depression is rectangular in shape and -40 to -50 mGal in amplitude. It extends also into the adjacent deep parts of the Cascadia Basin, where sediment thickness is 2.5 to 3 km. The gravity effect of thicker sediments in the Juno depression is probably offset by the effect of shallower ocean floor, which rises about 400 m. Along seismic line 85-02, gravity anomaly values are about -30 mGal between SP 1700 and 1300, where a bathymetric plateau lies at a water depth of 1300-1400 m. Unfortunately, free-air gravity profiles inboard from the Juno depression not so much reflect the geologic structure as mimic the morphology of the continental slope. This bathymetric mimic makes finding the Juno depression's inboard boundary from gravity data impossible. Besides, in this area Juno gravity signatures are contaminated by the first appearance of N-S anomalies, which become increasingly prominent towards the Brooks-Estevan embayment.

Typical oceanic-crustal magnetic anomaly lineations continue into the Juno depression from Pacific regions outboard (Fig. 27). Thickness of the crystalline crust beneath the lower slope is only about 7 km (Waldron et al., 1990). The Juno depression is thus akin to the Queen Charlotte Trough further north. Both were formed by downdropping of fault-bounded blocks of oceanic crust of the Juan de Fuca plate.

Genetic links are suggested between the Cascadia Basin and the Juno depression. Both contain sediment packages with similar seismic facies (stratified over transparent), though in the Juno depression each facies is about 50% thicker. Drilling 7 km west of the edge of the depression (ODP site 888) encountered Upper Pleistocene to Recent sediments 567 m thick. Though dominated by silt, this section contains at all levels abundant coarser material, such as fine to coarse sand and even gravel. Wood fragments were found in Upper Pleistocene and Holocene strata (Carson et al., 1993). It appears that abundant sediment supply from Vancouver Island allowed sedimentation not only to keep pace with subsidence of the Juno depression, but also to overshoot it and flow over into adjacent parts of the Cascadia Basin.

Limited northward extent of subduction-related thrust faults and mélange along the continental margin

Sedimentary rocks deformed into broken formation or mélange have previously been described along the entire continental margin from Oregon to southern British Columbia. Snavely (1987) reported mélange beneath the cover of stratified sedimentary rocks from at least five deep wells on the shelf off Washington. On the Oregon and Washington margin, such a style of deformation can readily be linked with west-vergent thrusting and subduction of the Juan de Fuca plate. Off Oregon, thrusts have been detected by sea-floor mapping and seismic surveys (Snavely, 1987; Goldfinger et al., 1992; Cochrane et al., 1994), and by recent drilling on the lower slope during the ODP Leg 146 (Carson et al., 1993).

On the northern Washington margin, development of mélange and broken formation may have been less intense than supposed previously. Results of new field mapping on northwestern Olympic Peninsula cast doubt on the perception of Hoh Basin sedimentary rocks as a simple mélange (Orange, 1990). Snavely (1987) noted structural complexity in that area and proposed that these rocks might have been affected by processes not of subduction, but of obduction. Recent publications show many faults in western Olympic coastal areas are rather steep (Orange et al., 1993).

Many blocks of Eocene basalts in western Oregon and southwestern Washington onshore have been found by paleomagnetic studies to be rotated clockwise. These rotations have been attributed to obliquity of plate convergence during the Tertiary (Wells and Coe, 1985). The amount of block rotation decreases along the coast to the north, and on the Olympic Peninsula it virtually disappears (Babcock et al., 1992, 1994).

These observations are circumstantial, and by themselves do not preclude the existence of some accreted rocks off Vancouver Island and northern Washington. However, the possible extent of any such rocks off Vancouver Island is severaly constrained by detailed studies of the submerged continental margin and by geological observations onshore. No mélange has been found on the shelf or slope off southern Vancouver Island. On the shelf, stratified sedimentary rocks as old as Eocene have been drilled to a depth of almost 4 km. No mélange (except fractured rocks created by sediment slumping or diapirism) is known on the Vancouver Island continental slope.

Estimating the volume of material available for accretion suggests that continuous accretion of oceanic sediments and mafic rocks at conventionally assumed rates is "not probable" (Waldron et al., 1990, p. 112). The volume of material presumably scraped off the Juan de Fuca plate greatly exceeds the space available to accommodate it. Presence of continental crystalline crust off Vancouver Island as far outboard as the mid-slope bathymetric break, and of the fault-bounded Juno structural depression on the lower slope, greatly limits the space available for an accretionary prism. Estimates from various tectonic models of the volume of material that should have been accreted (Clowes et al., 1987; Davis and Hyndman, 1989; Waldron et al., 1990) differ from one another, but all exceed by far any remaining available space.

For example, Davis and Hyndman (1989) calculated that the Cascadia Basin sediments, if scraped off the Juan de Fuca plate, would alone have contributed 170 km3 for each kilometer of margin length during the last 1.8 Ma, or about 100 km3 per 1 Ma. Thickness of these sediments at the foot of the slope, where they supposedly enter the accretionary prism, is 2.5-3 km. The estimate of Waldron et al. (1990) that scraping the top 2 km of mafic material off the oceanic crystalline crust would add 400 km3 per 1 km margin length per 1 Ma seems exaggerated. Nonetheless, for a prism containing sediments and mafic material in equal proportion, as was supposed by Finn (1990), a total estimate of 200 km3 per 1 km per 1 Ma appears consistent with the current models. However, given the shortage of available space, this number is much too high to be realistic if constant rates of accretion are assumed for the Cenozoic.

The upper-slope body with velocities of 4.8 to about 5.5 km/s was modeled by Drew and Clowes (1990) to have an average thickness of only 3 km and a maximum length of about 50 km (Fig. 53). It was interpreted by Waldron et al. (1990) as an accretionary prism. With a cross-section area around 150 km2, this space would have been filled in less than 1 Ma. Even if the entire space between the sea floor and the top of the subducted slab in this model were a mélange, the cross-section area of about 1200 km2 (Tofino Basin sedimentary rocks subtracted) would have been filled in just 6 Ma.

With such rates, assumption of continuous accretion would produce space problems even in Washington, but emplacement of a mélange in that area in a short pulse seems more plausible. Drillhole penetrations on the Washington shelf indicate the existence of a mélange involving pre-Late Miocene Tertiary rocks of various ages, beneath undisturbed younger strata (Snavely, 1987). Depth extent of this mélange is uncertain, but it is probably underlain by attenuated continental crystalline crust. The accretionary pulse that created this mélange was probably related to the mid-Miocene tectonic episode that also caused sudden contractional deformation in the Hoh Basin.

No such contractional episode is recognized on Vancouver Island, and models of an accretionary complex under the shelf and upper slope are inconsistent with the available evidence. The Pacific Rim "terrane" of Hyndman et al. (1990) and Dehler and Clowes (1992) encompasses rocks of different origins and types. From field mapping, metamorphic rocks of the Leech River complex are now clearly distinguished from slices of the Pacific Rim complex and its equivalent Pandora Peak unit (Rusmore and Cowan, 1985; Brandon et al., 1989a,b). These slices were probably emplaced by strike-slip movements, not thrusting.

Eocene basalts of the Crescent Formation (another presumed terrane in the accretionary-complex model) are now recognized to have erupted from many local volcanic centers in situ, probably in a rift setting (Brandon and Vance, 1992; Babcock et al., 1992, 1994). On the shelf off Vancouver Island, magnetic anomalies related to such basaltic bodies have only a limited extent. The distribution of basalts is reflected in narrow magnetic highs trending WNW and NNW. In the Prometheus H-68 well in the area of the anomalies, these basalts are about 500 m thick, but absence of correlative strong gravity highs suggests their total volume is small. Crescent basalts are underlain, with a hot contact, by sedimentary rocks onshore (Babcock et al., 1994). Similarly, in seismic line 89-06 on the shelf (Fig. 51), below the volcanics at

1900 ms lie short seismic events (SP 1840) which may be due to deep stratified rocks.

By analogy with the interpretation of similar seismic and magnetic anomalies in the Winona Basin (Chapter 5), and consistent with the distribution and stratigraphic position of Crescent basalts onshore, local magnetic anomalies on the Vancouver Island shelf are probably caused by local volcanic bodies. No single Crescent "terrane" was accreted to North America in Washington or British Columbia. Instead, individual igneous massifs of various sizes were produced in situ, by local eruptions.

Neither does geological evidence confirm the presence of crustal-scale thrust faults which are the cornerstone of the subduction-complex models of the Vancouver Island margin. Thrusts have been mapped on the east side of Vancouver Island, and local thrust splays are found within the OWSZ (see previous chapters). But no thrust-sheet structures are evident on western Vancouver Island or on the submerged margin. Tectonic slices of the Pacific Rim complex were displaced along steep strike-slip faults. The 15-km-wide slice in the Ucluet area on the west coast of the island wedges out to the north and south, where the Westcoast and Tofino faults that bound it are thought to merge (Muller, 1977a,b; Brandon, 1989a,b). On southern Vancouver Island, small slices of Pandora Peak rocks are found in the San Juan and Survey Mountain fault zones (Rusmore and Cowan, 1985), which belong to the OWSZ.

Large right-lateral movements on the Westcoast fault in the Late Cretaceous and/or early Tertiary (Brandon, 1989a) confirm that this fault is steep. The same is indicated by its measured dips

(Muller et al., 1974, 1981), and its trace is remarkably straight along the west coast of Vancouver Island. On Hesquiat Peninsula, it strikes N50°W, dips steeply to the east, and offsets Late Eocene and Oligocene strata of the Carmanah Group in a normal (east-side-down) sense. On Nootka Island, this fault is not exposed, but Carmanah Group strata maintain a N50°W strike. Such a straight fault is unlikely to be a low-angle thrust. Right-lateral movements in the Late Cretaceous or early Tertiary were only one of the many episodes in the history of this long-lived, crustal-scale, steep fault.

Unmetamorphosed Eocene and Oligocene granitoid plutons of the Catface suite in many parts of Vancouver Island are aligned along steep faults (Woodsworth et al., 1991). One such stock, dated at 52 Ma, intrudes the Pacific Rim complex (Brandon, 1989a). This rules out the supposed accretion of the Pacific Rim oceanic-crust "terrane" by thrusting around 42 Ma as was postulated by Clowes et al. (1987) and Hyndman et al. (1990).

No big thrusts are apparent in the Tofino Basin on the shelf. Almost 4 km of drilled Eocene to Plio-Pleistocene rocks are stratified and show no signs of significant thrusting. No thrusts appear in seismic reflection profiles. Differences in the stratigraphy between parts of the basin, combined with geophysical evidence for transverse faults which continue on the shelf from Vancouver Island, indicate that the basin evolved by alternating differential uplift and subsidence of fault-bounded blocks.

Along the egde of the southern and central shelf off Vancouver Island, seismic reflection profiles show a buried positive

structure over which stratified sedimentary rocks thin almost to
zero (Tiffin et al., 1972). Such a basin-bounding structural high
is observed in lines 85-01 (Fig. 52; SP 1300-1400) and 85-02 (Fig.
57; SP 1050-1150). Gravity anomalies over the outer shelf and
upper slope are relatively positive, whereas a strong gravity low
is found over the Tofino graben at mid-shelf.

Off Oregon and Washington, similar gravity lows on the shelf have
been interpreted as sediment-filled depressions (Couch and
Riddihough, 1989). An old gravity model (R.W. Couch in: Muller,
1977a) showed the Tofino Basin to be 8-9 km thick, but Thybo
(1990) interpreted seismic data to suggest a thickness of 5 km.
In another seismic line (89-06; Fig. 51), the stratified package
at SP 1700 and 2050 also continues to 3-3.5 s, or about 5 km.

Poor seismic-signal penetration hinders interpretation all along
the submerged margin of western North America (Bruns and Carlson,
1987; Lyatsky, 1991b). For example, the modern seismic line 85-05
resolved only the top 1.5 s (about 2 km) in the Fuca Basin which
is as much as 8 km deep (Niem and Snavely, 1991). In the Tofino
Basin, the basement is usually ill-defined. Bedding planes,
diapirs, folds and faults scatter acoustic energy and greatly
degrade seismic images, leading to observations by many
geophysicists that the structure of the shelf is difficult to
interpret (Iwasaki and Shimamura, 1990; Thybo, 1990).

Only locally is the base of penetration in the Tofino Basin
associated with clear events reasonably interpretable as volcanic-
related. In Line 85-01, SP 2050 to 2320 (Fig. 52), it is
associated with characteristic diffractions due to surface

roughness and internal inhomogeneity of flows, which were indeed penetrated by the Prometheus H-68 well at SP 2280. However, these events do not mark the true crystalline basement. Crescent basalts on the Olympic Peninsula are diachronous, and they pass into overlying Eocene sedimentary sequences gradationally. The Crescent Formation is laterally discontinuous, made up of many volcanic bodies. It is interbedded with marine sedimentary rocks, and it overlies them with a hot contact (Babcock et al., 1994). Seismic reflections below the volcanics on the shelf were noted in Line 89-06 (Fig. 51), suggesting Tofino Basin sedimentary rocks may also be present beneath the volcanics.

The mud-dominated Tofino Basin was found by drilling to be overpressured (Shouldice, 1971). Overpressuring led to sediment flowage, which was most common in the deep parts of the basin, such as off central Vancouver Island (Tiffin et al., 1972). Farther south, anticlines probably caused by mud flowage are observed at SP 1400-1800 and SP 2600-2700 in Line 85-01 (Fig. 52), and a piercement diapir around SP 200 in Line 85-02 (Fig. 57). In places, strata are draped over deeper structures (e.g., Line 85-01, SP 1500 to 2000).

A set of small grabens lies at SP 650-800 in Line 85-02. The half-graben in Line 85-01, SP 2320-2450, is bounded by steep normal faults which disturb shallow and deep strata. In keeping with the accretionary-complex model, this structure was previously interpreted as a "fossil trench" (Hyndman et al., 1990). However, grabens and half-grabens are normally associated with extensional, not compressional, tectonic regimes. No low-angle thrust faults are apparent in seismic reflection profiles on the shelf.

Thus, along the continental margin from Oregon to southern British Columbia, examination of Tertiary geology reveals a gradual northward decrease in the abundance of phenomena attributable to subduction.

Zoning in the distribution of continental and oceanic crust on Vancouver Island continental slope

Structure of the Insular Belt is dominated by several fault systems of crustal or even lithospheric scale. One such system, detected with seismic refraction data, lies beneath the Strait of Georgia and separates the Insular and Coast belts (White and Clowes, 1984). The Alberni-Cowichan Lake fault system lies between the Eastern and Western Vancouver Island blocks. The latter is separated from the Tofino Basin by the inner strand of the OWSZ - the Westcoast fault. The outer strand of the OWSZ - the Calawah fault - separates crustal blocks on the inner shelf from the Tofino graben.

Similarly, structural zoning occurs in the continental slope area. The mid-slope bathymetric break, between the relatively steep upper slope from the relatively gentle lower slope off southern Vancouver Island, also marks a structural boundary. Attenuated continental crust lies inboard, oceanic crust outboard.

A different zoning is found north of the Brooks-Estevan embayment, where the entire lower slope is underlain by attenuated continental crust and the boundary with oceanic crust follows the Revere-Dellwood fault.

The previous idea that the Winona Basin off northern Vancouver Island is underlain by a block of oceanic crust which subsided some 8 km during the last 1.5 Ma (Davis and Riddihough, 1982) was shown in Chapter 6 to be unrealistic. Yuan et al. (1992, p. 1516) stated that the basement seismic reflection under the Winona Basin in line 88-02 (Fig. 36) "undoubtedly represents the top of the subducting oceanic Explorer plate". Davis and Currie (1993) believed this basin is underlain by an independent, small oceanic Winona plate whose subduction has now stopped. However, magnetic stripes are entirely absent in this area and are only observed west of the Revere-Dellwood fault (Fig. 27). For 8 km of sediments to have been laid down in just 1.5 Ma requires unrealistically high rates of sedimentation. High seismic velocities of deep strata in this basin are consistent with presence of old sedimentary rocks. The underlying crystalline crust, whose thickness reaches 15 km, is probably continental.

Off southern Vancouver Island, by contrast, the lower slope is underlain by oceanic crust of the Juno depression. The outer boundary of this depression coincides roughly with the foot of the slope, and both its outer and inner boundaries are approximately in line with the Revere-Dellwood fault, if it were projected under the Brooks-Estevan embayment.

This embayment is about 70 km long. In its northwestern part, an extension of the Revere-Dellwood fault might be one of the buried faults near SP 400 in seismic reflection line 85-04 (Fig. 37). In the southern part of the embayment, however, coincident gravity and magnetic anomalies run N-S, and no clear extension of the Revere-Dellwood fault is apparent in seismic line 89-09 (Fig. 42).

But the plate-boundary structural zone continues for thousands of kilometers along the continental margin of western North America. Compared with that distance, a gap of a few tens of kilometers is minor. Alignment between the Revere-Dellwood fault and the boundary faults of the Juno depression is consistent with overall continuity of this system's outer strands. The SE extension of the Revere-Dellwood fault splits into two branches, which bound the Juno depression. The inner branch is of most importance, as it separates the continental and oceanic crust.

The inner strand of the NNW-trending plate-boundary structural zone controls the position of the shelf edge, which remains fairly straight all along Vancouver Island. Local disruptions in its position can be attributed to transverse faults, such as Estevan (Fig. 50). Just south of the Estevan fault, across which the shelf widens to the south by about 10 km, on the outer shelf lies the NNW-trending Apollo anticline (Shouldice, 1971; Tiffin et al., 1972; Yorath, 1980). This elongated anticline is on trend with the shelf edge to the north, which may indicate propagation of NNW-oriented faults onto the shelf. Presence of other NNW-oriented faults is suggested by the elongation and alignment of other diapirs in the Tofino Basin. The volcanic body that causes the elongated positive magnetic anomaly off Clayquot Sound also trends NNW. Like many other Crescent volcanic bodies, it may have formed along a fault.

The OWSZ, trending WNW, also has a regional extent, and it meets the plate-boundary fault system off Vancouver Island. The Westcoast fault is actually Mesozoic, but it was reactivated and incorporated into the OWSZ in the Tertiary. It continues along Vancouver Island, as far as the transverse Esperanza fault and possibly beyond (Muller et al., 1974, 1981). The Calawah fault also extends on the shelf, continuing as far as Barkley Sound. Another WNW-oriented fault is the one which controls the volcanic body and the magnetic anomaly just south of Hesquiat Peninsula. However, OWSZ-related faults are mostly confined to the shelf, and only a few WNW bathymetric trends are found on the slope in the Brooks-Estevan embayment.

The amount of Cenozoic compression increases along the western North America continental margin from north to south. No evidence of tectonic compression is found off southeastern Alaska, the Queen Charlotte Islands and northern Vancouver Island (see Bruns and Carlson, 1987; von Huene, 1989; and Chapter 6). In contrasts, thrusts are fully developed on the submerged continental margin off southern Washington and especially Oregon. Steep, NNW-trending strands of the plate-boundary fault system off southeastern Alaska and British Columbia run into the N-S-trending Cascadia subduction zone off Washington and Oregon.

The gradual northward decay in the intensity of compression along the margin may be related to non-rigid behavior of the heavily fragmented northern Juan de Fuca plate. Such unusual characteristics of the Cascadia subduction zone and of the Juan de Fuca plate are discussed in the next chapter.

CHAPTER 9 - INTERLOCKING OF CONTINENTAL AND OCEANIC CRUSTAL BLOCKS ALONG THE CONTINENTAL MARGIN AND NON-RIGID BEHAVIOR OF NORTHERN JUAN DE FUCA PLATE

Place of block interlocking in the plate-boundary zone

Vancouver Island has mostly been uplifted in the Cenozoic, providing sediments for the basins on its sides (Muller, 1977a). Sea-floor dredging and analysis of magnetic anomalies show that Mesozoic basement rocks similar to those on Vancouver Island also underlie Tertiary strata on the Kyuquot and Northern blocks on the shelf. To the south, also suggesting presence of underlying continental crystalline crust, felsic stocks and dikes cut Eocene volcanic massifs of the Crescent Formation including Metchosin (Muller, 1977c; Snavely, 1987), as well as the Pacific Rim complex (Brandon, 1989a). Crustal blocks on the Vancouver Island shelf and in the Strait of Juan de Fuca were delineated from various lines of evidence in a previous chapter.

Seismic refraction (Taber and Lewis, 1986; Drew and Clowes, 1990; Mereu, 1990) and gravity (Finn, 1990) models across the submerged margin off southern British Columbia and Washington show under the shelf and upper slope material with velocities and densities similar to those in continental crystalline crust. That material has traditionally been interpreted as an accretionary mélange containing both sedimentary and volcanic rocks (Finn, 1990; Waldron et al., 1990), but the analysis in the previous chapters shows that at least some of this material is probably continental crust. Off Washington, this crust forms a slab on top of an oceanic-crustal slab of the subducting Juan de Fuca plate (Couch and Riddihough, 1989). Off southern Vancouver Island, the amount of underthrusting is minimal, and on the inner side of the Juno depression blocks of continental and oceanic crust are juxtaposed.

Off northern Vancouver Island, blocks of continental and oceanic crust are separated by the Revere-Dellwood fault (Chapter 5). Further north, off the Queen Charlotte Islands and southeastern Alaska, other faults in the outer strand of the plate-boundary structural zone separate continental and oceanic crust. Local disruptions in this otherwise continuous structural zone were noted off Queen Charlotte Sound and in the Brooks-Estevan embayment off central Vancouver Island. The latter area is proposed here as the tectonotype for interlocking of continental and oceanic crustal blocks at continental margins.

Geomorphological expression of block interlocking

The bathymetry in the Brooks-Estevan embayment (Fig. 28) is less regular than to the north and south. The NW alignment of the continental slope is disrupted and the slope narrows to only 35-40 km. To the northwest and southeast, the slope is more regular and its NW orientation more clearly expressed. Off southern Vancouver Island, it is about 60 km wide and gentler than in the embayment. Transverse canyons are better defined and more isolated. By comparison, the slope in the embayment appears chaotic.

On closer inspection, however, many bathymetric trends in the embayment are distinctly N-S, NE and WNW. Most N-S trends occur in the southeastern part of the embayment, and most NE and WNW trends in the northwestern part. The NE trends are aligned with the Brooks fracture zone and the Estevan fault, which bound the embayment. The WNW trends are local and may be ascribed to the continuity into this area of the OWSZ. Most interesting are the N-S trends, which are aligned with magnetic and gravity anomalies that continue into the embayment from oceanic regions.

Magnetic and gravity expression of block interlocking

The N-S alignment of magnetic stripes over the northeastern Pacific Ocean has been recognized since the early work of Raff and Mason (1961). Off northern Vancouver Island, these stripes stop abruptly at the Revere-Dellwood fault, and inboard a blank magnetic zone marks foundered blocks of attenuated continental crust (Fig. 27). Northwest of the Brooks-Estevan embayment, the Winona Basin overlies such a block of continental crust.

In contrast, N-S magnetic stripes persist from outboard areas into the Juno block on the lower slope off southern Vancouver Island. This block is made of oceanic crust which originally was in continuity with the Juan de Fuca plate. Seismic profiles show it is downdropped along steep faults and covered by some 4 km of sediments, but magnetic stripes are nevertheless clear.

The Juno depression was delineated in the previous chapter largely from gravity maps: because it is downdropped and covered by low-density sediments, it is associated with a free-air low of -40 to -50 mGal (Fig. 53). Its main elongation is NW, parallel to faults of the plate-boundary structural zone along which it is downdropped. The southeastern end of the Juno depression lies on the NE-trending Nitinat fault zone. However, its northwestern end is not expressed clearly: around longitude 127°W, many gravity anomalies as low as -50 mGal run N-S as far as the edge of the shelf. The Juno low ends in that broad zone of N-S anomalies.

Magnetic anomalies in the southeastern part of the Brooks-Estevan embayment also trend N-S, continuing from the outboard oceanic regions (Fig. 41). Particularly well expressed is the linear anomaly around 127°40'W. Despite local breaks, it reaches the latitude of 49°40'N.

This relative high-low pair has amplitudes between <+150 nT and >+200 nT. Continuity of magnetic stripes from oceanic areas is usually thought to indicate continuity of oceanic crust. Off Washington, Finn (1990, 1991) noted that magnetic stripes, though faded, persist on the continental slope to the shelf, probably due to underthrusting of oceanic crust beneath this part of North America. Magnetic stripes in the Juno depression are much clearer, because the oceanic crystalline crust in this area is not underthrusted but merely covered by sediments. In southeastern Brooks-Estevan embayment, magnetic stripes also remain strong, suggesting the sediments are underlain by oceanic crust.

Coincident with the large N-S magnetic high-low pair in southeastern Brooks-Estevan embayment is a free-air relative gravity high of -30 mGal, which lies between relative lows of -60 to -70 mGal (Fig. 40). Such large variations in gravity anomaly values are probably caused by dip-slip offsets on faults trending N-S. Suggesting that movements on these faults also strongly influenced the ocean-floor morphology, in the same area N-S trends characterize the lower slope including the toe (Fig. 28). The origin of such a curious correlation between oceanic magnetic stripes and large faults is discussed in the following sections.

The northwestern part of the Brooks-Estevan embayment lacks N-S magnetic stripes. In this area begins the blank magnetic domain which is so well expressed over the Winona Basin.

The negative free-air gravity anomaly over the Winona Basin, which at its minimum drops below -130 mGal, does not continue onto the embayment. However, it does not stop at any strong NE-trending gradient zone that could be correlated with a large transverse fault. Rather, the Winona low breaks up into several "tongues" that wedge out to the SE. Anomaly values decline in the embayment gradually, from mostly below -70 mGal in the northwest to mostly above -70 mGal in the southeast. These tongues in the northwestern part of the embayment trend mostly NW-SE, in line with the Winona low, suggesting that the Winona Basin breaks up into smaller depocenters. Faults in this area trend NW-SE and are aligned with the plate-boundary structural zone. Like in the Winona Basin, the crust is probably continental. Thus, two types of crust underlie the Brooks-Estevan embayment - continental in the northwest, oceanic in the southeast.

Earthquake seismicity in the zone of block interlocking
The Scott Islands fracture zone, which belongs to the plate-boundary fault system, is shown by marine seismic reflection data to be a steep, west-dipping normal fault (Fig. 36). South of Brooks Peninsula, it meets the OWSZ. Structural complexity in this area is compounded by presence of the broad, NE-trending Brooks fracture zone. This zone transects the margin, bounding the Brooks-Estevan embayment and accounting for numerous NE-oriented bathymetric features on the continental slope.

No direct extensions of this zone into the deep oceanic areas outboard have been detected (e.g., Au and Clowes, 1982), but closer to shore, several NE-trending faults - Brooks fracture zone, Esperanza and Estevan faults - run onto the shelf and slope from Vancouver Island. There is no evidence that they continue onto the abyssal plain directly, though strong disruptions of magnetic stripes in that area suggests extensive faulting in the oceanic crust (cp. Davis and Currie, 1993).

Previously, assuming that plates in this area are rigid, other explanations were proposed for the structural complexity around the Brooks-Estevan embayment. A plate boundary was proposed to run across this area, from northern Juan de Fuca Ridge to Vancouver Island, along a broad, NE-oriented band of earthquake epicenters (e.g., Barr and Chase, 1974). Initially, it was believed to separate the Juan de Fuca plate from a triagnular oceanward protrusion of the North America plate, which was suggested to lie off northern Vancouver Island and Queen Charlotte Sound (Barr, 1974; Barr and Chase, 1974). Existence of such a protrusion was hard to reconcile with the usual understanding of plate kinematics, and anyway the supposed fault configuration was not confirmed by contemporary seismic data (D.L. Tiffin, pers. comm. to Barr, 1974, p. 1197).

Kinematically more plausible were the later ideas that the presumed NE-trending boundary separates the Juan de Fuca and Pacific plates, or fragments of the Juan de Fuca plate. Following regional studies of magnetic anomalies (Riddihough, 1977), two independent oceanic-lithosphere fragments - the Juan de Fuca and Explorer plates - were postulated to lie, respectively, south and north of the NE-trending band of earthquakes (Figs. 3, 4; Hyndman et al., 1979; Keen and Hyndman, 1979). The time of splitting of the Explorer plate from the larger Juan de Fuca plate has been estimated variously at 8 to 3.5 Ma (Hyndman et al., 1979; Riddihough, 1984; Riddihough and Hyndman, 1991).

The structure and even the location of their presumed boundary are unclear to this day. Barr (1974, her Fig. 10) showed it as a left-lateral fault from the northern end of the Juan de Fuca Ridge to the southern tip of Nootka Island. Hyndman et al. (1979), who named the presumed plate-boundary fault zone Nootka, showed it running from the junction of the Juan de Fuca Ridge with the Sovanco fracture zone, towards northern Nootka Island. They showed it as being 15-25 km wide, but having no particular gravity, magnetic or bathymetric expression. Recently, Davis and Currie (1993) showed the fault boundary between the Juan de Fuca and Explorer plate fragments beginning 50-55 km south of the junction of the Juan de Fuca Ridge with the Sovanco fracture zone. In another interpretation, the western part of Nootka fault zone has migrated southward so as to "disappear", and a new fault has been initiated at the northern end of the Juan de Fuca Ridge (Riddihough and Hyndman, 1991, p. 445). The eastern part of the Nootka fault zone in this new model has during the last 5 Ma been migrating along the coast northward.

Thus, confusion surrounds the definition of a plate boundary along the broad, NE-oriented band of earthquakes off northern Vancouver Island. Also misleading is the early supposition (Hyndman et al., 1979) that no faults related to this band are present on the island. Despite several attempts to constrain the width of this earthquake band (presumably to a single fault zone), it remains broad (Milne et al., 1978; Wahlström and Rogers, 1992). Borders of the Explorer plate are now believed to be "diffuse rather than single faults" (Wahlström and Rogers, 1992, p. 960).

Studies of regional seismicity around the Brooks-Estevan embayment (Fig. 59) have led to the now-common conclusion that modern strain in this area is diffuse. The band of earthquakes that runs NE from northern Juan de Fuca Ridge towards Nootka Island and beyond is about 40 km wide, much too broad to represent a single fault. Focal mechanisms throughout this band - in the pelagic oceanic areas, on the submerged margin and in adjacent parts of Vancouver Island - suggest dextral movements on NW-SE faults and/or sinistral movements on NE-SW faults (Rogers, 1979; Wahlström and Rogers, 1990). Faults with both these orientations are abundant on western Vancouver Island (Muller et al., 1974, 1981), and they are also common in the submerged Brooks-Estevan embayment.

Onshore, mapping shows the ages of many such faults to be pre-Neogene, or even pre-Tertiary. Faults oriented NE in many parts of the Insular Belt were inherited from Mesozoic time and reactivated in the Tertiary (Lyatsky, 1991a, 1993a). Earthquakes on Vancouver Island and the adjacent submerged margin indicate some of these faults are active still. Earthquakes farther outboard, however, reflect deformation in oceanic lithosphere.

On the shelf, earthquakes seem to be most common near the boundary between the Northern and Kyuquot blocks, the boundary between the Kyuquot and Cove blocks, and within the northern part of the Cove block (Fig. 59). The amount of seismicity declines drastically

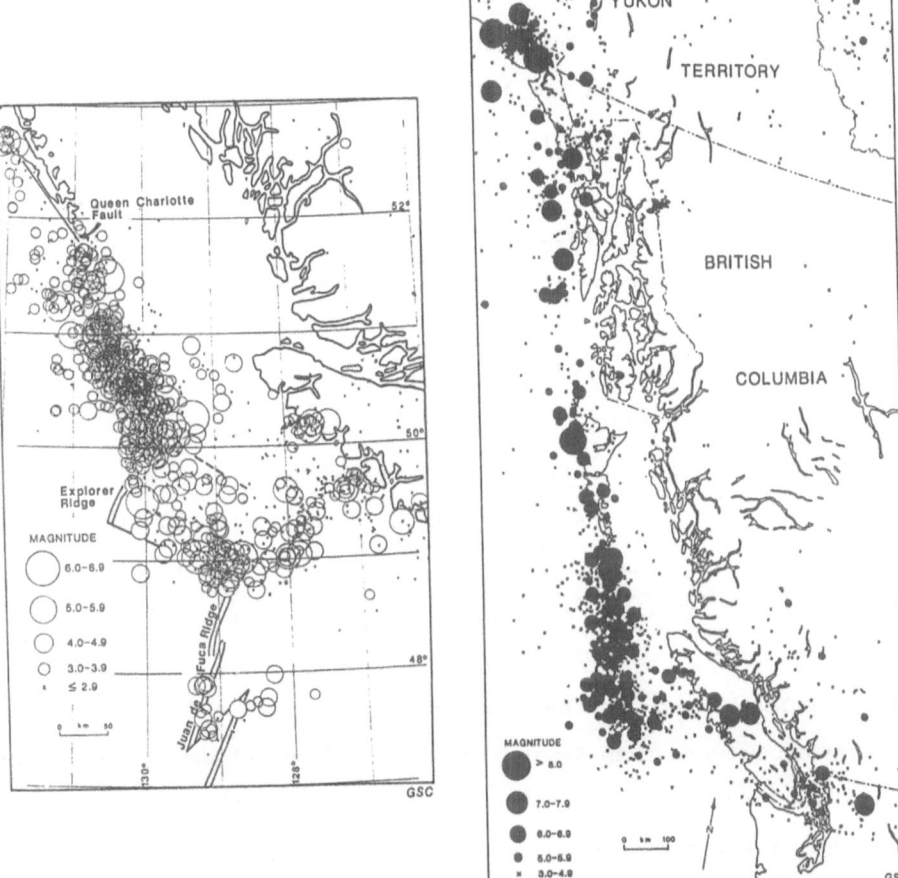

Figure 59. Distribution of earthquake epicenters along the British Columbia and southeastern Alaska continental margin (modified from Riddihough and Hyndman, 1991). The seismicity pattern off Vancouver Island is significantly different from the conventionally assumed plate configuration.

off southern Vancouver Island, and remains low off Washington (McCrumb et al., 1989a,b; Acharya, 1992). Such a pattern is hard to explain in terms of interaction of rigid plates, but other evidence indicates that the behavior of the Juan de Fuca plate off western North America is largely non-rigid.

Evidence for lack of rigidity of northern Juan de Fuca plate
From diffuse seismicity and curved and broken magnetic stripes, intraplate deformation has previously been inferred in the southern (Gorda) segment of the Juan de Fuca plate (Wilson, 1986; Stoddard, 1987, 1991). Subduction there appears to have stopped (Couch and Riddihough, 1989). Similar phenomena in the northern part of the Juan de Fuca plate also point to intraplate deformation, which has been confirmed by detailed local studies off the mouth of Queen Charlotte Sound (see Allan et al., 1993).

The Explorer and northern Juan de Fuca ridges and the Sovanco transform between them demarcate the boundary between the northern Juan de Fuca plate and the oceanic Pacific plate to the west. However, the Explorer Ridge and the Sovanco zone are remarkably aseismic and lacking many other indications of ongoing tectonic activity (e.g., Chase et al., 1975; Riddihough et al., 1983; McCrumb et al., 1989a). The Explorer Ridge has some geochemical characteristics uncommon at spreading centers (Michael et al., 1989; Allan et al., 1993). The Sovanco zone is a belt, several tens of kilometers wide, of fault-bounded oceanic-crustal blocks variously rotated and tilted (Cowan et al., 1986; Lister, 1991; Davis and Currie, 1993).

Discordant with the presumed plate boundaries off western Canada, a broad band of earthquakes runs NNW from northern Juan de Fuca Ridge towards the southern tip of the Queen Charlotte Islands (Fig. 58; also Milne et al., 1978; Wahlström and Rogers, 1990, 1992). Focal mechanisms in this band are NE-SW sinistral or NW-SE dextral, similar to those around the Brooks-Estevan embayment. As noted by Wahlström and Rogers (1992, p. 960), "lack of correlation of significant seismicity with some mapped ocean-bottom structures is evidently a physical reality". It indicates that intraplate deformation is occurring in this area.

It has also been reported (Au and Clowes, 1982, 1984) that under the northern Juan de Fuca plate, the P-wave and S-wave seismic velocity anisotropy in the oceanic upper mantle has the fast direction perpendicular to the Juan de Fuca Ridge. The P- and S-wave velocities perpendicular to the ridge are 8.3 km/s and 4.6 km/s, respectively, but only 7.5 km/s and 4.5 km/s parallel to the ridge. Such anisotropy might be explained if the upper mantle is decoupled from the overlying oceanic crust, adding a third dimension to the fragmentation of the Juan de Fuca plate.

If so, the N-S magnetic anomalies in northern Juan de Fuca plate may reflect the motion of only the oceanic crust, rather than of the entire lithosphere. In southeastern Brooks-Estevan embayment, correspondence of magnetic and gravity anomalies suggests that the oceanic crust is broken into blocks, which in that area are interlocked with blocks of continental crust.

Seismic refraction data revealed considerable variations in the velocity structure of the oceanic crust in the northern Juan de Fuca plate (Malecek and Clowes, 1978; Au and Clowes, 1982).

Thickness of the upper lithospheric layer distinguished as oceanic crust was found to vary from 6-7 km to as much as 10-11 km in some places. The crust was also noted to be thickening eastward, away from the Juan de Fuca Ridge. Most of these variations are accommodated in the interpteted lower-crustal layer 3 (beneath the sediments of layer 1 and the basalts and sheeted dikes of layer 2), whose thickness varies from 3.1 to 7.4 km (op. cit).

Au and Clowes (1982) alluded to unspecified processes of "crustal maturing" to account for thickness and velocity variations in their layer 3. On the other hand, Malecek and Clowes (1978) had ascribed thickening of the oceanic crust to its compressional "bunching-up" against the continental buttress. They had also reported large steep faults in the oceanic crust near the Explorer Ridge. Many other steep faults were noted in seismic reflection profiles between the northern Juan de Fuca Ridge and the Brooks-Estevan embayment (Hyndman et al., 1979).

Recent reflection profiles across northern Juan de Fuca plate also reveal thickness variations in layer 2 (Rohr et al., 1988). The top of oceanic-crust basalts under the Cascadia Basin sediments has repeatedly been found in seismic reflection profiles to have many hundreds of meters of relief (Carlson and Nelson, 1987; Calvert et al., 1990; Hasselgren et al., 1992; Rohr, 1994; see also the data of Hasselgren and Clowes, 1995). Off Vancouver Island, local east-side-down offsets (Fig. 52, SP 180, 400, 500, 550, 700; Fig. 56, SP 2000, 2400) cause the oceanic basement to deepen towards the continent stepwise.

The Cascadia Basin covers virtually the entire Juan de Fuca plate.

From magnetic anomalies, the age of the underlying oceanic crust is estimated at no more than 10 Ma (Riddihough, 1984), and the age of the Cascadia Basin is therefore inferred to be Late Miocene to Recent (see also Riddihough and Hyndman, 1991). Sparse data from deep-ocean drilling (such as DSDP site 174 off Oregon) and seismic reflection profiles, as summarized by Carlson and Nelson (1987) and Duncan and Kulm (1989), show the basin contains two units. Only Pliocene to Recent sediments have been drilled. The bottom unit, fairly transparent in seismic sections, consists of thinly bedded silty hemi-pelagic clay. The seismically reflective upper unit contains mostly thick- to thin-bedded, medium and fine sand of Upper Pleistocene to Holocene age.

The Cascadia Basin generally thickens eastward, from the Juan de Fuca Ridge towards the continent. At the foot of the slope, it is 2.5-3 km thick. Water depth increases gradually from 2400-2600 m off Vancouver Island to as much as 3000 off northern California. Modern channels on the abyssal plain distribute sediments across the Juan de Fuca plate mostly from north to south. This plate thus seems to have a slight but persistent southward tilt.

Variations in the thickness of extrusive basalts in layer 2A, considered to be the main source of oceanic magnetic stripes, by several hundred meters in the vicinity of the Juan de Fuca Ridge have been found to have a noticeable effect on the character of magnetic anomalies (Tivey and Johnson, 1993; Tivey, 1994). In the northern part of the Juan de Fuca plate, magnetic anomalies are disrupted by many faults (Figs. 28, 41), and therefore do not always provide a reliable framework for plate reconstructions off western Canada (Davis and Currie, 1993).

In the southeastern Brooks-Estevan embayment, magnetic and gravity anomalies coincide with each other and with major bathymetric features on the continental slope. Off central Washington, gravity anomalies over the continental slope and the adjacent parts of the abyssal plain correspond to magnetic anomalies of opposite polarity. The two profiles in that area are almost mirror images of one another (Fig. 54), but farther outboard the antisymmetry is less well expressed. Such a correspondence between gravity and magnetic anomalies may reflect keyboard-style shuffling of elongated blocks or warping of weak oceanic crust. If so, the faulting or warping affect the arrangement of magnetic stripes. Plate reconstructions may therefore suffer from the effects of structural deformation of oceanic crust.

Constraints on the timing of intraplate deformation in the Juan de Fuca plate

A fundamental assumption in plate-motion reconstructions from oceanic magnetic lineations and hot-spot tracks is that plates are coherent, rigid lithospheric entities. Conventional models of interactions at plate boundaries, which flow largely from such reconstructions, usually also assume plates to be rigid. However, during the Cenozoic the oceanic Farallon plate has been undergoing fragmentation since about 55 Ma (Atwater, 1989; Stock and Lee, 1994). Its modern remnant, the small Juan de Fuca plate, is still being fragmented into ever smaller pieces, particularly on its northern and southern ends. For a crumbling plate, generalized models assuming rigid-plate behavior may be inappropriate.

Probably because the Juan de Fuca plate is young, it is thin (30-

35 km lithospheric thickness at the Washington margin; Finn, 1990) and weak. Heavy shearing, vertical and horizontal, has affected the upper part of this plate's lithosphere.

Parts of the Juan de Fuca plate are still coherent enough to support a flexural bulge, which has been inferred along the continental margin off Oregon and southern Washington from marine gravity data (Jachens et al., 1989). No such bulge is apparent in the northern and southern parts of this plate. Due to intraplate deformation, magnetic stripes are curved and earthquakes broadly distributed in the southern part of the Juan de Fuca plate off northern California (Wilson, 1986; Stoddard, 1987, 1991). Similar phenomena occur in the northern part of this plate off Vancouver Island. The broad plate-boundary structural zone continues to the Vancouver Island margin from the margin off southeastern Alaska and the Queen Charlotte Islands, past the mouth of Queen Charlotte Sound (also Allan et al., 1993). A diffuse boundary between the Juan de Fuca and Pacific plates has been proposed to run through the "Explorer plate" (Furlong et al., 1994).

The timing when intraplate deformation began is hard to pinpoint: fragmentation of the Farallon plate went on through most of the Cenozoic. Magnetic stripes have been used to estimate the age of the present-day Juan de Fuca lithosphere as no more than 10 Ma at the foot of the slope (Riddihough, 1984); DSDP drilling some 80 km outboard confirms the age of the sediments is at least Pliocene (Duncan and Kulm, 1989).

Seismic reflection profiles show relief in the basaltic basement, usually in the hundreds of meters, throughout the basin (Carlson and Nelson, 1987; Calvert et al., 1990; Hasselgren et al., 1992; Rohr, 1994). However, though Pliocene sediments are strongly affected, Pleistocene beds are less disturbed. This suggests that some intraplate deformation occurred in the Pliocene in many parts of the Juan de Fuca plate. At present, strong active deformation is taking place in the plate's northern and southern parts.

Genetic aspects of the geology of western North America continental margin

Continuity of continental-crust structures on the submerged continental margin

North of the Olympic Mountains, seismic refraction data suggest the continental crust on Vancouver Island is about 40 km thick (McMechan and Spence, 1983; Drew and Clowes, 1990). Father north in the Insular Belt, its thickness varies considerably: 23 km under Queen Charlotte Sound; 29 km under the Queen Charlotte Islands; 27-29 km under Hecate Strait; 25-30 km under Dixon Entrance and southeastern Alaska (Johnson, 1972; Mackie et al., 1989; Sweeney and Seemann, 1991; Brew et al., 1991; Yuan et al., 1992; Spence and Asudeh, 1993; Hole et al., 1993). Apart from varying along the Insular Belt, in many of these areas the crust thins towards the continental margin stepwise (see Chapters 6, 7).

In an E-W seismic refraction profile across west-central Washington, the crust, with velocities not exceeding 7.1 km/s, is 20 km thick at the coast and more than 30 km thick under the Puget lowlands (Taber and Lewis, 1986). The composition of the deep crust in western Oregon and Washington south of the Olympic Peninsula is still a topic of discussion (Keach et al., 1989; Finn, 1990), but thr Moho is smooth and shallows towards the continental margin gradually (Mooney and Weaver, 1989).

Still, thickness of the crystalline crust in these areas varies abruptly, as thickness of the sedimentary cover varies. Localized depressions are apparently filled with many kilometers of Tertiary sediments alone. Several such very deep depressions, 100-200 km across and about 200 km apart, on the shelf off Oregon and Washington, might have been created by large-scale shearing along the plate boundary (Couch and Riddihough, 1989). Smaller fault-bounded Tertiary depressions, some 3 km deep and tens of kilometers across, lie on the shelf off southeastern Alaska (Bruns and Carlson, 1987). Such depressions in the continental crust on the exterior shelf were initiated before the latest Tertiary, as the many kilometers of sediments in them must have taken millions of years to accumulate.

Regional structural analysis suggests that areas inboard from the exterior shelf were not significantly affected by shearing propagating from the plate-boundary structural zone (Lyatsky, 1993a,b). Modern earthquakes, with right-lateral and in places reverse focal mechanisms, along the southeastern Alaska and Queen Charlotte Islands margin, are concentrated on this zone's inner strand (Rogers, 1983, 1986; Bérubé et al., 1989). In the past, however, the outer strand was the most active, and no more than a small amount of shearing might have propagated inboard. It is the outer strand that separates the continental and oceanic crust.

Seismicity is less intense along the margin off southern Vancouver Island (Fig. 58) as well as off Washington and Oregon (Riddihough et al., 1983; Taber and Smith, 1985; Crosson and Owens, 1987;

McCrumb et al., 1989a,b). Most of the seismicity in western Washington is clustered around Puget Sound, where the OWSZ interacts with the eastern side of the crustal-scale, rectangular North Olympic block (see Chapter 5).

Various parts of the OWSZ have manifested themselves differently throughout the Cenozoic, though apparently without large strike-slip movements. East of Puget Sound, the OWSZ is intruded by the Snoqualmie batholith whose age is 17 to 19.7 Ma (Reidel et al., 1994). On the Olympic Peninsula and southern Vancouver Island, strands of the OWSZ show no evidence of consistently displacing in a strike-slip sense the associated massifs of Eocene Crescent basalts (Babcock et al., 1994). The suspected strike-slip offset by the Calawah fault of Eocene and even late Pliocene sedimentary rocks on northwestern Olympic Peninsula, if real, is local and probably minor (see MacLeod et al., 1977, p. 277). This offset indicates, nonetheless, that this fault was active recently.

The Calawah fault is the southern strand of the OWSZ. Both it and the OWSZ's northern strand - the Westcoast fault - continue WNW onto the Vancouver Island exterior shelf. Whereas the Calawah fault does not continue directly past Barkley Sound, the Westcoast fault reaches northern Vancouver Island (Figs. 16, 50). These faults bound a number of crustal blocks on the shelf, which rose and subsided differentially since at least the Eocene. Between these blocks lie transcurrent faults trending NE and N-S. They continue on the shelf from Vancouver Island, and some of them, such as the Nitinat fault zone, cut the entire continental margin.

This NE-oriented fault zone continues to the lower continental slope. There it marks the southeastern boundary of the Juno depression, which was formed by downdropping of a rectangular block of oceanic crust. South of this fault zone, the gravity low at the foot of the slope off Washington is less regular and lacks a clear rectangular shape (see Finn, 1990; Finn et al., 1991). Unlike the fault-bounded Juno depression, the along-slope depression off Washington may simply be an undulation of the Juan de Fuca plate.

North of Juno depression, blocks of continental and oceanic crust are juxtaposed complexly. In the Brooks-Estevan embayment and off the mouth of Queen Charlotte Sound, they are interlocked. Blocks of oceanic crust were probably emplaced in these parts of the plate-boundary zone along faults.

Regional seismicity and questions about the subduction megathrust

To maintain space balance, oceanic lithosphere moving away from a spreading ridge must be subducted at another plate boundary and consumed into the mantle. At such a boundary, a great thrust fault must develop between the overriding and subducting plates, dipping under the overriding plate. Such a classical structural configuration is best illustrated at the Japan-Kurile and Alaska-Aleutian margins. There, many strong earthquakes due to friction of slabs at megathrusts occur in Wadati-Benioff seismicity zones, which can continue to depths of hundreds of kilometers. Many such earthquakes have thrust focal mechanisms.

Among the subduction zones worldwide, the Cascadia zone is "unusual" in many respects (Acharya, 1992). It lacks a strong deep earthquakes or earthquakes with thrust mechanisms. At one time, this was considered evidence against ongoing subduction of the Juan de Fuca plate beneath North America, and subduction was thought to have ceased (see McCrumb et al., 1989b). In the 1970s that attitude was rejected, but the tectonic models established at that time assumed the Juan de Fuca plate was rigid (Riddihough, 1977; Keen and Hyndman, 1979). Despite the large subsequent effort to prove that the Cascadia zone fits the typical subduction-zone model, the absence of some essential attributes remains. Intraplate seismicity offshore is incompatible with generalized models of rigid-plate behavior, and the Cascadia subduction zone is abnormally aseismic.

But still, evidence for Holocene submarine volcanism suggests that the Juan de Fuca Ridge, though also largely aseismic, is producing oceanic lithosphere (Johnson and Holmes, 1989). Addition of that lithosphere to the Juan de Fuca plate is now accommodated partly by intraplate deformation.

Large mismatches in Pacific plate-tectonic reconstructions for the late Cenozoic have been well publicized (Stock and Molnar, 1988; DeMets et al., 1990). The assumption that all oceanic plates are rigid has pushed the search for explanation in the direction of the North American continent, where internal deformation is undeniable. However, another part of the answer may be that the oceanic Juan de Fuca plate is not rigid.

Intraplate deformation occurs elsewhere, and it complicates the behavior of the Indian plate (e.g., DeMets et al., 1990). The Juan de Fuca plate is deformed more penetratively - by faulting, warping, crust-mantle delamination. That the rigid-plate assumption is inappropriate for the Gorda area off northern California is now recognized (Wilson, 1986; Stoddard, 1987, 1991), and a similar conclusion has been reached for the area off the mouth of Queen Charlotte Sound (Carbotte et al., 1989; Allan et al., 1993). The data presented here suggest that the rigid-plate assumption fails for much of northern Juan de Fuca plate.

The atypical character of the Juan de Fuca plate was acknowledged since the early plate-tectonic interpretations of the Pacific region, but was not explained with the data available at the time (e.g., McManus et al., 1972). Later, abnormalities of plate behavior were still not comprehensively explained, because tectonic models for this area relied on the rigid-plate assumption (Riddihough and Hyndman, 1989, 1991). However, the Cascadia subduction zone "is unique among other subduction zones the world over" (McCrumb et al., 1989b, p. 605). It is remarkable by the absence of seismogenic megathrust, a bathymetric trench, or a coherent magmatic arc.

Absence of megathrust earthquakes

The lack of seismicity related to a subduction megathrust is the most striking abnormality of the Cascadia subduction zone. Earthquakes near the continental margin are few, and their magnitudes are low. Epicenters occur in clusters, like the one in Puget Sound, and no earthquakes are deeper than 80 km. No thrust earthquakes have been recorded along the Cascadia subduction zone. Hypocenters of such earthquakes as do occur lie in the North America plate and in the inferred descending slab – but not at their supposed east-dipping megathrust boundary (Riddihough et al., 1983; Taber and Smith, 1985; Taber and Lewis, 1986; Crosson and Owens, 1987; Mooney and Weaver, 1989; Couch and Riddihough, 1989; Rogers et al., 1990).

Normal-fault focal mechanisms are typical at the lower seismicity level (Taber and Smith, 1985). Above, shallow earthquakes in the continental crust suggest dextral movements on NNW- to WNW-striking faults and reverse movements on faults with various orientations (e.g., Riddihough et al., 1983; Weaver et al., 1990). Consistent with other data, such as late Cenozoic structures in the interior of the Cordillera as well as drillhole breakouts and recent ground movements, this indicates that the principal stress in Washington and western British Columbia during the late Cenozoic has been N-S compression (Dragert, 1987; McCrumb et al., 1989a; Werner et al., 1991; Reidel et al., 1994).

Without recorded thrust earthquakes, conflicting models hold that either ongoing subduction is aseismic or the stress accumulates slowly and is released in periodic cataclysms (see Rogers, 1988). Both these explanations seem artificial.

That an active translithospheric fault should be completely aseismic is improbable. Cited in support of the currently prevailing second option is indirect evidence such as drowned coastal forests and peat swamps, sea-floor turbidites, slumps. These phenomena have been interpreted as triggered by sudden movements on the megathrust (e.g., Atwater, 1987, 1994).

However, a connection between occasional turbidity currents or local drowning of forests, and past earthquakes on the megathrust, is tenuous. Such phenomena occur in coastal areas worldwide, with no relation to subduction zones. Their causes may be various: sliding of rain-drenched sediments, small local earthquakes, sea storms, tsunamis, etc. (Acharya, 1992). Raised Holocene terraces are generally absent along the coast of Oregon, whereas Pleistocene wave-cut platforms are common. This probably indicates slow and gradual upwarping of the Oregon margin (West and McCrumb, 1988). On the other hand, the generally irregular distribution of subsidence and uplift zones along the entire coastal belt of Oregon, Washington and British Columbia is inconsistent with regional buildup of compressive stress.

Data from leveling surveys and geodetic strain measurements, as well as tidal data and observations of raised or subsided paleoshorelines, are sometimes linked with the presumed kinematics in the Cascadia subduction zone. Reliability of these data varies (Crosson, 1986), and their tectonic interpretations are ambiguous (e.g., Dragert, 1987). Discrimination between contributions from post-gracial isostatic rebound in western British Columbia and northwestern Washington, local block movements and eustatic sea-level fluctuations remains problematic (e.g., Riddihough, 1982b).

Complex, irregular vertical ground movements (Fig. 59), not easily explicable by two-dimensional flexure models, have been recorded along the Oregon, Washington and British Columbia margin (see the data presented by Lisowski, 1985; Dragert, 1987; Riddihough and Hyndman, 1991; Dragert et al., 1994; Mitchell et al., 1994). Local blocks are sinking, rising and tilting in different directions. These blocks are separated by narrow gradient zones trending mostly NE but in places also NW. A broad, E-W-oriented zone of subsidence near latitude 45°N is closed on the shelf but

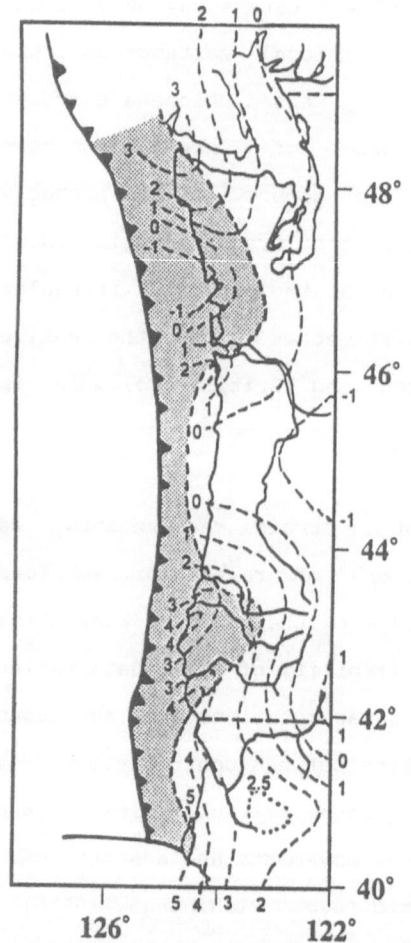

Figure 60. Present-day rates of coastal subsidence and uplift in western Washington and Oregon (modified from Mitchell et al., 1994). Subsidence and uplift rates (contoured) are in mm/year. The stippled area represents the conventionally assumed region of elastic-strain accumulation, based on the rigid-plate model. However, the irregular pattern of subsidence and uplift is not consistent with simple 2-D subduction models.

open landward. In contrast, a similar subsidence zone near 47°N
is closed landward but open to the shelf. These two downwarps are
bounded by uplift zones near 46°N and 48°N. Patterns of
subsidence and uplift are also irregular on the Western Canada
Archipelago. On Vancouver Island, crustal blocks separated by a
NE-oriented gradient zone running through the central part of the
island are at present being tilted at dissimilar rates. These
rates have varied during the last several decades (see Lisowski,
1985; Dragert, 1987).

Other evidence, discussed in previous chapters, suggests that
there is no extensive subducted slab or megathrust under the
continental margin of British Columbia. In the Cascadia
subduction zone farther south, no thrust seismicity is occurring,
and there is no clear indication that large earthquakes occurred
on the megathrust in the recent past. Thus, the idea that
accumulation of elastic stress in this entire region will soon
lead to a gigantic earthquake is unfounded.

Segmentation of volcanic chains in western Cordillera
Cenozoic volcanism in the western Washington and Oregon Cordillera
was subdivided by Armstrong (1978) into four episodes, beginning
at 55 Ma. However, volcanism before the Oligocene is no longer
thought to represent subduction-related arc magmatism, which is
believed to have begun in this region only around 36 Ma (Brandon
and Vance, 1992). Eocene magmatism in coastal areas of Washington
and Oregon is now thought to be a result of continental rifting
(also Babcock et al., 1992, 1994).

Late Cenozoic volcanoes in the western Cordillera can be grouped

into four main belts. The Cascade belt runs N-S from northwestern
California to northwestern Washington. Form there, the NNW-
trending Garibaldi belt runs along the southern Coast Belt orogen
and loses definition gradually around 50°30'N. Inland from Queen
Charlotte Sound, the ENE-oriented Anahim belt crosses much of the
Canadian Cordillera. In northwestern British Columbia, near the
border with southeastern Alaska, lies the N-S-oriented Stikine
volcanic belt.

The Cascade Mountains area contains the Late Pliocene to Recent
(post-2 Ma) volcanic rocks of the High Cascade suite, as well as
older, Oligocene and Miocene volcanics. Discontinuous character
of volcanism in the late Cenozoic is indicated by a jump in the
location of volcanic centers and a structural unconformity at the
base of the High Cascade suite. The previous volcanic sequence,
dated at 19 to 2 Ma, has a wide regional distribution (Cheney,
1994). In the Cordilleran interior, it includes basalts of the
Columbia River province, which erupted in an intracontinental
setting unrelated to the continental margin (Cheney, 1994; Reidel
et al., 1994).

Attributes of a subduction-related magmatic arc are most apparent
in the post-2 Ma High Cascade volcanic suite (e.g., Taylor, 1990),
but in northwestern Washington and southwestern British Columbia
the mode of volcanism and the composition of lavas change (Sherrod
and Smith, 1990; Green, 1990). Under western Washington and
Oregon, segmentation of the subducted slab is inferred from
teleseismic data (Michaelson and Weaver, 1986). Fault-controlled
segmentation of the High Cascade volcanic chain has been inferred
in part from gravity maps (Blakely and Jachens, 1990). Different

segments of this chain (Fig. 10) are characterized by dissimilar timing of volcanism and composition of erupted lavas (Guffanti and Weaver, 1988; Scott, 1990).

The High Cascade arc proper, in Oregon and southern Washington, is built largely of overlapping, mafic lava fields. Lavas on its northern end are more felsic. Quaternary eruption rates decline along the High Cascade chain from south to north. They decline further, dramatically, across the OWSZ (Sherrod and Smith, 1990), which separates the High Cascade and Garibaldi volcanic chains.

North of the OWSZ, volcanoes are generally aligned along faults parallel to the grain of the Coast Belt orogen. The NNW trend of the Garibaldi chain is accentuated by a series of hot springs (Green, 1990; Read, 1990; Souther, 1976, 1990). Green (1990) noted that the rock assemblage of the Garibaldi volcanic chain has geochemical signatures closely resembling magmatic associations in regions of extensional continental tectonism. Abundance of andesite, dacite and rhyolite, besides basalts, indicates presence in magmas of material from continental crust. Other volcanic belts in the British Columbia Cordillera, such as Anahim and Stikine, developed far away from the continental margin and are also unrelated to processes of subduction. Whereas large eruptions occurred recently at Mt. St. Helens in western Washington, volcanoes in the Garibaldi chain are currently inactive (Souther, 1990).

Drilling tests of geophysical models of continental-margin structure

Leg 146 of the Ocean Drilling Program was planned to test the accretionary-prism models off Oregon and southern Vancouver Island. On the Oregon continental slope, low-angle thrusts dipping to the east are apparent in seismic reflection profiles (e.g., Snavely, 1987; Goldfinger et al., 1992). A detachment is clearly imaged within the sedimentary section, 1.5 s above the basaltic oceanic-crustal basement (Cochrane et al., 1994). The overlying sediments are cut by distinctive, low-angle thrust faults which flatten out into this detachment. The sediments between the detachment and the basement are not affected and maintain subhorizontal stratification. In contrast, no such extensive, low-angle thrust structures are apparent in seismic lines off Vancouver Island (see Chapters 6, 8).

Existence of thrust faults on the Oregon slope was confirmed by ODP drilling. No thrust faults were penetrated off Vancouver Island (Carson et al., 1993; MacKay et al., 1994).

On the lower slope off Oregon, fault zones were encountered at depths of 260, 308, 375 and 440 m below seabed at the ODP site 891. These faults cut clayey silt and silty clay with subordinate sand layers. At site 892 nearby, a fault zone was penetrated at a depth of 52 m, followed by shear bands and stratal disruptions interpreted as another fault zone at 62.5-67 m; yet another interval with strongly developed fault-zone fabric (including mélange) was encountered at 106-175 m. These faults probably serve as conduits for fluids expelled from sediments in the growing accretionary wedge, and local hydrogeochemical anomalies are associated with the penetrated fault zones.

No such phenomena were encountered off southern Vancouver Island.

Some 7 km west of the foot of the slope, at ODP site 888, drilling to a depth of 567 m found the Cascadia Basin sediments to be undisturbed. The bedded Upper Pleistocene and Holocene sediments, less than 780,000 years in age, are laminated clayey silt intercalated with fine to coarse sand and even gravel. These sediments, derived from adjacent land areas, were deposited on the ocean floor by turbidity currents. The upper part of the section contains pieces of wood.

Wells at site 889/890 were drilled on the middle slope, at a water depth of 1322 m. This site sits on a bathymetric ledge, in an area which which may or may not be part of northern Juno depression. Upper Pliocene to Quaternary sediments, drilled to a depth of 345 m below seabed, are silt, clayey silt and fine sand. The bedding is subhorizontal in the top 104 m. Between 104 and 127 m, beds dip $40°-70°$ to the west. Below 127 m, rocks are fractured. The fractures are pervasive, interlocking and occasionally slickensided below 150 m.

Significantly, in sharp contrast to the accretionary prism off Oregon, no faults have been penetrated on the slope off Vancouver Island. Without fault-plane conduits, the flow of fluids escaping from compacting sediments is diffuse (Carson et al., 1993; MacKay et al., 1994).

Deformation of sediments on the continental slope off southern Vancouver Island was caused primarily by downslope slumping. Drilling results support the interpretation of seismic data from this area that only sparse, steep reverse faults, one of which is a reactivated normal fault, accommodate no more than minor tectonic shortening.

The pros and cons of dying subduction

The idea that subduction of the Juan de Fuca plate is stopping in an ongoing plate reorganization is voiced at times (Srivastava, 1973; Lister, 1991). However, it has received little attention (see McCrumb et al., 1989b), as the rigid-plate assumption continues to be applied (Riddihough and Hyndman, 1991).

Simply to argue that the subduction of the entire Juan de Fuca plate has stopped would indeed be difficult. No comprehensive alternative yet exists to the general principles of plate tectonics, which require that oceanic lithosphere created at spreading ridges must be disposed of in subduction zones to avoid space problems. In the worldwide practice, these principles have met considerable success. In the study area, unusual as it is, many geological and geophysical observations are consistent with continuing sea-floor spreading at the Juan de Fuca Ridge and underthrusting at parts the western North America margin. In Oregon in particular, continentward-dipping thrusts have been found to cut Eocene sedimentary rocks along the coast (Snavely, 1987). Existence of such thrusts on the Oregon continental slope is confirmed by seismic data (Cochrane et al., 1994) and by ODP drilling (Carson et al., 1993). The volcanic arc along the margin in Oregon and southern Washington is still extant.

It has been suggested that subduction may no longer be occurring in the southern (Gorda) part of the Juan de Fuca plate (Couch and Riddihough, 1989), where plate convergence is probably taken up by internal deformation of oceanic lithosphere (Wilson, 1986;

Stoddard, 1987, 1991). Disrupted magnetic stripes and diffuse seismicity are also found in the northern part of the Juan de Fuca plate off British Columbia (Atwater and Severinghaus, 1989; Wahlström and Rogers, 1992; Davis and Currie, 1993; Furlong et al., 1994). Detailed studies off the mouth of Queen Charlotte Sound have found that to apply the standard rigid-plate assumption in that part of the plate is awkward (Carbotte et al., 1989; Allan et al., 1993).

Rather than a comprehensive stoppage of subduction, it seems more realistic that the Cascadia subduction zone is shrinking gradually from the north and south. The Juan de Fuca plate increasingly accommodates convergence by internal deformation.

Slow changes in tectonic regime at the continental margin
Early doubts whether the Juan de Fuca plate and the Cascadia subduction zone are behaving in a "typical" manner (McManus et al., 1972; Stacey, 1973; Srivastava, 1973) were soon dismissed (Riddihough and Hyndman, 1976), in part by being ascribed to presumably less-than-wholehearted acceptance by some of the doubters of plate tectonics (Riddihough et al., 1983). The Cascadia subduction zone was extended as far as Queen Charlotte Sound (Hyndman et al., 1979; Riddihough, 1984). Big geophysical, mostly reflection seismic, surveys were used to corroborate this model (Keen and Hyndman, 1979; Clowes et al., 1984, 1987; Hyndman et al., 1990).

In the scope of this model, the seismic profiles were interpreted in terms of subduction-related structures. Low-angle events were readily interpreted as thrust faults.

However, these events could also be related to lithologic contacts in the thick Phanerozoic volcano-sedimentary succession, intrusive igneous bodies, metamorphic fronts, etc. Reprocessing of seismic data showed that many of the events initially assigned a structural significance are in fact diffractions (Milkereit et al., 1990) or off-line noise (Hawthorne, 1990; Levato et al., 1990). Geologic mapping shows that most faults on western and southern Vancouver Island are steep (Muller et al., 1974, 1981; Fairchild and Cowan, 1982; Rusmore and Cowan, 1985).

Faults in the plate-boundary structural zone, which continues to the Vancouver Island margin from Alaska, and in the OWSZ, which continues there from Idaho, are deep-seated and steep. Both these zones begin in the continental interior. The plate-boundary structural zone begins as the Fairweather fault in the Cordilleran interior in Alaska. Farther south, it separates the continental and oceanic plates off southeastern Alaska and British Columbia. Off Vancouver Island, it meets the OWSZ, which is a long-lived zone of crustal weakness that transects the principal geologic grain of the Washington and British Columbia Cordillera and separates large Cordilleran geologic provinces. Such a composite, huge structural zone, thousands of kilometers long and running on both ends into the continent, has no relation to subduction.

From Vancouver Island, the boundary between the North America and Juan de Fuca plates runs southward, and the Cascadia subduction zone is found there. The OWSZ part of the composite structural zone trends ESE and runs into the continental interior.

On the upper continental slope off southern Vancouver Island and the Olympic Peninsula, integrated analysis of geophysical data suggests the presence of continental crust. Off northern Vancouver Island, attenuated continental crust underlies both the upper and lower slope (Chapters 5-8).

On the Olympic Peninsula, extensive deformation in the Miocene might have been induced by convergence of North America with the Farallon/Juan de Fuca plate. In the Central Olympic Basin and in the Olympic Mountains, compressional structures with a regional westerly vergency accommodate considerable shortening. However, in the Hoh Basin to the west, many shear zones are steep and separated by areas of coherent bedding; some of the faults may be related to obduction rather than subduction (Snavely, 1987; Orange, 1990; Orange et al., 1993). An accretionary sedimentary pile, which developed largely in the Late Miocene, lies farther outboard, on the submerged margin; in part it overlies attenuated continental crust.

On the lower slope off the Olympic Peninsula, some thrusting went on till the Quaternary (Line 76-19; Fig. 56). However, the age of the mélange drilled on the Olympic Peninsula shelf beneath stratified younger sediments, and the estimates of likely rates of mélange growth (Chapter 8), suggest that most of the prism developed in the Late Miocene - around the time of uplift and deformation in the Hoh Basin. Since then, no major compression is known to have occurred in the Hoh Basin. On the lower slope, continentward-dipping thrust faults cut sediments as young as Late Pliocene, but only one, at the foot of the slope, cuts Quaternary sediments (Fig. 56; Snavely and Wagner, 1981). This indicates the intensity of compressional tectonism at the Olympic Peninsula continental margin has probably declined since the Late Miocene.

Many bathymetric ridges on the Washington continental slope are caused by thrust faults. A number of these faults are dipping oceanward, rather than continentward as expected in a typical accretionary prism (Snavely, 1987). Laboratory experiments indicate that landward-verging thrusts may indeed form due to variations in local conditions in a thick, semi-consolidated, compressed sediment pile (Seely, 1977).

Aside from thrusting, overpressuring may lead to diapirism, and diapirs on the Washington shelf deform and pierce the Upper Miocene to Recent sedimentary blanket (Snavely, 1987). More diapirs, probably caused by a combination of overpressuring and movements on underlying steep faults, occur on the shelf off Vancouver Island (Shouldice, 1971; Tiffin et al., 1972). Such diapirs are found worldwide in mud-dominated basins that have no relation to subduction.

Also unrelated to subduction are the young sediment slumps that have been imaged with sonar and seismic techniques on the upper and lower continental slope off northern and southern Vancouver Island (Tiffin et al., 1972; Figs. 5 and 6 of Davis and Hyndman, 1989). Buried slumps, like those on the inboard side of the Juno depression in seismic line 85-01 (Fig. 52), probably account for the penetrative yet diffuse deformation of sediments encountered at ODP drilling site 889/890 at mid-slope.

Seismic profiles show that on the slope off southern Vancouver

Island reverse faults are sparse and steep, and no such faults are observed off northern Vancouver Island (see Chapters 6, 8). No accretionary mélange was penetrated in the Cygnet J-100 well on the southern Vancouver Island outer shelf. As was discussed above, little room is available for such a mélange off southern Vancouver Island, and none at all off central and northern Vancouver Island.

Structural and geophysical evidence from the Olympic Peninsula and the adjacent submerged margin indicates that manifestations of subduction in that area have decreased gradually in time, since the Late Miocene. They also decrease gradually in space, from the Washington margin northward.

Importance of geological paradigm as a guide for geophysical interpretation

Because interpretation of geophysical data generally does not yield a unique solution, selection between various geophysically permissible options must be guided by geological reasoning. Based on the principles of tectonics of rigid plates, a switch of regime along a continental margin from subduction to transform cannot occur gradually: a well-defined triple junction is required between these tectonic domains. Such a model has previously guided the interpretation of geophysical data along the continental margin of western North America (Riddihough and Hyndman, 1989, 1991). It is internally logical as it seeks to avoid space problems in reconstructing plate motions - as long as plate rigidity is accepted. But if the oceanic lithosphere is not rigid, rethinking of the model is required.

Consistent with the gradual northward loss of stiffness of the Juan de Fuca plate, manifestations of subduction also decrease northward. Seismic data show the continental slope off Oregon and southern Washington is cut by many continentward-dipping thrust faults (Snavely, 1987; Carlson and Nelson, 1987; Goldfinger et al., 1992) coalescing into a detachment (Cochrane et al., 1994). No such thrust detachment is apparent in seismic profiles off Vancouver Island. The idea that the Tofino Basin on the shelf is cut by low-angle thrusts (Yorath, 1980, 1987) is contradicted by both seismic profiles and well information.

Thrust-sheet structure has been inferred from deep seismic data on Vancouver Island and the adjacent submerged margin (Yorath et al., 1985a,b; Clowes et al., 1987; Hyndman et al., 1990). However, extremely poor signal penetration and chaotic character of events on the middle and upper continental slope make reliable interpretation difficult. Besides, whereas seismic data are good at imaging low-angle discontinuities in the subsurface, steep discontinuitues often remain undetected. Geologic mapping (Muller et al., 1974, 1981; Muller, 1977a-c; Fairchild and Cowan, 1982; Rusmore and Cowan, 1985) and detailed seismic studies (Mayrand et al., 1987) show the major faults on the island to be steep. As well, no objective criteria have yet been developed to interpret deep seismic events as thrusts, rather than as metamorphic fronts (Hyndman, 1988), pluton tops, sills, stratigraphic contacts, etc.

All this leaves much room for speculation. In interpreting seismic data across the continental margin, it has been customary since the 1970s to concentrate on continentward-dipping events assumed to represent subduction-related thrusts.

Recently, however, some of the deep seismic events on Vancouver Island previously interpreted as thrust structures have been shown by data-processing tests to be diffractions (Milkereit et al., 1990). The event "F" in Line 84-01 (Fig. 46), interpreted variously as the top (Hyndman et al., 1990) or the base (Clowes et al., 1987) of the undergoing oceanic crust, has been shown to probably be off-line noise (Hawthorne, 1990; Levato et al., 1990). Along the main LITHOPROBE transect through the island and adjacent submerged margin, very different interpretations of the same reflection and refraction data have been produced by different workers (see papers in Green, ed., 1990).

On southernmost Vancouver Island, in the NW half of the short, isolated reflection line 84-04 which crosses the steep San Juan fault (Figs. 14, 23), events dipping NW are observed from 2 to 5 s. In keeping with the thrust-complex model, the shallowest of them ("B") was postulated to be linked with low-angle events on the SE end of this profile and interpreted as a reflection from the Leech River "thrust" (Clowes et al., 1987). But field mapping shows the Leech River fault is steep (e.g., Muller, 1977c). In any case, Line 84-04 does not cross this fault, so projecting any events to its surface trace is conjectural. As well, the event "B" is not continuous. A break in it occurs in the middle of the profile at the San Juan fault, which was shown by detailed analysis of these data to be subvertical (Mayrand et al., 1987).

The Leech River complex adjoins the Eocene Metchosin igneous massif. This massif is regarded in some models as part of an accreted oceanic-crustal Crescent "terrane" (Hyndman et al., 1990) which under Vancouver Island forms a 4-km-thick, east-dipping panel (Dehler and Clowes, 1992). But field mapping onshore shows the Crescent Formation comprises many discrete igneous massifs of various sizes, which erupted from multiple centers in a rift setting, in situ. Structural evidence from the Olympic Peninsula argues against their accretion (Brandon and Vance, 1992; Babcock et al., 1992, 1994). Inconsistent with oceanic-crustal origin, felsic rocks are found in these massifs on the Olympic Peninsula (Snavely, 1987) and southern Vancouver Island (Muller, 1977c).

Crescent basalts are in stratigraphic contact with sedimentary rocks of both the Central Olympic and Fuca basins on the Olympic Peninsula, and no fossil subduction megathrust is recognized between them (Babcock et al., 1994; Chapter 5).

Magnetic anomalies over three Eocene basaltic bodies on the Vancouver Island exterior shelf are elongated and narrow. On northern Olympic Peninsula, other elongated Crescent bodies are associated with faults of the OWSZ, along which they seem to have erupted. On the shelf, an extension of one of these faults (Calawah) bounds the Prometheus basaltic body, suggesting in-situ eruption in that area as well. Absence of gravity anomalies over these bodies confirms that their size is small.

South of the Olympic Peninsula, other Crescent massifs which cause strong gravity anomalies were modeled by Finn (1990) to be up to 30 km thick. The gravity high over the Metchosin massif is also strong and centrally symmetrical, and remains so when gravity data are upward continued to 20 km. There is no indication of lateral offset in a thrust slice, but rather a suggestion that a frozen mafic magma chamber underlies this massif at mid-crustal level.

Another presumed thrust-bounded exotic "terrane" is the Pacific Rim complex on western and southern Vancouver Island (Hyndman et al., 1990; Dehler and Clowes, 1992). However, field mapping shows it is confined to narrow fault-bounded slices which were probably dispersed into these areas by strike-slip faulting from the San Juan Islands (Rusmore and Cowan, 1985; Brandon, 1989a,b). One of the faults bounding a Pacific Rim slice is the Westcoast fault. It is long-lived, having acted during its Mesozoic and Cenozoic history in a normal, reverse and strike-slip sense (Muller et al., 1974, 1981; Brandon, 1989a,b). During the Tertiary, it was incorporated into the OWSZ.

South of the OWSZ, the Hoh sedimentary basin is made up of three distinct inconformity-bounded packages of Eocene, Upper Oligocene to Middle Miocene, and Pliocene to Recent ages. In the Late Miocene, it was deformed and uplifted, around the time when most of the accretionary mélange was created outboard. However, plate interactions had less direct effect on the development of other deep basins in this region. The Tofino Basin was controlled by differential uplift and subsidence of fault-bounded crustal blocks, and the Fuca Basin lies within the OWSZ. Subsidence of the Puget and Georgia basins occurred in conjunction with uplift of the North Cascade and Coast Mountains.

Fragmentation of the northern Juan de Fuca plate is indicated by various lines of evidence. The supposed boundaries of the Explorer fragment with the Pacific plate - the Explorer Ridge and the Sovanco fracture zone - are not active seismically (Barr,

1974; Barr and Chase, 1974; Wahlström and Rogers, 1992). Basalts along the Explorer Ridge have geochemical characteristics unexpected at spreading centers (Michael et al., 1989; Allan et al., 1993). The Sovanco fracture zone is tens of kilometers wide and contains many rotated blocks (Cowan et al., 1986; Lister, 1991; Davis and Currie, 1993). No active spreading centers or plate triple junction exist off Queen Charlotte Sound. The northern boundary between the Pacific and Juan de Fuca (or Explorer) plates is still undefined and may be diffuse (Furlong et al., 1994). The assumption that oceanic lithosphere behaves rigidly off Queen Charlotte Sound has been found to be unrealistic (Carbotte et al., 1989; Allan et al., 1993).

The broad plate-boundary fault system continues along the western North America continental margin from southeastern Alaska, past the Queen Charlotte Islands, to northern Vancouver Island. In the Brooks-Estevan embayment, small blocks of continental and oceanic crust are juxtaposed along faults expressed in gravity and magnetic maps. From the Brooks-Estevan embayment, the plate-boundary structural zone continues to the submerged margin off southern Vancouver Island, where it meets the OWSZ.

Ongoing subduction off Vancouver Island is inconsistent with the observed pattern of modern ground movements. From leveling surveys, Dragert (1987) reported dissimilar elevation changes and rates of tilting of different blocks on the island. Upwarping in the Comox and Campbell River areas, on the island's east side, continued at about 1 mm/year between 1946 and 1977. Between 1977 and 1984, the Comox area rose at the same rate, but uplift near Campbell River just to the north accelerated to 5 mm/year. The boundary between these two blocks, on central Vancouver Island, trends NE. To Dragert, such a pattern of modern block movements suggested regional N-S compression, which has also been inferred for the entire western British Columbia and Washington (Riddihough et al., 1983; Reidel et al., 1994). It is "inconsistent with simple two-dimensional subduction" (Dragert, 1987, p. 695).

Gradual change of tectonic regime along the margin occurs off northern Washington and southern British Columbia. An end-member regime, subduction, is evident off Oregon and southern Washington. From detailed studies of the structure of the submerged margin and patterns of coastal subsidence and uplift involving Quaternary deposits, McCrory (1994) concluded that the northern Washington margin is an area of transition between "subduction-dominated tectonic processes" to the south and unspecified "margin-dominated processes" to the north.

Along the Western Canada Archipelago, no direct signs of ongoing or recent subduction are apparent from geological or geophysical data. Unlike in western Oregon and Washington, the continental crust thins towards the margin stepwise. The boundary between juxtaposed blocks of attenuated continental and oceanic crust runs along the middle slope off southern Vancouver Island and even farther outboard, along the Revere-Dellwood fault, off northern Vancouver Island. Off the Queen Charlotte Islands, this boundary follows the outer scarp of the Queen Charlotte Terrace. The inner strand of the plate-boundary fault system off northern Vancouver Island is a west-dipping normal fault (Fig. 36).

No compressional structures have yet been found to accommodate the presumed very oblique convergence between the North America and Pacific plates modeled off the Queen Charlotte Islands. The uplift and continentward tilt of the islands, supposedly in the Pliocene, has been cited as evidence of convergence (Yorath and Hyndman, 1983; Riddihough and Hyndman, 1991). Geologic mapping, however, shows that uplift of the entire Western Canada Archipelago began much earlier. On the Queen Charlotte Islands, it was accommodated by many steep faults and occurred without significant eastward tilting (Thompson et al., 1991; Hickson, 1991). Many more faults, normal and reverse, are found on the interior shelf in Hecate Strait (Lyatsky, 1993a). Most of Vancouver Island was raised throughout the Tertiary. Its uplift is continuing at present (Dragert et al., 1994), but in a complex fashion suggestive of N-S compression (Dragert, 1987).

Along the plate-boundary fault system off southeastern Alaska, seismicity and offsets of sea-floor bathymetric features suggest that late Cenozoic displacements have mostly been strike-slip. The inner strand of the fault system is active at present, but outboard strands are thought to have been active in the past (Bruns and Carlson, 1987; von Huene, 1989). These faults are linked directly with the faults bounding the Queen Charlotte Terrace. North of Cross Sound, the Fairweather fault component of this system runs into the interior of southern Alaska.

Thus, no simple separation division can be made between the subduction and strike-slip regimes along the continental margin. Such a division was made previously on the assumption that the behavior of oceanic plate in this region is rigid (Riddihough and Hyndman, 1991). Instead, the subduction regime, well developed off Oregon and southern Washington, decays northward gradually. A transitional zone lies off northern Washington and perhaps southern Vancouver Island.

Other evidence for non-rigid behavior of the Juan de Fuca plate and gradual change of tectonic regime along the continental margin
<u>Absence of bathymetric trench</u>

For decades, the absence of a bathymetric trench along the Cascadia zone of subduction has remained a mystery. Trenches are typical at other subduction zones in the circum-Pacific region and elsewhere, and absence of a trench makes the Cascadia zone highly unusual (McCrumb et al., 1989a).

All the more strange, then, is the presence outboard, off Oregon and southern Washington, of a fore-trench bulge. Such bulges are typically caused by downward bending of oceanic plates entering subduction zones, and thus are intimately associated with trenches. But the Cascadia bulge is only weakly developed, identified mostly from gravity data (e.g., Jachens et al., 1989) rather than from relief of the ocean floor or of the top of the basalts beneath the Cascadia Basin. Presence of a bulge is not indicated by gravity data farther north.

Such abnormal phenomena are easier to explain by gradual northward loss of the strength of oceanic lithosphere and by general non-rigid behavior of the Juan de Fuca plate.

<u>Isometric geoid anomaly off the western North America continental margin</u>

Active subduction zones around the perimeter of the Pacific Ocean

- Japan, Kurile, Aleutian, Chile, etc. - are marked by pronounced geoid lows which run along subduction-related trenches. These degree-10 residual anomalies have been found to be narrow (200-400 km), long (thousands of kilometers) and distinctly negative (-20 m on average; Bowin, 1983).

A different geoid anomaly lies off western North America, from California to southeastern Alaska. It is nearly isometric, with a huge diameter of about 2500 km. It is concentric and only about -10 m in amplitude. Unrelated to subduction, positive and negative anomalies of similar dimensions are the norm all across the Pacific.

Absence along the Cascadia subduction zone of an elongated geoid low typical of other subduction zones might be related to the weakness of the Juan de Fuca plate: the flexure it supports is too small to produce a trench or an associated low in geoid maps. Weakness of the subducting plate might also be the reason for the lack of earthquakes on the megathrust.

Deformation of the basaltic basement in the abyssal Cascadia Basin

Drillhole and seismic data show that the Cascadia Basin contains thinly bedded, continent-derived, pelagic and turbiditic siltstone and sandstone with an average total thickness of about 500 m. However, the top of underlying ocean-crust basalts is rugged, with hundreds of meters of relief (Carlson and Nelson, 1987; Calvert et al., 1990; Hasselgren et al., 1992; Rohr, 1994) caused by buried mounds and faults. Depressions in this relief are filled with sediments of the lower, seismically transparent, sedimentary unit. The overlying upper unit consists of turbiditic, well-bedded deposits. Near the continental slope, mostly in the Astoria and Nitinat fans, the Cascadia Basin is up to 2.5-3 km thick.

Along the lower continental slope off British Columbia, faults of the plate-boundary structural zone accommodated subsidence of blocks and formation of grabens in both the oceanic (e.g., Juno depression) and continental (e.g., Winona Basin) crust. In the Brooks-Peninsula embayment, fault-related gravity anomalies coincide with magnetic linear anomalies continuing into this area from outboard parts of the northern Juan de Fuca plate. If reactivation of older faults in the oceanic crust was involved, origin of magnetic anomalies in parts of the northern Juan de Fuca plate might be related to local faulting. Interpretation of such anomalies as isochrons related to sea-floor spreading alone would contaminate local plate reconstructions.

Presence of a broad belt of earthquakes running from northern Juan de Fuca Ridge NNW (e.g., Wahlström and Rogers, 1992) suggests considerable in-plate deformation. If late-stage faulting caused dike magmatism, resulting magnetic anomalies might deceptively resemble those conventionally related to sea-floor spreading.

Shearing of the Juan de Fuca plate in the third dimension
Earthquake data, as well as studies of young geologic structures, modern ground motion, well break-outs, etc., indicate that the stress field in Washington varies with depth: N-S compression in the continental crust, vs. extension at deeper levels presumably associated with the subducted slab. Causes of this contrast are unclear. The continental crust in western North America is probably thick enough, and rich enough in sources of radiogenic

heat, to be capable of tectono-magmatic and tectono-metamorphic self-development. At any rate, the N-S compression has never been fully explained in terms of ongoing subduction (McCrumb et al., 1989a). Presence of such stresses in the Cordilleran interior, where they manifested themselves by creating late Cenozoic structures in the Yakima belt (Reidel et al., 1994), casts further doubt on a relationship to subduction.

Probably, these stresses are related to regional adjustment as plate interactions are changing. Jostling of fault-bounded blocks of continental crust in a changing stress field is reflected in the irregular current ground movements.

Reflecting ongoing change in plate interactions, the amount of recent compression at the continental slope and the intensity of current arc magmatism decrease from south to north. Most of the change takes place in northern Washington and perhaps southernmost British Columbia. The stoppage of subduction is encroaching on the Cascadia subduction zone gradually, from the north, and the plate interactions slowly continue to change.

Jostling of blocks adjusting to a new regime is also apparent in the northern Juan de Fuca plate. Earthquakes there occur in broad bands. Due to structural readjustments and perhaps some related deep-seated processes, velocity structure of the oceanic crust is variable, and its thickness locally seems to reach 11 km (Malecek and Clowes, 1978; Au and Clowes, 1982, 1984). New seismic reflection profiles contain dipping events within the oceanic crust and reveal many other variations in crustal structure which have been ascribed to magmatic underplating (Hasselgren et al.,

1992). Alternatively, thickening of the oceanic crust might be structural (the "bunching-up" of Malecek and Clowes, 1978), in which case the dipping reflections may indicate low-angle zones of intracrustal structural delamination.

Gradual change in tectonic regime along the continental margin and deep structure of the margin region from teleseismic data

Deep, if coarse, images of continental-margin structure obtained with teleseismic data show that the subducted Juan de Fuca slab beneath western Oregon and Washington is deteriorating. Tears perpendicular to the margin divide the slab into segments with different dips (Michaelson and Weaver, 1986). Overall, though, the dip of the subducted slab is remarkably steep, up to 60° just a short distance from the Cascadia subduction zone (Rasmussen and Humphreys, 1988; Humphreys and Dueker, 1994a,b). A tear propagating through the subducted slab in a N-S direction, at a depth of 150-300 km, has also been suggested (VanDecar et al., 1990; VanDecar, 1991).

The rate of subduction in the entire Cascadia zone has decreased in the last several million years (Riddihough, 1984). Though the decline of the subduction rate seems to be greater in the north than in the south, degradation of the underthrusted slab seems to be proceeding from south to north. In the north, the entire structure of the continental-margin region, from the crustal levels to a depth of several hundred kilometers, appears to be different than in the south (Humphreys and Dueker, 1994a,b).

CHAPTER 10 - CONCLUDING REMARKS

Continental-crust structures in boundary zones between continental and oceanic plates have received little attention in the last several decades. Perhaps because the northeastern Pacific is the cradle of the plate-tectonic theory, a perception has evolved that this region, above all, should fit perfectly the generalized global models of plate interactions. As a result, many geological observations deviating from the expected "typical" behavior have been underemphasized. The temptation has developed even for the geologists who make field observations to fit their findings to pre-conceived, simplistic tectonic scenarios.

But hiding from mismatches between facts and models neither resolves the problem nor makes it go away. Room for improvement of our knowledge about the nature is endless, and no level of understanding we achieve is likely to be final.

The contacts between continental and oceanic crust off western North America are of two main types: along steep faults, or along thrusts. In the case of thrust emplacement, depending on which type of crust overrides which, a distinction can be made between subduction and obduction. The former predominates, and various lines of evidence indicate that a subduction megathrust is well developed off Oregon and southern Washington. Juxtaposition along steep faults may also be of two types: a single, long bounding fault may clearly separate oceanic and continental crust, or many blocks detached or semi-detached from their mother plates may be interlocked along many local faults.

No clear break is observed along the continental margin between the tectonic domains where juxtaposition and thrust relationships predominate. The lack of a sharp tectonic-domain break can be explained if the Juan de Fuca plate is not perfectly rigid. This plate itself and the mode of its subduction beneath North America are now acknowledged to be "unusual". Particularly remarkable are the high degree of intraplate deformation, the absence of Wadati-Benioff seismicity or of a bathymetric trench, and the gradual change along the continental margin between the regimes of juxtaposition and underthrusting of oceanic crust.

The common explanation for the Juan de Fuca plate's weakness - that it is young, warm, thin and soft - accounts for some of these phenomena but largely leaves unexplained the interlocking between oceanic and continental crust. Comparing the behavior of this plate with other oceanic plates worldwide, including those which are young and warm, Acharya (1992) still concluded that the Juan de Fuca plate is unusual even against that background. Perhaps it should be considered a member of a special class.

Whatever the bahavior of the Juan de Fuca plate, the role of continental crust in shaping the structure of the margin was also considerable. Attenuated continental crust underlies parts of the shelf and upper slope. The capacity for self-development of continental crust, which contains its own sources of radiogenic heat, complicates the relationship between the structure of the continental margin and the behavior of oceanic plates outboard.

Fault networks found in the continental crust on land also continue far offshore. A huge structural zone, comprising the

OWSZ, the Scott Islands fracture zone, the Queen Charlotte fault zone, the Chichagof-Baranof fault zone, and the Fairweather fault, runs from the interior of the Pacific Northwest Cordillera, to the submerged margin, and back into the Cordilleran interior in Alaska. This fault system has at least in the late Cenozoic controlled the position of the plate boundary off northern Washington, British Columbia and southeastern Alaska.

Continental crust does more than react passively to motions of adjacent oceanic plates. The dynamic, deep thermal structure of continental plates, though only sketchily understood, may nevertheless be related to large vertical movements of continental regions in the past (e.g., Forte et al., 1993). Potential influences on the development of continental crust are many, and they may have nothing to do with the motion of neighboring plates.

The global tectonic framework from plate reconstructions may help unravel the evolution of specific regions. Like any scientific tool, however, these reconstructions can be misused. Misuses are to rely on this one tool whilst neglecting other lines of inquiry, or to use it uncritically, applying a rather blunt instrument where sophistication and finesse are in order.

In the early days of plate tectonics, most of the scientific attention was paid not so much to local details as to erecting the main pillars of the new global thinking. Two general assumptions have emerged from that effort: that oceanic magnetic stripes reflect the creation and motion not just of the oceanic crust but of the entire lithosphere; and that each plate, oceanic or continental, behaves as a stiff, rigid lithospheric entity.

In the first approximation, these assumptions have performed well. The early plate reconstructions flowing from them seemed, by and large, internally coherent, and dealing with any remaining discrepancies and improving the resolution of models was expected to be a matter of mopping up. However, refinements of plate reconstructions have failed to eliminate the discrepancies. Ill-defined plate boundaries and misties in global and regional plate circuits have received a lot of attention in the recent years, because to proper reconstructions they are unacceptable (Stock and Molnar, 1988; DeMets et al., 1990).

Intraplate deformation is sometimes invoked to explain the discrepancies: Basin and Range extension in the interior of the western North American continent; or warping of oceanic lithosphere in the Indian plate. If a plate expands, contracts, or is delaminated internally, magnetic stripes correspondingly become less reliable as indicators of motion of the entire lithospheric plate. Applying erroneous reconstructions might lead to errors in local tectonic interpretations.

To guard against this danger, it pays to test tectonic models for specific regions through geology-based local studies. Such studies, of course, are easiest in continental areas onshore. The usual advantage of detailed local investigations is the possibility of incorporating into a single interpretation the results of various geological and geophysical surveys.

No one method of geoscience inquiry can lead to a unique solution. Plate-motion reconstructions relying only on oceanic magnetic stripes and hot-spot tracks are limited because these methods of reconstruction are limited. In continental regions, in contrast, many more lines of evidence are available: paleontological, sedimentological, geochemical, etc. For example, burial histories of sedimentary basins may be deduced from patterns of thermal maturation, or from fission-track analysis of rocks. If rocks are metamorphosed, studies of their metamorphic history offer clues to the history of their deep burial and exhumation. Styles of magmatism may be related to the composition of the underlying lithosphere. Structures in rocks reflect the history of their deformation. The list goes on.

A model of a region's evolution involving all such considerations is better constrained than a model based on just one or two lines of evidence. The more is known, the less need there is to make assumptions about what is not. Global plate reconstructions can provide useful constraints on models of regional tectonics, but they should be checked by independent multidisciplinary analysis of specific regions. A region should be treated as a complex geodynamical system.

The natural system examined here is a small element of the globe - a part of the North American continent. The planet Earth as a whole is a much bigger system, of a higher hierarchical order. The nature - or its parts - is complex, made up of more bits than we can ever know, interacting in more ways than we could possibly imagine. One can only conceive of something that complex, that limitless, in a simplified form.

So we conjure up our inherently limited models. They help us to understand the internal and external structure of natural systems and their evolution. But abstraction, indispensable as it is, can also be misleading if it takes one too far away from reality. Modeling is a means to an end, not an end in itself.

But what is a "realistic" model? When is a model too simplistic? How accurate is accurate enough? One check on a model's reliability is its internal logical consistency, but that alone does not prove the model is "true". The model must also be testable (Oreskes et al., 1994; Lyatsky, 1994).

Building a model requires carefully defining the system and examining its observable phenomenological manifestations. Only then, keeping in mind the intended objective of the study, can one begin to develop a suitable simulation. No one step is more important than another, and all assumptions and conclusions need to be tested by observable facts.

REFERENCES

Acharya, H., 1992. Comparison of seismicity parameters in different subduction zones and its implications for the Cascadia subduction zone; Journal of Geophysical Research, v. 97, p. 8831-8842.

Adair, M.L., Talmage, R.H., Crosby, T.W., and Testa, S.M., 1989. Geology and seismicity of the Skagit nuclear power plant site; in Galster, R.W. (ed.), Engineering Geology in Washington, v. I; Washington Division of Geology and Earth Resources, Bulletin 78, p. 607-624.

Allan, J.F., Chase, R.L., Cousens, B., Michael, P.J., Gorton, M.P., and Scott, S.D., 1993. The Tuzo Wilson volcanic field, NE Pacific: alcaline volcanism at a complex, diffuse, transform-trench-ridge triple junction; Journal of Geophysical Research, v. 98, p. 22,367-22,387.

Anders, M.H. and Christie-Blick, N., 1994. Is the Sevier Desert reflection of west-central Utah a normal fault?; Geology, v. 22, p. 771-774.

Anderson, R.G. and Greig, C.J., 1989. Jurassic and Tertiary plutonism in the Queen Charlotte Islands, British Columbia; in Current Research, Part H; Geological Survey of Canada, Paper 89-1H, p. 95-104.

Anderson, R.G. and Reichenbach, I., 1991. U-Pb and K-Ar framework for Middle to Late Jurassic (172-158 Ma) and Tertiary (46-27 Ma) plutons in Queen Charlotte Islands, British Columbia; in G.J. Woodsworth (ed.), Evolution and Hydrocarbon Potential of the Queen Charlotte Basin, British Columbia; Geological Survey of Canada, Paper 90-10, p. 59-87.

Andrew, A., Armstrong, R.L., and Runkle, D., 1991. Neodymium-strontium-lead isotopic study of Vancouver Island igneous rocks; Canadian Journal of Earth Sciences, v. 28, p. 1744-1752.

Andrew, A. and Godwin, C.I., 1989a. Lead- and strontium-isotope geochemistry of Paleozoic Sicker Group and Jurassic Bonanza Group volcanic rocks and island Intrusions, Vancouver Island, British Columbia; Canadian Journal of Earth Sciences, v. 26, p. 894-907.

Andrew, A. and Godwin, C.I., 1989b. Lead- and strontium-isotope geochemistry of the Karmutsen Formation, Vancouver Island, British Columbia; Canadian Journal of Earth Sciences, v. 26, p. 908-919.

Andrew, A. and Godwin, C.I., 1989c. Lead- and strontium-isotope geochemistry of the Tertiary Catface intrusions and related mineralization, Vancouver Island, British Columbia; Canadian Journal of Earth Sciences, v. 26, p. 920-926.

Andrew, A. and Godwin, C.I., 1989d. Galena lead isotope model for Vancouver Island; in Geological Fieldwork 1988, British Columbia Ministry of Energy, Mines and Petroleum Resources, Paper 1989-1, p. 75-79.

Archibald, D.A. and Nixon, G.T., 1995. 40Ar/39Ar geochronometry of igneous rocks in the Quatsino-Port McNeill map area, northern Vancouver Island (92L/12,11); in Geological Fieldwork 1994, British Columbia Ministry of Energy, Mines and Petroleum Resources, Paper 1995-1, p. 49-59.

Arkani-Hamed, J. and Strangway, D.W., 1988. Interpretation of the aeromagnetic anomalies of southern Vancouver Island; Canadian Journal of Earth Sciences, v. 25, p. 801-809.

Armstrong, R.L., 1978. Cenozoic igneous history of the U.S. Cordillera from lat. 42˘ to 49˘N; in R.B. Smith and G.P. Eaton (eds.), Cenozoic Tectonics and Regional Geophysics of the Western

Cordillera; Geological Society of America, Memoir 152, p. 265-282.

Armstrong, R.L., 1988. Mesozoic and early Cenozoic magmatic evilution of the Canadian Cordillera; in S.P. Clark, B.C. Burchfiel, and J. Suppe (eds.), Processes in Continental Lithospheric Deformation; Geological Society of America, Special Paper 218, p. 55-91.

Armstrong, R.L., Muller, J.E., Harakal, J.E., and Muelenbachs, K., 1985. The Neogene Alert Bay volcanic belt of northern Vancouver Island, Canada: descending-plate-edge volcanism in the arc-trench gap; Journal of Volcanology and Geothermal Research, v. 26, p. 387-399.

Atwater, B.F., 1987. Evidence for great Holocene earthquakes along the outer coast of Washington State; Science, v. 236, p. 942-944.

Atwater, B.F., 1994. Prehistoric earthquakes in western Washington; in R. Lasmanis and E.S. Cheney (eds.), Regional Geology of Washington State; Washington Division of Geology and Earth Resources, Bulletin 80, p. 219-222.

Atwater, T., 1970. Implications of plate tectonics for the Cenozoic tectonic evolution of western North America; Bulletin of Geological Society of America, v. 81, p. 3518-3536.

Atwater, T., 1989. Plate tectonic history of the northeast Pacific and western North America; in E.L. Winterer, D.M. Hussong, and R.W. Decker (eds.), The Eastern Pacific Ocean and Hawaii; Geological Society of America, The Geology of North America, v. N, p. 21-72.

Atwater, T. and Severinghaus, J., 1989. Tectonic maps of the northeast Pacific; in E.L. Winterer, D.M. Hussong, and R.W. Decker (eds.), The Eastern Pacific Ocean and Hawaii; Geological Society of America, The Geology of North America, v. N, p. 15-20.

Au, D. and Clowes, R.M., 1982. Crustal structure from an OBS survey of the Nootka fault zone off western Canada; Geophysical Journal of Royal Astronomical Society, v. 68, p. 27-47.

Au, D. and Clowes, R.M., 1984. Shear-wave velocity structure of the oceanic lithosphere from ocean bottom seismometer studies; Geophysical Journal of Royal Astronomical Society, v. 77, p. 105-123.

Baars, D.L., 1978. The Olymipic-Wichita Lineament: a continental-scale basement fracture system; Global Tectonics and Metallogeny, v. 1/1, p. 83-87.

Babcock, R.S., Burmester, R.F., Engebretson, D.C., Warnock, A.C., and Clark, K.P., 1992. A rifted margin origin for the Crescent basalts and related rocks in the northern Coast Range volcanic province, Washington and British Columbia; Journal of Geophysical Research, v. 97, p. 6799-6821.

Babcock, R.S., Suszek, C.A., and Engebretson, D.C., 1994. The Crescent "terrane", Olympic Peninsula and southern Vancouver Island; in R. Lasmanis and E.S. Cheney (eds.), Regional Geology of Washington State; Washington Division of Geology and Earth Resources, Bulletin 80, p. 141-157.

Bailey, R.C., 1990. Trapping of aqueous fluids in the deep crust; Geophysical Research Letters, v. 17, p. 1129-1132.

Barker, F., Sutherland Brown, A., Budahn, J.R., and Plafker, G., 1989. Back-arc with frontal-arc component origin of Triassic Karmutsen basalt, British Columbia, Canada; Chemical Geology, v. 75, p. 81-102.

Barnard, W.D., 1978. The Washington continental slope: Quaternary tectonics and sedimentation; Marine Geology, v. 27, p. 79-114.

Barr, S.M., 1974. Structure and tectonics of the continental slope west of southern Vancouver Island; Canadian Journal of Earth Sciences, v. 11, p. 1187-1199.

Barr, S.M. and Chase, R.L., 1974. Geology of the northern end of Juan de Fuca Ridge and seafloor-spreading; Canadian Journal of Earth Sciences, v. 11, p. 1384-1406.

Bérubé, J., Rogers, G.C., Ellis, R.M., and Hasselgren, E.O., 1989. A microseismicity study of the Queen Charlotte Islands region; Canadian Journal of Earth Sciences, v. 26, p. 2556-2566.

Blakely, R.J. and Connard, G.G., 1989. Crustal studies using magnetic data; in L.C. Pakiser and W.D. Mooney (eds.), Geophysical Framework of the Continental United States; Geological Society of America, Memoir 172, p. 45-60.

Blakely, R.J. and Jachens, R.C., 1990. Volcanism, isostatic residual gravity, and regional tectonic setting of the Cascade volcanic province; Journal of Geophysical Research, v. 95, p. 19,439-19,451.

Bowin, C., 1983. Depth of principal mass anomalies contributing to the Earth's geoidal undulations and gravity anomalies; Marine Geodesy, v. 7, p. 61-100.

Brandon, M.T., 1989a. Origin of igneous rocks associated with mélanges of the Pacific Rim Complex, western Vancouver Island, Canada; Tectonics, v. 8, p. 1115-1136.

Brandon, M.T., 1989b. Deformation styles in a sequence of olistostromal mélanges, Pacific Rim Complex, western Vancouver Island, Canada; Bulletin of Geological Society of America, v. 101, p. 1520-1542.

Brandon, M.T. and Calderwood, A.R., 1990. High-pressure metamorphism and uplift of the Olympic subduction complex; Geology, v. 18, p. 1252-1255.

Brandon, M.T., Cowan, D.S., and Vance, J.A., 1988. The Late Cretaceous San Juan Thrust System, San Juan Islands, Washington; Geological Society of America, Special Paper 221, 81 p.

Brandon, M.T. and Vance, J.A., 1992. Tectonic evolution of the Cenozoic Olympic subduction complex, Washington State, as deduced from fission track ages for detrital zircons; American Journal of Science, v. 292, p. 565-636.

Brew, D.A. and Ford, A.B., 1983. Comment on 'Tectonic accretion and the origin of the two major metamorphic and plutonic welts in the Canadian Cordillera'; Geology, v. 11, p. 427-428.

Brew, D.A., Karl, S.A., Barnes, D.F., Jachens, R.C., Ford, A.B., and Horner, R., 1991. A northern Cordilleran ocean-continent transect: Sitka Sound, Alaska, to Atlin Lake, British Columbia; Canadian Journal of Earth Sciences, v. 28, p. 840-853.

Broome, J., Simard, R., and Teskey, D., 1985. Presentation of magnetic anomaly data by stereo projection of magnetic shadowgrams; Canadian Journal of Earth Sciences, v. 22, p. 311-314.

Brown, E.H., Cary, J.A., Dougan, B.E., Dragovich, J.D., Fluke, S.M., and McShane, D.P., 1994. Tectonic evolution of the Cascades Crystalline Core in the Cascade River area, Washington; in R. Lasmanis and E.S. Cheney (eds.), Regional Geology of Washington State; Washington Division of Geology and Earth Resources, Bulletin 80, p. 93-113.

Bruns, T.R. and Carlson, P.R., 1987. Geology and petroleum potential of the southeast Alaska continental margin; in D.W. Scholl, A. Grantz, and J.G. Vedder (eds.), Geology and Resource Potential of the Continental Margin of Western North America and Adjacent Ocean Basins - Beaufort Sea to Baja California; Circum-Pacific Council for Energy and Mineral Resources, Earth Science Series 6, p. 269-282.

Burchfiel, B.C., Cowan, D.S., and Davis, G.A., 1992. Tectonic overview of the Cordilleran orogen in the western United States; in B.C. Burchfiel, P.W. Lipman, and M.L. Zoback (eds.), The Cordilleran Orogen: Conterminous U.S.; Geological Society of America, The Geology of North America, v. G-3, p. 407-479.

Burchfiel, B.C., Lipman, P.W., and Zoback, M.L. (eds.), 1992. The Cordillaran Orogen: Conterminous U.S.; Geological Society of America, The Geology of North America, v. G-3, 724 p.

Calvert, A.J. and Clowes, R.M., 1990. Deep, high-amplitude reflections from a major shear zone above the subducting Juan de Fuca plate; Geology, v. 18, p. 1091-1094.

Calvert, A.J. and Clowes, R.M., 1991. Seismic evidence for the migration of fluids within the accretionary complexes of western Canada; Canadian Journal of Earth Sciences, v. 28, p. 542-556.

Calvert, A.J., Hasselgren, E.O., and Clowes, R.M., 1990. Oceanic rift propagation - a cause of crustal underplating and seamount volcanism; Geology, v. 18, p. 886-889.

Cameron, B.E.B. and Tipper, H.W., 1985. Jurassic Stratigraphy of the Queen Charlotte Islands, British Columbia; Geological Survey of Canada, Bulletin 365, 49 p.

Campbell, N.P., 1989. Structural and stratigraphic interpretation of rocks under the Yakima fold belt, Columbia Basin, based on recent surface mapping and well data; in S.P. Reidel and P.R. Hooper (eds.), Volcanism and Tectonism in the Columbia River Flood-Basalt Province; Geological Society of America, Special Paper 239, p. 209-222.

Carbotte, S.M., Dixon, J.M., Farrar, E., Davis, E.E., and Riddihough, R.P., 1989. Geological and geophysical characteristics of the Tuzo Wilson Seamounts: implications for plate geometry in the vicinity of the Pacific-North America-Explorer triple junction; Canadian Journal of Earth Sciences, v. 26, p. 2365-2384.

Carlson, P.R. and Nelson, C.H., 1987. Marine geology and resource potential of Cascadia Basin; in D.W. Scholl, A. Grantz, and J.G. Vedder (eds.), Geology and Resource Potential of the Continental Margin of Western North America and Adjacent Ocean Basins - Beaufort Sea to Baja California; Circum-Pacific Council for Energy and Mineral Resources, Earth Science Series 6, p. 523-535.

Carson, B., Westbrook, G., and Musgrave, R., 1993. Cascadia margin; JOIDES Journal, v. 19/2, p. 11-16.

Carter, L., 1974. An evaluation of the provenance of terrigenous sediments from offshore Vancouver Island; Canadian Journal of Earth Sciences, v. 11, p. 664-677.

Catchings, R.D. and Mooney, W.D., 1988. Crustal structure of the Columbia Plateau: evidence for continental rifting; Journal of Geophysical Research, v. 93, p. 459-474.

Chase, R.L., 1977. J. Tuzo Wilson Knolls: Canadian hotspot; Nature, v. 266, p. 344-346.

Chase, R.L. and Tiffin, D.L., 1972. Queen Charlotte Fault-Zone, British Columbia; in Marine Geology and Geophysics, Section 8, 24th International Geological Congress, Montreal, Proceedings, p. 17-27.

Chase, R.L., Tiffin, D.L., and Murray, J.W., 1975. The western Canadian continental shelf; in C,J, Yorath, E.R. Parker, and D.J. Glass (eds.), Canada's Continental Margins and Offshore Petroleum Exploration; Canadian Society of Petroleum Geologists, Memoir 4, p. 701-721.

Cheney, E.S., 1994. Cenozoic unconformity-bounded sequences of central and eastern Washington; in R. Lasmanis and E.S. Cheney (eds.), Regional Geology of Washington State; Washington Division of Geology and Earth Resources, Bulletin 80, p. 115-139.

Christensen, N.I. and Mooney, W.D., 1995. Seismic velocity structure and composition of the continental crust: a global view; Journal of Geophysical Research, in press.

Christiansen, R.L. and Yeats, R.S., 1992. Post-Laramide geology of the U.S. Cordilleran region; in B.C. Burchfiel, P.W. Lipman, and M.L. Zoback (eds.), The Cordilleran Orogen: Conterminous U.S.; Geological Society of America, The Geology of North America, v. G-3, p. 261-406.

Clauser, C. and Huenges, E., 1993. KTB thermal regime and heat transport mechanism - current knowledge; Scientific Drilling, v. 3, p. 271-281.

Cloos, M., 1993. Lithosphere buoyancy and collisional orogenesis: subduction of oceanic plateaus, continental margins, island arcs, spreading ridges, and seamounts; Bulletin of Geological Society of America, v. 105, p. 715-737.

Clowes, R.M., Brandon, M.T., Green, A.G., Yorath, C.J., Sutherland Brown, A., Kanasewich, E.R., and Spencer, C., 1987. LITHOPROBE - southern Vancouver Island: Cenozoic subduction complex imaged by deep seismic reflections; Canadian Journal of Earth Sciences, v. 24, p. 31-51.

Clowes, R.M., Green, A.G., Yorath, C.J., Kanasewich, E.R., West, G.F., and Garland, G.D., 1984. Lithoprobe - a national program for studying the third dimension of geology; Journal of the Canadian Society of Exploration Geophysicists, v. 20, p. 23-39.

Clowes, R.M., Thorleifson, A.J., and Lynch, S., 1981. Winona Basin, west coast Canada: crustal structure from marine seismic studies; Journal of Geophysical Research, v. 86, p. 225-242.

Cochrane, G.R., Moore, J.C., MacKay M.E., and Moore, G.F., 1994. Velocity and inferred porosity model of the Oregon accretionary prism from multichannel seismic reflection data: implications on sediment dewatering and overpressure; Journal of Geophysical Research, v. 99, p. 7033-7043.

Coles, R.L. and Currie, R.G., 1977. Magnetic anomalies and rock magnetizations in the southern Coast Mountains, British Columbia: possible relation to subduction; Canadian Journal of Earth Sciences, v. 14, p. 1753-1770.

Cordell, L.E. and Grauch, V.J.S., 1985. Mapping basement magnetization zones from aeromagnetic data in the San Juan basin, New Mexico; in W.J. Hinze, M.F. Kane, and N.W. O'Hara (eds.), The Utility of Regional Gravity and Magnetic Anomaly Maps; Society of Exploration Geophysicists, p. 181-197.

Couch, R. and Chase, R., 1973. Site survey of Paul Revere Ridge west of northern Vancouver Island; in L.F. Musich and O.E. Weser (eds.), Initial Reports of the Deep Sea Drilling Project, v. XVIII; U.S. Government Printing Office, p. 987-995.

Couch, R.W. and Riddihough, R.P., 1989. The crustal structure of the western continental margin of North America; in L.C. Pakiser and W.D. Mooney (eds.), Geophysical Framework of the Continental United States; Geological Society of America, Memoir 172, p. 103-128.

Cousens, B.L., Chase, R.L., and Schilling, J.-G., 1984. Basalt geochemistry of the Explorer Ridge area, northeast Pacific Ocean; Canadian Journal of Earth Sciences, v. 21, p. 157-170.

Cousens, B.L., Chase, R.L., and Schilling, J.-G., 1985. Geochemistry and origin of volcanic rocks from Tuzo Wilson and Bowie seamounts, northeast Pacific Ocean; Canadian Journal of Earth Sciences, v. 22, p. 1609-1617.

Cowan, D.S., Botros, M., and Johnson, H.P., 1986. Bookshelf tectonics: rotated crustal blocks within the Sovanco Fracture Zone; Geophysical Research Letters, v. 13, p. 995-998.

Cowan, D.S. and Bruhn, R.L., 1992. Late Jurassic to early Late Cretaceous geology of the U.S. Cordillera; in B.C. Burchfiel, P.W. Lipman, and M.L. Zoback (eds.), The Cordilleran Orogen: Conterminous U.S.; Geological Society of America, The Geology of North America, v. G-3, p. 169-203.

Crawford, M.L., Hollister, L.S., and Woodsworth, G.J., 1987. Crustal deformation and regional metamorphism across a terrane boundary, Coast Plutonic Complex, British Columbia; Tectonics, v. 6, p. 343-361.

Crosson, R.S., 1986. Comment on 'Geodetic strain measurements in Washington'; Journal of Geophysical Research, v. 91, p. 7555-7557.

Crosson, R.S. and Owens, T.J., 1987. Slab geometry of the Cascadia subduction zone beneath Washington from earthquake hypocenters and teleseismic converted waves; Geophysical Research Letters, v. 14, p. 824-827.

Currie, R.G., Seemann, D.A., and Riddihough, R.P., 1983a. Total field magnetic anomaly offshore British Columbia; Geological Survey of Canada, Open File 828, scale 1:1,000,000.

Currie, R.G., Cooper, R.V., Riddihough, R.P., and Seemann, D.A., 1983b. Multiparameter geophysical surveys off the west coast of Canada; in Current Research, Part A; Geological Survey of Canada, Paper 83-1A, p. 207-212.

Currie, R.G. and Muller, J.E., 1976. Magnetic susceptibility as a diagnostic parameter of Vancouver Island volcanic rocks; in Report of Activities, Part B; Geological Survey of Canada, Paper 76-1B, p. 97-98.

Currie, R.G. and Teskey, D.J., 1988. Magnetics component of the Frontier Geoscience Program on the west coast of Canada; in Current Research, Part E; Geological Survey of Canada, Paper 88-1E, p. 287.

Davis, E.E., 1982. Evidence for extensive basalt flows on the sea floor; Bulletin of Geological Society of America, v. 93, p. 1023-1029.

Davis, E.E. and Clowes, R.M., 1986. High velocities and seismic anisotropy in Pleistocene turbidites off Western Canada; Geophysical Journal of Royal Astronomical Society, v. 84, p. 381-399.

Davis, E.E. and Currie, R.G., 1993. Geophysical observations of the northern Juan de Fuca Ridge system: lessons in sea-floor spreading; Canadian Journal of Earth of Sciences, v. 30, p. 278-300.

Davis, E.E. and Hyndman, R.D., 1989. Accretion and recent deformation of sediments along the northern Cascadia subduction zone; Bulletin of Geological Society of America, v. 101, p. 1465-1480.

Davis, E.E. and Riddihough, R.P., 1982. The Winona Basin: structure and tectonics; Canadian Journal of Earth Sciences, v. 19, p. 767-788.

Davis, E.E. and Seemann, D.A., 1981. A Compilation of Seismic Reflection Profiles Across the Continental Margin of Western Canada; Geological Survey of Canada, Open File 751.

Dehler, S.A., 1991. Integrated Geophysical Modelling of the Northern Cascadia Subduction Zone; Ph.D. thesis, Dept. of Geophysics & Astronomy, University of British Columbia, Vancouver, 151 p.

Dehler, S.A. and Clowes, R.M., 1988. The Queen Charlotte

Islands refraction project: Part I - the Queen Charlotte Fault zone; Canadian Journal of Earth Sciences, v. 25, p. 1857-1870.

Dehler, S.A. and Clowes, R.M., 1992. Integrated geophysical modelling of terranes and other structural features along the western Canadian margin; Canadian Journal of Earth Sciences, v. 29, p. 1492-1508.

Dehlinger, P., Couch, R.W., McManus, D.A., and Gemperle, M., 1970. Northeast Pacific Structure; in A.E. Maxwell (ed.), The Sea, v. 14/2, p. 133-189.

DeMets, C., Gordon, R.G., Argus, D.F., and Stein, S., 1990. Current plate motions; Geophysical Journal International, v. 101, p. 425-478.

Desrochers, A., 1989. Depositional history of Upper Triassic carbonate platforms on Wrangellia Terrane, western British Columbia, Canada (abs.); Bulletin of American Association of Petroleum Geologists, v. 73, p. 349-350.

Dickinson, W.R., 1973. Widths of modern arc-trench gaps proportional to past duration of igneous activity in associated magmatic arcs; Journal of Geophysical Research, v. 78, p. 3376-3389.

Dickinson, W.R., 1976. Sedimentary basins developed during evolution of Mesozoic-Cenozoic arc-trench system in western North America; Canadian Journal of Earth Sciences, v. 13, p. 1268-1287.

Dietz, R.S., 1952. Geomorphic evolution of continental terrace (continental shelf and slope); Bulletin of American Association of Petroleum Geologists, v. 36, p. 1802-1819.

Dietz, R.S., 1964. The origin of continental slopes; American Scientist, v. 52, p. 50-59.

Digel, S. and Ghent, E.D., 1994. Fluid-mineral equilibria in prehnite-pumpellyite to greenschist facies metabasites near Flin Flon, Manitoba, Canada: implications for petrogenetic grids; Journal of Metamorphic Geology, v. 12, p. 467-477.

Douglas, R.J.W. (ed.), 1970. Geology and Economic Minerals of Canada; Geological Survey of Canada, Economic Geology Report No. 1, 838 p.

Dragert, H., 1987. The fall (and rise) of central Vancouver Island: 1930-1985; Canadian Journal of Earth Sciences, v. 24, p. 689-697.

Dragert, H., Hyndman, R.D., Rogers, G.C., and Wang, K., 1994. Current deformation and the width of the seismogenic zone of the northern Cascadia subduction zone; Journal of Geophysical Research, v. 99, p. 653-668.

Drew, J.J. and Clowes, R.M., 1990. A re-interpretation of the seismic structure across the active subduction zone of western Canada; in A.G. Green (ed.), Studies of Laterally Heterogeneous Structures Using Seismic Refraction and Reflection Data; Geological Survey of Canada, Paper 89-13, p. 115-132.

Duncan, R.A., 1982. A captured island chain in the Coast Range of Oregon and Washington; Journal of Geophysical Research, v. 87, p. 10,827-10,837.

Duncan, R.A. and Kulm, L.D., 1989. Plate tectonic evolution of the Cascades arc-subduction complex; in E.L. Winterer, D.M. Hussong, and R.W. Decker (eds.), The Eastern Pacific Ocean and Hawaii; Geological Society of America, The Geology of North America, v. N, p. 413-438.

Egger, A. and Ansorge, J., 1990. Interpretation of seismic refraction data-CCSS data set I: Vancouver Island - continental margin; in A.G. Gereen (ed.), Studies of Laterally Heterogeneous Structures Using Seismic Refraction and Reflection Data; Geological Survey of Canada, paper 89-13, p. 133-150.

Engebretson, D.C., Cox, A., and Gordon, R.G., 1985. Relative Motions Between Oceanic and Continental Plates in the Pacific Basin; Geological Society of America, Special Paper 206, 59 p.

England, T.D.J., 1991. Late Cretaceous to Paleogene structural and stratigraphic evolution of Georgia Basin, southwestern British Columbia: implications for hydrocarbon potential; Washington Geology, v. 19/4, p. 10-11.

England, T.D.J. and Calon, T.J., 1991. The Cowichan fold and thrust system, Vancouver Island, southwestern British Columbia; Bulletin of Geological Society of America, v. 103, p. 336-362.

England, T.D.J. and Hiscott, R.N., 1992. Lithostratigraphy and deep-water setting of the upper Nanaimo Group (Upper Cretaceous), outer Gulf Islands of southwestern British Columbia; Canadian Journal of Earth Sciences, v. 29, p. 574-595.

Evans, J.E., 1994. Depositional history of the Eocene Chumstick Formation: implications of tectonic partitioning for the history of the Leavenworth and Entiat-Eagle Creek fault systems, Washington; Tectonics, v. 13, p. 1425-1444.

Evarts, R.C., 1990. Before Mount St. Helens: the Eocene to Miocene Cascade volcanic arc in southern Washington (abs.); Geoscience Canada, v. 17, p. 126.

Fairchild, L.H. and Cowan, D.S., 1982. Structure, petrology, and tectonic history of the Leech River complex northwest of Voctoria, Vancouver Island; Canadian Journal of Earth Sciences, v. 19, p. 1817-1835.

Finn, C., 1990. Geophysical constraints on Washington convergent margin structure; Journal of Geophysical Research, v. 95, p. 19,533-19,546.

Finn, C., 1991. Comment on 'U.S. west coast revisited: an aeromagnetic perspective'; Geology, v. 19, p. 950.

Finn, C., Phillips, W.M., and Williams, D.L., 1991. Gravity anomaly and terrain maps of Washington, scales 1:1,000,000 and 1:500,000; U.S. Geological Survey, Map GP-988.

Forte, A.M., Peltier, W.R., Dziewonski, A.M., and Woodward, R.L., 1993. Dymanic surface topography: a new interpretation based upon mantle flow models derived from seismic tomography; Geophysical Research Letters, v. 20, p. 225-238.

Fowler, C.M.R. and Pandit, B.I., 1990. Analysis of CCSS data set I: reflection refraction data from the Vancouver Island continental margin of western Canada; in A.G. Green (ed.), Studies of Laterally Heterogeneous Structures Using Seismic Refraction and Reflection Data; Geological Survey of Canada, Paper 89-13, p. 79-90.

Friedman, G.M. and Sanders, J.E., 1978. Principles of Sedimentology; Wiley & Sons, New York.

Frizzell, V.A., Tabor, R.W., Booth, D.B., Ort, K.M., and Waitt, R.B., 1984. Preliminary Geologic Map of the Snoqualmie Pass Quardangle, Washington; U.S. Geological Survey, Open-File Map 84-693, 43 p., scale 1:100,000.

Fuis, G.S. and Clowes, R.M., 1993. Comparison of deep structure along three transects of the western North American continental margin; Tectonics, v. 12, p. 1420-1435.

Furlong, K.P., Rohr, K.M.M., and Lowe, C., 1994. Evolution of the Pacific-Juan de Fuca-North America triple junction (abs.); American Geophysical Union, Fall Meeting, San Francisco, Abstracts; Supplement to EOS, p. 620.

Gabrielse, H., Monger, J.W.H., Wheeler, J.O., and Yorath, C.J., 1991. Morphogeological belts, tectonic assemblages, and terranes; in H. Gabrielse and C.J. Yorath (eds.), Geology of the Cordilleran Orogen in Canada; Geological Society of America, The Geology of North America, v. G-2, p. 15-28.

Gabrielse, H. and Yorath, C.J. (eds.), 1991. Geology of the Cordilleran Orogen in Canada; Geological Society of America, The Geology of North America, v. G-2.

Galster, R.W., Coombs, H.A., and Waldron, H.H., 1989. Engineering geology in Washington - an introduction; in R.W. Galster (ed.), Engineering Geology in Washington, v. I; Washington Division of Geology and Earth Resources, Bulletin 78, p. 3-12.

Gardner, M.C., Bergman, S.C., Cushing, G.W., MacEvett, E.M., Plafker, G., Campbell, R.B., Dodds, C.J., McLelland, W.C., and Mueller, P.A., 1988. Pennsylvanian pluton stitching of Wrangellia and the Alexander Terrane, Wrangell Mountains, Alaska; Geology, v. 16, p. 967-971.

Garver, J.I. and Brandon, M.T., 1994. Erosional denudation of the British Columbia Coast Ranges as determined from fission-track ages of detrital zircon from the Tofino Basin, Olympic Peninsula, Washington; Bulletin of Geological Society of America, v. 106, p. 1398-1412.

Gehrels, G.E., 1990. Late Proterozoic-Cambrian metamorphic basement of the Alexander terrane on Long and Dall Islands, southeast Alaska; Bulletin of Geological Society of America, v. 102, p. 760-767.

Gehrels, G.E. and Saleeby, J.B., 1987a. Geology of southern Prince of Wales Island, Southeastern Alaska; Bulletin of Geological Society of America, v. 98, p. 123-137.

Gehrels, G.E. and Saleeby, J.B., 1987b. Geologic framework, tectonic evolution, and displacement history of the Alexander Terrane; Tectonics, v. 6, p. 151-173.

Gehrels, G.E., Saleeby, J.B., and Berg, H.C., 1987. Geology of Annette, Gravina, and Duke islands, southeastern Alaska; Canadian Journal of Earth Sciences, v. 24, p. 866-881.

Goldfinger, C., Kulm, L.D., Yeats, R.S., Applegate, B., MacKay, M.E., and Moore, G.F., 1992. Transverse structural trends along the Oregon convergent margin: implications for Cascadia earthquake potential and crustal rotations; Geology, v. 20, p. 141-144.

Goodacre, A.K., Grieve, R.A.F., and Halpenny, J.F., 1987a. Observed gravity values of Canada; Geological Survey of Canada, Canadian Geophysical Atlas, Map 1, scale 1:10,000,000.

Goodacre, A.K., Grieve, R.A.F., and Halpenny, J.F., 1987b. Free air gravity anomaly map of Canada; Geological Survey of Canada, Canadian Geophysical Atlas, Map 2, scale 1:10,000,000.

Goodacre, A.K., Grieve, R.A.F., and Halpenny, J.F., 1987c. Bouguer gravity anomaly map of Canada; Geological Survey of Canada, Canadian Geophysical Atlas, Map 3, scale 1:10,000,000.

Goodacre, A.K., Grieve, R.A.F., and Halpenny, J.F., 1987d. Isostatic gravity anomaly map of Canada; Geological Survey of Canada, Canadian Geophysical Atlas, Map 4, scale 1:10,000,000.

Goodacre, A.K., Grieve, R.A.F., Halpenny, J.F., and Sharpton, V.L., 1987e. Horizontal gradient of the Bouguer gravity anomaly map of Canada; Geological Survey of Canada, Canadian Geophysical Atlas, Map 5, scale 1:10,000,000.

Grant, A.C., 1980. Problems with plate tectonics: the Labrador Sea; Bulletin of Canadian Petroleum Geology, v. 28, p. 252-278.

Grant, A.C., 1987. Inversion tectonics on the continental margin east of Newfoundland; Geology, v. 15, p. 845-848.

Grant, F.C. and West, G.F., 1965. Interpretation Theory in Applied Geophysics; McGraw-Hill, Toronto, 583 p.

Green, A.G. (ed.), 1990. Studies of Laterally Heterogeneous Structures Using Seismic Refraction and Reflection Data; Geological Survey of Canada, Paper 89-13, 224 p.

337

Green, A.G., Berry, M.J., Spencer, C.P., Kanasewich, E.R., Chiu, S., Clowes, R.M., Yorath, C.J., Stewart, D.B., Unger, J.D., and Poole, W.H., 1986. Recent seismic reflection studies in Canada; in M. Barazangi and L. Brown (eds.), Reflection Seismology: Global Perspectives; American Geophysical Union, Geodynamics Series, v. 13, p. 85-97.

Green, A.G., Clowes, R.M., and Ellis, R.M., 1990. Crustal studies across Vancouver Island and adjacent offshore margin; in A.G. Green (ed.), Studies of Laterally Heterogeneous Structures Using Seismic Refraction and Reflection Data; Geological Survey of Canada, Paper 89-13, p. 3-25.

Green A.G., Milkereit, B., Mayrand, L., Spencer, C., Kurtz, R., and Clowes, R.M., 1987. Lithoprobe seismic reflection profiling across Vancouver Island: results from reprocessing; Geophysical Journal of Royal Astronomical Society, v. 89, p. 85-90.

Green, N.L., 1990. Late Cenozoic volcanism in the Mount Garibaldi and Garibaldi Lake volcanic fields, Garibaldi Volcanic Belt, southwestern British Columbia; Geoscience Canada, v. 17, p. 171-175.

Gretener, P.E., 1986. General comments on listric normal faults with particular reference to growth faults and their role in hydrocarbon trapping; Bulletin of Swiss Association of Petroleum-Geologists and Engineers, v. 52, No. 122, p. 21-34.

Grocott, J., Brown, M., Dallmeyer, R.D., Taylor, G.K., and Treloar, P.J., 1994. Mechanisms of continental growth in extensional arcs: an example from the Andean plate-boundary zone; Geology, v. 22, p. 391-394.

Guffanti, M. and Weaver, C.S., 1988. Distribution of late Cenozoic volcanic vents in the Cascade Range: volcanic arc segmentation and regional tectonic considerations; Journal of Geophysical Research, v. 93, p. 6513-6529.

Haggart, J.W., 1991. A synthesis of Cretaceous stratigraphy, Queen Charlotte Islands, British Columbia; in G.J. Woodsworth (ed.), Evolution and Hydrocarbon Potential of the Queen Charlotte Basin, British Columbia; Geological Survey of Canada, Paper 90-10, p. 253-278.

Haggart, J.W., 1993. Latest Jurassic and Cretaceous paleogeography of the northern Insular Belt, British Columbia; in G.C. Dunne and K.A. McDougall (eds.), Mesozoic Paleogeography of the Western United States-II; Society of Economic Paleontologists and Mineralogists, Pacific Section, Book 71, p. 463-475.

Hasselgren, E.O. and Clowes, R.M., 1995. Crustal structure of northern Juan de Fuca plate from multichannel reflection data; Journal of Geophysical Research, v. 100, p. 6469-6486.

Hasselgren, E.O., Clowes, R.M., and Calvert, A.J., 1992. Propagating rift pseudofaults - zones of crustal underplating imaged by multichannel seismic reflection data; Geophysical Research Letters, v. 19, p. 485-488.

Haugerud, R.A., 1989. Geology of the metamorphic core of the North Cascades; in N.L. Joseph (ed.), Geologic Guidebook for Washington and Adjacent Areas; Washington Division of Geology and Earth Resources, Information Circular 86, p. 119-136.

Hawthorne, R., 1990. Reprocessing Vancouver Island LITHOPROBE data; in A.G. Green (ed.), Studies of Laterally Heterogeneous Structures Using Seismic Refraction and Reflection Data; Geological Survey of Canada, Paper 89-13, p. 175-190.

Heller, P.L., Tabor, R.W., O'Neil, J.R., Pevear, D.R., Shafiqullah, M., and Winslow, N.S., 1992. Isotopic provenance of Paleogene sandstones from the accretionary core of the Olympic Mountains, Washington; Bulletin of Geological Society of America, v. 104, p. 140-153.

Hesthammer, J., Indrelid, J., Lewis, P.D., and Orchard, M.J., 1991. Permian strata on the Queen Charlotte Islands, British Columbia; in Current Research, Part A; Geological Survey of Canada, Paper 91-1A, p. 321-329.

Hickson, C.J., 1991. The Masset Formation on Graham Island, Queen Charlotte Islands, British Columbia; in G.J. Woodsworth (ed.), Evolution and Hydrocarbon Potential of the Queen Charlotte Basin, British Columbia; Geological Survey of Canada, Paper 90-10, p. 295-304.

Higgs, R., 1991. Sedimentology, basin-fill architecture and petroleum geology of the Tertiary Queen Charlotte Basin; in G.J. Woodsworth (ed.), Evolution and Petroleum Potential of the Queen Charlotte Basin, British Columbia; Geological Survey of Canada, Paper 90-10, p. 295-304.

Hole, J.A., Clowes, R.M., and Ellis, R.M., 1993. Interpretation of three-dimensional seismic refraction data from western Hecate Strait, British Columbia: structure of the crust; Canadian Journal of Earth Sciences, v. 30, p. 1440-1452.

Hollister, L.S., 1993. The role of melt in the uplift and exhumation of orogenic belts; Chemical Geology, v. 108, p. 31-48.

Hooper, P.R. and Conrey, R.M., 1989. A model for the tectonic setting of the Columbia River basalt eruptions; in S.P. Reidel and P.R. Hooper (eds.), Volcanism and Tectonism in the Columbia River Flood Basalt Province; Geological Society of America, Special Paper 239, p. 293-306.

Horn, J.R., Clowes, R.M., Ellis, R.M., and Bird, D.N., 1984. The seismic structure across an active oceanic/continental transform fault zone; Journal of Geophysical Research, v. 89, p. 3107-3120.

Humphreys, E.D. and Dueker, K.G., 1994a. Western U.S. upper mantle structure; Journal of Geophysical Research, v. 99, p. 9615-9634.

Humphreys, E.D. and Dueker, K.G., 1994b. Physical state of the western U.S. upper mantle; Journal of Geophysical Research, v. 99, p. 9635-9650.

Hutchison, W.W., 1982. Geology of the Prince Rupert-Skeena Map Area, British Columbia; Geological Survey of Canada, Memoir 394, 116 p.

Hyndman, R.D., 1988. Dipping seismic reflectors, electrically conductive zones, and trapped water in the crust over a subducting plate; Journal of Geophysical Research, v. 83, p. 13,391-13,405.

Hyndman, R.G. and Ellis, R.M., 1981. Queen Charlotte fault zone: microearthquakes from a temporary array of land stations and ocean bottom seismographs; Canadian Journal of Earth Sciences, v. 18, p. 776-788.

Hyndman, R.D. and Hamilton, T.S., 1993. Queen Charlotte area tectonics and volcanism and their association with relative plate motions along the Northeast Pacific margin; Journal of Geophysical Research, v. 98, p. 14,257-14,277.

Hyndman, R.D., Lewis, T.J., Wright, J.A., Burgess, M., Chapman, D.S., and Yamano, M., 1982. Queen Charlotte fault zone: heat flow measurements; Canadian Journal of Earth Sciences, v. 19, p. 1657-1669.

Hyndman, R.D., Riddihough, R.P., and Herzer, R., 1979. The Nootka Fault Zone - a new plate boundary off western Canada; Geophysical Journal of Royal Astronomical Society, v. 58, p. 667-683.

Hyndman, R.D. and Rogers, G.C., 1981. Seismicity surveys with ocean bottom seismographs off western Canada; Journal of Geophysical Research, v. 86, p. 3867-3880.

Hyndman, R.D. and Wang, K., 1993. Thermal constraints on the zone of major thrust earthquake failure: the Cascadia subduction zone; Journal of Geophysical Research, v. 98, p. 2039-2060.

Hyndman, R.D., Yorath, C.J., Clowes, R.M., and Davis, E.E., 1990. The northern Cascadia subduction zone at Vancouver Island: seismic structure and tectonic history; Canadian Journal of Earth Sciences, v. 27, p. 313-329.

Indrelid, J., 1991. Stratigraphy, Structural Geology and Petroleum Potential of Cretaceous and Tertiary Rocks in the Central Graham Island Area, Queen Charlotte Islands, British Columbia; M.Sc. thesis, Dept. of Geological Sciences, University of British Columbia, Vancouver, 176 p.

Irving, E., Souther, J.G., and Baker, J., 1992. Tertiary extension and tilting in the Queen Charlotte Islands, evidence from dyke swarms and their paleomagnetism; Canadian Journal of Earth Sciences, v. 29, p. 1878-1898.

Isachsen, C.E., 1987. Geology, geochemistry, and cooling history of the Westcoast Crystalline Complex and related rocks, Meares Island and vicinity, Vancouver Island, British Columbia; Canadian Journal of Earth Sciences, v. 24, p. 2047-2064.

Iwasaki, T. and Shimamura, H., 1990. Velocity structure model determined from onshore-offshore seismic profiling across Vancouver Island and adjacent continental margin; in A.G. Green (ed.), Studies of Laterally Heterogeneous Structures Using Seismic Refraction and Reflection Data; Geological Survey of Canada, Paper 89-13, p. 91-103.

Jachens, R.C., Simpson, R.W., Blakely, R.J., and Saltus, R.W., 1989. Isostatic residual gravity and crustal geology of the United States; in L.C. Pakiser and W.D. Mooney (eds.), Geophysical Framework of the Continental United States; Geological Society of America, Memoir 172, p. 405-424.

Jarchow, C.M., Catchings, R.D., and Lutter, W.J., 1994. Large-explosive source, wide-recording aperture, seismic profiling on the Columbia Plateau, Washington; Geophysics, v. 59, p. 259-271.

Jeletzky, J.A., 1976. Mesozoic and ?Tertiary Rocks of Quatsino Sound, Vancouver Island, British Columbia; Geological Survey of Canada, Bulletin 242, 243 p.

Johnson, D.W., 1919. Shore Processes and Shoreline Development; John Wiley & Sons, 584 p.

Johnson, H.P. and Holmes, M.L., 1989. Evolution in plate tectonics; the Juan de Fuca Ridge; in E.L. Winterer, D.M. Hussong, and R.W. Decker, The Eastern Pacific Ocean and Hawaii; Geological Society of America, The Geology of North America, v. N, p. 73-91.

Johnson, P.R., Zietz, I., and Bond, K.R., 1990. U.S. west coast revisited: and aeromagnetic perspective; Geology, v. 18, p. 332-335.

Johnson, S.H., Couch, R.W., Gemperle, M., and Banks, E.R., 1972. Seismic refraction measurements in southeast Alaska and western British Columbia; Canadian Journal of Earth Sciences, v. 9, p. 1756-1765.

Johnson, S.Y., 1984. Evidence for margin-truncating transcurrent fault (pre-late Eocene) in western Washington; Geology, v. 12, p. 538-541.

Jones, A.G., Bailey, R.C., and Mareschal, M., 1994. High-resolution electromagnetic images of conducting zones in an upthrust crustal block; Geophysical Research Letters, v. 21, p. 1807-1810.

Jones, D.L., Silberling, N.J., and Hillhouse, J., 1977. Wrangellia - a displaced terrane in northwestern North America;

Canadian Journal of Earth Sciences, v. 14, p. 2566-2577.

Keach, W.R., Oliver, J.E., Brown, L.D., and Kaufman, S., 1989. Cenozoic active margin and shallow Cascades structure: COCORP results from western Oregon; Bulletin of Geological Society of America, v. 101, p. 783-794.

Keen, C.E. and Hyndman, R.D., 1979. Geophysical review of the continental margins of eastern and western Canada; Canadian Journal of Earth Sciences, v. 16, p. 712-747.

Kenyon, C. and Bickford, C.G.C., 1989. Vitrinite reflectance study of Nanaimo Group coals of Vancouver Island; in Geological Fieldwork 1988, British Columbia Ministry of Energy, Mines and Petroleum Resources, Paper 1989-1, p. 543-558.

King, P.B., 1969. Tectonic Map of North America, scale 1:5,000,000; U.S. Geological Survey.

Kozlovsky, Ye.A. (ed.), 1984 (translated 1987). The Superdeep Well of the Kola Peninsula; Springer-Verlag, 558 p.

Kulm, L.P., von Huene, R., Duncan, J.R., Ingle, J.C., Kling, S.A., Musich, L.F., Piper, D.J.W., Pratt, R.M., Schrader, H., Weser, O., and Wise, S.W., 1973. Site 177; Initial Reports of the Deep Sea Drilling Project, No. 13; U.S. Government Printing Office, p. 233-243.

Kurtz, R.D., DeLaurier, J.M., and Gupta, J.C., 1986. A magnetotelluric sounding across Vancouver Island detects the subducting Juan de Fuca plate; Nature, v. 321, p. 596-599.

Kurtz, R.D., DeLaurier, J.M., and Gupta, J.C., 1990. The electrical conductivity distribution beneath Vancouver Island: a region of active plate subduction; Journal of Geophysical Research, v. 95, p. 10,929-10,946.

Lapp, D.B., Owens, T.J., and Crosson, R.S., 1990. P-waveform analysis for local subduction geometry south of Puget Sound, Washington; Pure and Applied Geophysics, v. 133, p. 349-365.

Leeman, W.P., Smith, D.R., Hildreth, W., Palacz, Z., and Rogers, N., 1990. Compositional diversity of late Cenozoic basalts in a transect across the southern Washington Cascades: implications for subduction zone magmatism; Journal of Geophysical Research, v. 95, p. 19,561-19,582.

Lees, J.M. and Crosson, R.S., 1990. Tomographic imaging of local earthquake delay times for three-dimensional velocity variation in western Washington; Journal of Geophysical Research, v. 95, p. 4763-4776.

Levato, L., Alioth, D., Olivier, R., and Wagner, J.-J., 1990. Reflection seismic data from Vancouver Island processed using Geovecteur software package on a Cray supercomputer (CCSS topic III); in A.G. Green (ed.), Studies of Laterally Heterogeneous Structures Using Seismic Refraction and Reflection Data; Geological Survey of Canada, Paper 89-13, p. 191-205.

Levi, S. and Riddihough, R., 1986. Why are marine magnetic anomalies suppressed over sedimented spreading centers?; Geology, v. 14, p. 651-654.

Lewis, P.D., Haggart, J.W., Anderson, R.G., Hickson, C.J., Thompson, R.I., Dietrich, J.R., and Rohr, K.M.M., 1991a. Triassic to Neogene geological evolution of the Queen Charlotte Basin; Canadian Journal of Earth Sciences, v. 28, p. 854-869.

Lewis, P.D., Thompson, R.I., Haggart, J.W., and Hickson, C.J., 1991b. Discussion of 'Sedimentology and tectonic implications of Cretaceous fan-delta conglomerates, Queen Charlotte Islands, Canada'; Sedimentology, v. 38, p. 1173-1179.

Lisowski, M., 1985. Geodetic Strain Measurements in Central Vancouver Island; M.Sc. thesis, Dept. of Geological Sciences, University of British Columbia, Vancouver, 100 p.

Lister, C.R.B., 1989. Plate tectonics at an awkward junction: rules for the evolution of Sovanco Ridge area, NE Pacific; Geophysical Journal, v. 96, p. 191-201.

Ludwin, R.S., Weaver, C.S., and Catchings, R.D., 1989. Apparent structural relations between crustal earthquakes and continental rifting in the Columbia Plateau, Washington (abs.); Seismological Research Letters, v. 60, p. 29.

Lutter, W.J., Catchings, R.D., and Jarchow, C.M., 1994. An image of the Columbia Plateau from inversion of high-resolution seismic data; Geophysics, v. 59, p. 1278-1289.

Lyatsky, H.V., 1991a. Regional geophysical constraints on crustal structure and geologic evolution of the Insular Belt, British Columbia; in G.J. Woodsworth (ed.), Evolution and Hydrocarbon Potential of the Queen Charlotte Basin, British Columbia; Geological Survey of Canada, Paper 90-10, p. 97-106.

Lyatsky, H.V., 1991b. Diachronous acoustic basement in seismic reflection data from the Queen Charlotte Basin, British Columbia; in Current Research, Part A; Geological Survey of Canada, Paper 91-1A, p. 401-407.

Lyatsky, H.V., 1993a. Basement-controlled structure and evolution of the Queen Charlotte Basin, west coast of Canada; Tectonophysics, v. 228, p. 123-140.

Lyatsky, H.V., 1993b. Geophysical methods in a multidisciplinary study of western Canadian continental margin (abs.); European Association of Exploration Geophysicists, 55th Meeting and Technical Exhibition, Stavanger, Norway; Extended Abstracts of Papers, paper D-030.

Lyatsky, H.V., 1994. Formation of non-compressional sedimentary basins on continental crust: limitations on modern models; Journal of Petroleum Geology, v. 17, p. 301-316.

Lyatsky, H.V. and Haggart, J.W., 1993. Petroleum exploration model for the Queen Charlotte Basin, offshore British Columbia; Canadian Journal of Earth Sciences, v. 30, p. 918-927.

Lyatsky, H.V., Haggart, J.W., Hickson, C.J., and Woodsworth, G.J., 1991. Diffuse continent-ocean boundary at the continental margin of western Canada (abs.); American Geophysical Union, 38th Annual Pacific Northwest Regional Meeting, Washington State University, Tri-Cities, Program and Abstracts, p. 23.

Lyatsky, H.V., Haynes, A.K., Brown, R.J., Thurston, J.B., and Lyatsky, V.B., 1992a. Aeromagnetic Horizontal-Gradient Vector Map of the Queen Charlotte Basin Area, British Columbia; Geological Survey of Canada, Open File 2436, scale 1:1,000,000.

Lyatsky, H.V., Thurston, J.B., Brown, R.J., and Lyatsky, V.B., 1992b. Hydrocarbon-exploration applications of potential-field horizontal-gradient vector maps; Recorder (Canadian Society of Exploration Geophysicists), v. XVII/9, p. 10-15.

Lyatsky, V.B. and Lyatsky, H.V., 1990. Integrated geological basin analysis as a method of hydrocarbon exploration on continental shelves; in 22nd Offshore Technology Conference, Houston, Proceedings, p. 237-242.

Mackie, D.J., Clowes, R.M., Dehler, S.A., Ellis, R.M., and Morel-à-l'Huissier, P., 1989. The Queen Charlotte Islands refraction project. Part II. Structural model for transition from Pacific plate to North American plate; Canadian Journal of Earth Sciences, v. 26, p. 1713-1725.

MacKay, M.E., Jarrard, R.D., Westbrook, G.K., Hyndman, R.D., and Shipboard Scientific Party of Ocean Drilling Program Leg 146; Origin of bottom-simulating reflectors: geophysical evidence from the Cascadia accretionary prism; Geology, v. 22, p. 459-462.

MacLeod, N.S., Tiffin, D.L., Snavely, P.D., and Currie, R.D.,

1977. Geologic interpretation of magnetic and gravity anomalies in the Strait of Juan de Fuca, U.S.-Canada; Canadian Journal of Earth Sciences, v. 14, p. 223-238.

Malecek, S.J. and Clowes, R.M., 1978. Crustal structure near Explorer Ridge from a marine deep seismic sounding survey; Journal of Geophysical Research, v. 83, p. 5899-5912.

Mann, G.M. and Meyer, 1993. Late Cenozoic structure and correlations to seismicity along the Olympic-Wallowa Lineament, northwest United States; Bulletin of Geological Society of America, v. 105, p. 853-871.

Massey, N.W.D., 1986. Metchosin Igneous Complex, southern Vancouver Island: ophiolite stratigraphy developed in an emegrent island setting; Geology, v. 14, p. 602-605.

Massey, N.W.D. and Friday, S.J., 1989. Geology of the Alberni-Nanaimo Lakes area, Vancouver island; in Geological Fieldwork 1988, British Columbia Ministry of Energy, Mines and Petroleum Resources, Paper 1989-1, p. 61-74.

Mayrand, L.J., Green, A.G., and Milkereit, B., 1987. A quantitative approach to bedrock velocity resolution and precision: the LITHOPROBE Vancouver Island experiment; Journal of Geophysical Research, v. 92, p. 4837-4845.

McClain, K.J., 1981. A Geophysical Study of Accretionary Processes on the Washington Continental Margin; Ph.D. thesis, University of Washington, Seattle, 141 p.

McCrory, P.A., 1994. Late Quaternary thrust faulting along the Cascadia margin, Washington: implications for partitioning of strain (abs.); American Geophysical Union, Fall Meeting 1994, Program and Abstracts; Published as a supplement to EOS, p. 662.

McCrumb, D.R., Galster, R.W., West, D.O., Crosson, R.S., Ludwin, R.S., Hancock, W.E., and Mann, L.V., 1989a. Tectonics, seismicity, and engineering seismology in Washington; in R.W. Galster (ed.), Engineering Geology in Washington, v. I; Washington Division of Geology and Earth Resources, Bulletin 78, p. 97-120.

McCrumb, D.S., West, D.O., and Kiel, W.A., 1989b. Geology and seismic considerations of the Satsop nuclear power plant site; in R.W. Galster (ed.), Engineering Geology in Washington, v. I; Washington Division of Geology and Earth Resources, Bulletin 78, p. 589-606.

McKenzie, D., 1978. Some remarks on the development of sedimentary basins; Earth and Planetary Science Letters, v. 40, p. 25-32.

McManus, D.A., Holmes, M.L., Carson, B., and Barr, S.M., 1972. Late Quaternary tectonics, northern end of Juan de Fuca Ridge (northeast Pacific); Marine Geology, v. 12, p. 141-164.

McMechan, G.A. and Spence, G.D., 1983. P-wave velocity structure of the Earth's crust beneath Vancouver Island; Canadian Journal of Earth Sciences, v. 20, p. 742-752.

McMillan, W.J., 1991. Overview of the tectonic evolution and setting of mineral deposits in the Canadian Cordillera; in Ore Deposits, Tectonics and Metallogeny in the Canadian Cordillera; British Columbia Ministry of Energy, Mines and Petroleum Resources, Paper 1991-4, p. 5-24.

Meissner, R., 1989. Rupture, creep, lamellae and crocodiles: happenings in the continental crust; Terra Nova, v. 1, p. 17-28.

Meissner, R. and Wever, T., 1992. The possible role of fluids for the structuring of the continental crust; Earth-Science Reviews, v. 32, p. 19-32.

Mereu, R.F., 1990. An interpretation of CCSS data set I using the triangular block model method; in A.G. Green (ed.), Studies of Laterally Heterogeneous Structures Using Seismic Refraction and

Reflection Data; Geological Survey of Canada, Paper 89-13, p. 53-63.

Michael, P.J., Chase, R.L., and Allan, J.F., 1989. Petrologic and geologic variations along the southern Explorer Ridge, northeast Pacifc Ocean; Journal of Geophysical Research, v. 94, p. 13,895-13,918.

Michaelson, C.A. and Weaver, C.S., 1986. Upper mantle structure from teleseismic P wave arrivals in Washington and northern Oregon; Journal of Geophysical Research, v. 91, p. 2077-2094.

Milkereit, B., Spencer, C., and Mayrand, L.J., 1990. Migration and amplitude analysis of deep seismic reflection data: processing results of CCSS data sets II and III; in A.G. Green (ed.), Studies in Laterally Heterogeneous Structures Using Seismic Refraction and Reflection Data; Geological Survey of Canada, Paper 89-13, p. 151-164.

Miller, D.M., Nilsen, T.H., and Bilodeau, W.L., 1992. Late Cretaceous to early Eocene geologic evolution of the U.S. Cordillera; in B.C. Burchfiel, P.W. Lipman, and M.L. Zoback (eds.), The Cordillera Orogen: Conterminous U.S.; Geological Society of America, The Geology of North America, v. G-3, p. 205-260.

Milne, W.G., Rogers, G.C., Riddihough, R.P., McMechan, G.A., and Hyndman, R.D., 1978. Seismicity of western Canada; Canadian Journal of Earth Sciences, v. 15, p. 1170-1190.

Minster, J.B. and Jordan, T.H., 1978. Present day plate motions; Journal of Geophysical Research, v. 83, p. 5331-5354.

Mitchell, C.E., Vincent, P., Weldon, R.J., and Richards, M.A., 1994. Present-day vertical deformation of the Cascadia margin, Pacific Northwest, United States; Journal of Geophysical Research, v. 99, p. 12,257-12,277.

Monger, J.W.H., 1991. Late Mesozoic to Recent evolution of the Georgia Strait-Puget Sound region, British Columbia and Washington; Washington Geology, v. 19/4, p. 3-7.

Monger, J.W.H., 1993. Canadian Cordilleran tectonics: from geosynclines to crustal collage; Canadian Journal of Earth Sciences, v. 30, p. 209-231.

Monger, J.W.H., Price, R.A., and Tempelman-Kluit, D.J., 1982. Tectonic accretion and the origin of the two major metamorphic and plutonic welts in the Canadian Cordillera; Geology, v. 10, p. 70-75.

Mooney, W.D. and Weaver, C.S., 1989. Regional crustal structure and tectonics of the Pacific Coastal States; California, Oregon, and Washington; in L.C. Pakiser and W.D. Mooney (eds.), Geophysical Framework of the Continental United States; Geological Society of America, Memoir 172, p. 129-161.

Moran, J.E. and Lister, C.R.B., 1987. Heat flow across Cascadia Basin near 47°N, 128°W; Journal of Geophysical Research, v. 92, p. 11,416-11,432.

Morgan, J. and Warner, M., 1990. Interpretation of a combined refraction and reflection profile across the western Canadian active margin; in A.G. Green (ed.), Studies of Laterally Heterogeneous Structures Using Seismic Refraction and Reflection Data; Geological Survey of Canada, Paper 89-13, p. 31-41.

Muller, J.E., 1977a. Evolution of the Pacific margin, Vancouver Island, and adjacent regions; Canadian Journal of Earth Sciences, v. 14, p. 2062-2085.

Muller, J.E., 1977b. Geology of Vancouver Island; Geological Survey of Canada, Open File 463, scale 1:250,000.

Muller, J.E., 1977c. Metchosin Volcanics and Sooke Intrusions of southern Vancouver Island; in Report of Activities, Part A;

Geological Survey of Canada, Paper 77-1A, p. 287-294.

Muller, J.E., 1980a. The Paleozoic Sicker Group of Vancouver Island, British Columbia; Geological Survey of Canada, Paper 79-30, 23 p.

Muller, J.E., 1980b. Chemistry and origin of the Eocene Metchosin Volcanics, Vancouver Island, British Columbia; Canadian Journal of Earth Sciences, v. 17, p. 199-209.

Muller, J.E., Cameron, B.E.B., and Northcote, K.E., 1981. Geology and Mineral Deposits of Nootka Sound Map-Area (92E) Vancouver Island, British Columbia; Geological Survey of Canada, Paper 80-16, 53 p.

Muller, J.E., Northcote, K.E., and Carlisle, D., 1974. Geology and Mineral Deposits of Alert-Cape Scott Map-Area (92L-102I) Vancouver Island, British Columbia; Geological Survey of Canada, Paper 74-8, 77 p.

Mustard, P.S., 1991. Stratigraphy and sedimentology of the Georgia Basin, British Columbia and Washington State; Washington Geology, v. 19/4, p. 7-9.

Niem A.R., Snavely, P.D., and Niem, W.A., 1992b. Olympic Mountains core rocks; in Christiansen, R.L. and Yeats, R.S., 1992 (full reference above), p. 278-281.

Niem, A.R. and Snavely, P.D., 1991. Geology and preliminary hydrocarbon evaluation of the Tertiary Juan de Fuca Basin, Olympic Peninsula, Northwest Washington; Washington Geology, v. 19/4, p. 27-34.

Niem, W.R., Niem, A.R., and Snavely, P.D., 1992a. Western Washington-Oregon coastal sequence; in Christiansen, R.I. and Yeats, R.S., 1992 (full reference above), p. 265-270.

Niem, W.R., Niem, A.R., and Snavely, P.D., 1992c. Sedimentary embayments of the Washington-Oregon coast; in Christaiansen, R.L. and Yeats, R.S., 1992 (full reference above), p. 314-319.

Nixon, G.T., Hammack, J.L., Payie, G.J., Snyder, L.D., Archibald, D.A., and Barron, D.J., 1995. Quatsino-San Josef map area, northern Vancouver Island (92L/12W, 102I/8,9); in Geological Fieldwork 1994, British Columbia Ministry of Energy, Mines and Petroleum Resources, Paper 1995-1, p. 9-21.

Orange, D.L., 1990. Criteria helpful in recognizing shear-zone and diapiric mélanges: examples from the Hoh accretionary complex, Olympic Peninsula, Washington; Bulletin of Geological Society of America, v. 102, p. 935-951.

Orange, D.L., Geddes, D.S., and Moore, J.C., 1993. Structural and fluid evolution of a young accretionary complex: the Hoh rock assemblage of the western Olympic Peninsula, Washington; Bulletin of Geological Society of America, v. 105, p. 1053-1075.

Oreskes, N., Shrader-Frechette, K., and Belitz, K., 1994. Verification, validation, and confirmation of numerical models in the earth sciences; Science, v. 263, p. 641-646.

Owens, T.J., Crosson, R.S., and Hendrickson, M.A., 1988. Constraints on the subduction geometry beneath western Washington from broadband teleseismic waveform modeling; Bulletin of Seismological Society of America, v. 78, p. 1319-1334.

Pacht, J.A., 1984. Petrologic evolution and paleogeography of the Late Cretaceous Nanaimo Basin, Washington and British Columbia: implications for Cretaceous tectonics; Bulletin of Geological Society of America, v. 95, p. 766-778.

Pakiser, L.C. and Mooney, W.D., eds., 1989. Geophysical Framework of the Continental United States; Geological Society of America, Memoir 172, 826 p.

Parrish, R.R., 1983. Cenozoic thermal evolution and tectonics of the Coast Mountains of British Columbia. 1. Fission track dating, apparent uplift rates, and patterns of uplift; Tectonics, v. 2, p. 601-631.

Pavlenkova, N.I., 1989. The Kola well and its significance for deep seismic sounding; Sovetskaia Geologia (Soviet Geology), 1989/6, p. 16-23 (in Russian).

Petford, N., Kerr, R.C., and Lister, J.R., 1993. Dike transport of granitoid magmas; Geology, v. 21, p. 845-848.

Plafker, G., 1987. Regional geology and petroleum potential of the northern Gulf of Alaska continental margin; in D.W. Scholl, A. Grantz, and J.G. Vedder (eds.), Geology and Resource Potential of the Continental Margin of Western North America and Adjacent Ocean Basins - Beaufort Sea to Baja California; Circum-Pacific Council for Energy and Mineral Resources, Earth Science Series, v. 6, p. 229-268.

Price, E.H. and Watkinson, A.J., 1989. Structural geometry and strain distribution within eastern Umtanum fold ridge, south-central Washington; in S.P. Reidel and P.R. Hooper (eds.), Volcanism and Tectonism in the Columbia River Flood-Basalt Province; Geological Society of America, Special Paper 239, p. 265-281.

Price, R.A. and Douglas, R.J.W. (eds.), 1972. Variations in Tectonic Styles in Canada; Geological Association of Canada, Special Paper No. 11, 688 p.

Raff, A.D. and Mason, R.G., 1961. Magnetic survey off the west coast of North America, 40˚N. latitude to 52˚N. latitude; Bulletin of Geological Society of America, v. 72, p. 1267-1270.

Raisz, E., 1945. The Olympic-Wallowa Lineament; American Journal of Science, v. 243-A, p. 479-485.

Rasmussen, J.R. and Humphreys, E.D., 1988. Tomographic image of the Juan de Fuca plate beneath Washington and western Oregon using teleseismic P-wave travel times; Geophysical Research Letters, v. 15, p. 1417-1420.

Read, P.B., 1990. Mount Meager Complex, Garibaldi Belt, southwestern British Columbia; Geoscience Canada, v. 17, p. 167-170.

Reidel, S.P. and Campbell, N.P., 1989. Structure of the Yakima Fold Belt, central Washington; in N.L. Joseph (ed.), Geologic Guidebook for Washington and Adjacent Areas; Washington Division of Geology and Earth Resources, Information Circular 86, p. 277-288.

Reidel, S.P., Campbell, N.P., Fecht, K.R., and Lindsey, K.A., 1994. Late Cenozoic structure and stratigraphy of south-central Washington; in R. Lasmanis and E.S. Cheney (eds.), Regional Geology of Washington State; Washington Division of Geology and Earth Resources, Bulletin 80, p. 159-180.

Reidel, S.P. and Hooper, P.R. (eds.), 1989. Volcanism and Tectonism in the Columbia River Flood-Basalt Province; Geological Society of America, Special Paper 239, 386 p.

Reidel, S.P., Tolan, T.L., Hooper, P.R., Beeson, M.H., Fecht, K.R., Bentley, R.D., and Anderson, J.L., 1989. The Grande Ronde Basalt, Columbia River Basalt Group; stratigraphic descriptions and correlations in Washington, Oregon, and Idaho; in S.P. Reidel and P.R. Hooper (eds.), Volcanism and Tectonism in the Columbia River Flood-Basalt Province; Geological Society of America, Special Paper 239, p. 21-53.

Reynolds, R.L., Rosenbaum, J.G., Hudson, M.R., and Fishman, N.S., 1990. Rock magnetism, the distribution of magnetic minerals in the Earth's crust, and aeromagnetic anomalies; in W.F. Hanna (ed.), Geologic Applications of Modern Aeromagnetic Surveys; U.S. Geological Survey, Bulletin 1924, p. 24-45.

346

Riddihough, R.P., 1977. A model for recent plate interactions off Canada's west coast; Canadian Journal of Earth Sciences, v. 14, p. 384-396.

Riddihough, R.P., 1979. Gravity and structure of an active margin - British Columbia and Washington; Canadian Journal of Earth Sciences, v. 16, p. 350-363.

Riddihough, R.P., 1982a. One hundred million years of plate tectonics in western Canada; Geoscience Canada, v. 9, p. 28-34.

Riddihough, R.P., 1982b. Contemporary movements and tectonics on Canada's west coast: a discussion; Tectoniphysics, v. 86, p. 319-341.

Riddihough, R.P., 1984. Recent movements of the Juan de Fuca plate system; Journal of Geophysical Research, v. 89, p. 6980-6994.

Riddihough, R.P., Beck, M.E., Chase, R.L., Davis, E.E., Hyndman, R.D., Johnson, S.H., and Rogers, G.C., 1983. Geodymanics of the Juan de Fuca plate; in Geodynamics of the Eastern Pacific Region, Carribean and Scotia Arcs; American Geophysical Union, Geodynamics Series, v. 9, p. 5-21.

Riddihough, R.P., Currie, R.G., and Hyndman, R.D., 1980. The Dellwood knolls and their role in triple junction tectonics off northern Vancouver Island; Canadian Journal of Earth Sciences, v. 17, p. 577-593.

Riddihough, R.P. and Hyndman, R.D., 1976. Canada's active margin - the case for subduction; Geoscience Canada, v. 3, p. 269-279.

Riddihough, R.P. and Hyndman, R.D., 1989. Queen Charlotte Islands margin; in E.L. Winterer, D.M. Hussong, and R.W. Decker (eds.), The Eastern Pacific Ocean and Hawaii; Geological Society of America, The Geology of North America, v. N, p. 403-411.

Riddihough, R.P. and Hyndman, R.D., 1991. Modern plate tectonic regime of the continental margin of western Canada; in H. Gabrielse and C.J. Yorath (eds.), Geology of the Cordilleran orogen in Canada; Geological Society of America, The Geology of North America, v. G-2, p. 437-455.

Rogers, G.C., 1979. Earthquake fault plane solutions near Vancouver Island; Canadian Journal of Earth Sciences, v. 16, p. 523-531.

Rogers, G.C., 1983. Seismotectonics of British Columbia; Ph.D. thesis, Dept. of Geophysics & Astronomy, University of British Columbia, Vancouver, 247 p.

Rogers, G.C., 1986. Seismic gaps along the Queen Charlotte fault; Earthquake Prediction Research, v. 4, p. 1-11.

Rogers, G.C., 1988. An assessment of the megathrust earthquake potential of the Cascadia subduction zone; Canadian Journal of Earth Sciences, v. 25, p. 844-852.

Rogers, G.C., Spindler, C., and Hyndman, R.D., 1990. Seismicity along the Vancouver Island Lithoprobe corridor; Lithoprobe, Southern Canadian Cordillera Transect Workshop, Calgary, p. 166-169.

Rohr, K.M.M., 1994. Increase of seismic velocities in upper oceanic crust and hydrothermal circulation in the Juan de Fuca plate; Geophysical Research Letters, v. 21, p. 2163-2166.

Rohr, K. and Dietrich, J., 1990. Deep Seismic Survey of Queen Charlotte Basin; Geological Survey of Canada, Open File 2258.

Rohr, K.M.M. and Dietrich, J.R., 1992. Strike-slip tectonics and development of the Tertiary Queen Charlotte Basin, offshore western Canada: evidence from seismic reflection data; Basin Research, v. 4, p. 1-19.

Rohr, K.M.M., Milkereit, B., and Yorath, C.J., 1988. Asymmetric deep crustal structure across the Juan de Fuca Ridge; Geology, v. 16, p. 533-537.

Rosendahl, B.R., Meyers, J., Groschel, H., and Scott, D., 1992. Nature of the transition from continental to oceanic crust and the meaning of reflection Moho; Geology, v. 20, p. 721-724.

Rusmore, M.E. and Cowan, D.S., 1985. Jurassic-Cretaceous rock units along the southern edge of the Wrangellia terrane on Vancouver Island; Canadian Journal of Earth Sciences, v. 22, p. 1223-1232.

Rusmore, M.E. and Woodsworth, G.J., 1991. Coast Plutonic Complex: a mid-Cretaceous contractional orogen; Geology, v. 19, p. 941-944.

Rusmore, M.E. and Woodsworth, G.J., 1994. Evolution of the eastern Waddington thrust belt and its relation to the mid-Cretaceous Coast Mountains arc, western British Columbia; Tectonics, v. 13, p. 1052-1067.

Saleeby, J.B. and Busby-Spera, C., 1992. Early Mesozoic tectonic evolution of the western U.S. Cordillera; in B.C. Burchfiel, P.W. Lipman, and M.L. Zoback (eds.), The Cordilleran Orogen: Conterminous U.S.; Geological Society of America, The Geology of North America, v. G-3, p. 107-168.

Saltus, R.W., 1993. Upper-crustal structure beneath the Columbia River Basalt Group, Washington: gravity interpretation controlled by borehole and seismic studies; Bulletin of Geological Society of America, v. 105, p. 1247-1259.

Scott, W.E., 1990. Patterns of volcanism in the Cascade arc during the past 15,000 years; Geoscience Canada, v. 17, p. 179-183.

Seely, D.R., 1977. The significance of landward vergence and oblique structural trends on trench inner slopes; in M. Talwani and W.C. Pitman (eds.), Island Arcs, Deep Sea Trenches and Back-Arc Basins; American Geophysical Union, Maurice Ewing Series, v. 1, p. 187-198.

Seemann, D.A., 1982. Bathymetry off the Coast of British Columbia; Earth Physics Branch, Open File 82-25, scale 1:1,000,000.

Sharpton, V.L., Grieve, R.A.F., Thomas, M.D., and Halpenny, J.F., 1987. Horizontal gravity gradient: an aid to the definition of crustal structure in North America; Geophysical Research Letters, v. 14, p. 808-811.

Sherrod, D.R. and Smith, J.G., 1990. Quaternary extrusion rates of the Cascade Range, northwestern United States and southern British Columbia; Journal of Geophysical Research, v. 95, p. 19,465-19,474.

Shouldice, D.H., 1971. Geology of the western Canadian continental shelf; Bulletin of Canadian Petroleum Geology, v. 19, p. 405-436.

Simpson, R.W. and Jachens, R.C., 1989. Gravity methods in regional studies; in L.C. Pakiser and W.D. Mooney (eds.), Geophysical Framework of the Continental United States; Geological Society of America, Memoir 172, p. 35-44.

Simpson, R.W., Jachens, R.C., Blakeley, R.J., and Saltus, R.W., 1986. A new isostatic residual gravity map of the conterminous United States with a discussion on the significance of isostatic residual anomalies; Journal of Geophysical Research, v. 91, p. 8348-8372.

Smithson, S.B. and Johnson, R.A., 1989. Crustal structure of the western U.S. based on reflection seismology; in L.C. Pakiser and W.D. Mooney (eds.), Geophysical Framework of the Continental United States; Geological Society of America, Memoir 172, p. 577-612.

Snavely, P.D., 1987. Tertiary geologic framework, neotectonics, and petroleum potential of the Oregon-Washington continental margin; in D.W. Scholl, A. Grantz, and J.G. Vedder (eds.), Geology and Resource Potential of the Continental Margin of Western North America and Adjacent Oceanic Basins - Beaufort Sea to Baja California; Circum-Pacific Council for Energy and Mineral Resources, Earth Science Series, v. 6, p. 305-335.

Snavely, P.D., Tiffin, D.L., and Tompkins, D.H., 1980. Seismic Reflection Profile Across the Queen Charlotte Fault Zone, Dixon Entrance, Canada-U.S.; U.S. Geological Survey, Open-File Report 80-1063.

Snavely, P.D. and Wagner, H.C., 1981. Geologic Cross Section Across the Continental Margin off Cape Flattery, Washington and Vancouver Island, British Columbia; U.S. Geological Survey, Open File Report 81-978.

Sobczak, L.W., 1988. Discussion of 'LITHOPROBE - southern Vancouver Island: Cenozoic subduction complex imaged by deep seismic reflections'; Canadian Journal of Earth Sciences, v. 25, p. 163.

Sobczak, L.W. and Halpenny, J.F., 1990. Isostatic and Enhanced Isostatic Gravity Map of the Arctic; Geological Survey of Canada Paper 89-16, 9 p.

Souther, J.G., 1976. Geothermal potential of western Canada; Proceedings, 2nd U.N. Symposium on the Development and Use of Geothermal Resources, San Francisco, v. 1, p. 259-267.

Souther, J.G., 1990. Volcano tectonics of Canada; in C.A. Wood and J. Kienle (eds.), Volcanoes of North America - United States and Canada; Cambridge University Press, p. 111-116.

Souther, J.G. and Jessop, A., 1991. Dyke swarms in the Queen Charlotte Islands, British Columbia, and implications for hydrocarbon exploration; in G.J. Woodsworth (ed.), Evolution and Hydrocarbon Potential of the Queen Charlotte Basin, British Columbia; Geological Survey of Canada, Paper 90-10, p. 465-487.

Spence, G.D. and Asudeh, I., 1993. Seismic velocity structure of the Queen Charlotte Basin beneath Hecate Strait; Canadian Journal of Earth Sciences, v. 30, p. 787-805.

Spence, G.D., Clowes, R.M., and Ellis, R.M., 1985. Seismic structure across the active subduction zone of western Canada; Journal of Geophysical Research, v. 90, p. 6754-6772.

Spence, G.D., Hyndman, R.D., Davis, E.E., and Yorath, C.J., 1991. Seismic structure of the northwestern Cascadia accretionary prism: evidence from new multichannel seismic reflection data; in R. Meissner, L. Brown, H.-J. Dürbaum, W. Franke, K. Fuchs, and F. Seifert (eds.), Continental Lithosphere: Deep Seismic Reflections; American Geophysical Union, Geodynamics Series, Vol. 22, p. 257-263.

Srivastava, S.P., 1973. Interpretation of gravity and magnetic measurements across the continental margin of British Columbia, Canada; Canadian Journal of Earth Sciences, v. 10, p. 1664-1677.

Srivastava, S.P., Barrett, D.L., Keen, C.E., Manchester, K.S., Shih, K.G., Tiffin, D.L., Chase, R.L., Thomlinson, A.G., Davis, E.E., and Lister, C.R.B., 1971. Preliminary analysis of geophysical measurements north of Juan de Fuca Ridge; Canadian Journal of Earth Sciences, v. 8, p. 1265-1281.

Stacey, R.A., 1973. Gravity anomalies, crustal structure and plate tectonics in the Canadian Cordillera; Canadian Journal of Earth Sciences, v. 10, p. 615-628.

Stacey, R.A., 1975. Structure of the Queen Charlotte Basin; in C.J. Yorath, E.R. Parker, and D.J. Glass (eds.), Canada's

Continental Margins and Offshore Petroleum Exploration; Canadian Society of Petroleum Geologists, Memoir 4, p. 723-741.

Stacey, R.A. and Stephens, L.E., 1969. An interpretation of gravity measurements on the West Coast of Canada; Canadian Journal of Earth Sciences, v. 6, p. 463-474.

Stock, J.M. and Lee, J., 1994. Do microplates in subduction zones leave a geological record?; Tectonics, v. 13, p. 1472-1487.

Stock, J.M. and Molnar, P., 1988. Uncertainties and implications of the Late Cretaceous and Tertiary position of North America relative to the Farallon, Kula, and Pacific plates; Tectonics, v. 6, p. 1339-1384.

Stoddard, P.R., 1987. A kinematic model for the evolution of the Gorda plate; Journal of Geophysical Research, v. 92, p. 11,524-11,532.

Stoddard, P.R., 1991. A comparison of brittle deformation models for the Gorda plate; Tectonophysics, v. 187, p. 205-214.

Sutherland Brown, A., 1968. Geology of the Queen Charlotte Islands, British Columbia; British Columbia Department of Mines and Petroleum Resources, Bulletin 54, 226 p.

Sweeney, J.F. and Seemann, D.A., 1991. Crustal density structure, Queen Charlotte Islands and Hecate Strait; in G.J. Woodsworth (ed.), Evolution and Hydrocarbon Potential of the Queen Charlotte Basin, British Columbia; Geological Survey of Canada, Paper 90-10, p. 89-96.

Taber, J.J. and Lewis, B.T.R., 1986. Crustal structure of the Washington continental margin from refraction data; Bulletin of Seismological Society of America, v. 76, p. 1011-1024.

Taber, J.J. and Smith, S.W., 1985. Seismicity and focal mechanisms associated with the subduction of the Juan de Fuca plate beneath the Olympic Peninsula, Washington; Bulletin of Seismological Society of America, v. 75, p. 237-249.

Tabor, R.W., 1972. Age of the Olympic metamorphism, Washington - K-Ar dating of low-grade metamorphic rocks; Bulletin of Geological Sociery of America, v. 83, p. 1805-1816.

Tabor, R.W. and Cady, W.M., 1978. The Structure of the Olympic Mountains, Washington - Analysis of a Subduction Zone; U.S. Geological Survey, Professional Paper 1033, 38 p.

Taylor, E.M., 1990. Volcanic history and tectonic development of the central High Cascade Range, Oregon; Journal of Geophysical Research, v. 95, p. 19,611-19,622.

Teskey, D.J., Hood, P.J., and Dods, S.D., 1989a. Magnetic Anomaly Map of Canada - Upward Continued to 40 km; Geological Survey of Canada, Canadian Geophysical Atlas, Map 13, scale 1:10,000,000.

Teskey, D.J., Hood, P.J., and Dods, S.D., 1989b. Shaded Relief Presentation of the Magnetic Anomaly Map of Canada; Geological Survey of Canada, Canadian Geophysical Atlas, Map 14, scale 1:10,000,000.

Thompson, G.A., Catchings, R., Goodwin, E., Holbrook, S., Jarchow, C., Mann, C., Mccarthy, J., and Okaya, D., 1989. Geophysics of the western Basin and Range province; in L.C. Pakiser and W.D. Mooney (eds.), Geophysical Framework of the Continental United States; Geological Society of America, Memoir 172, p. 177-203.

Thompson, R.I., Haggart, J.W., and Lewis, P.D., 1991. Late Triassic through early Tertiary evolution of the Queen Charlotte Basin, British Columbia, with a perspective on hydrocarbon potential; in G.J. Woodsworth (ed.), Evolution and Hydrocarbon Potential of the Queen Charlotte Basin, British Columbia; Geological Survey of Canada, Paper 90-10, p. 3-29.

Thybo, H., 1990. Interpretation of coincident seismic reflection and refraction profiles across the active subduction zone of western Canada; in A.G. Green (ed.), Studies of Laterally Heterogeneous Structures Using Sesimic Refraction and Reflection Data; Geological Survey of Canada, Paper 89-13, p. 65-77.

Tiffin, D.L., Cameron, B.E.B., and Murray, J.W., 1972. Tectonics and depositional history of the continental margin off Vancouver Island, British Columbia; Canadian Journal of Earth Sciences, v. 9, p. 280-296.

Tipper, H.W. and Richards, T.A., 1976. Jurassic Stratigraphy and History of North-Central British Columbia; Geological Survey of Canada, Bulletin 270, p. 73 p.

Tivey, M.A., 1994. Fine-scale magnetic anomaly field over the southern Juan de Fuca Ridge: axial magnetization low and implications for crustal structure; Journal of Geophysical Research, v. 99, p. 4833-4855.

Tivey, M.A. and Johnson, H.P., 1993. Variations in oceanic crustal structure and implications for the fine-scale magnetic anomaly signal; Geophysical Research Letters, v. 20, p. 1879-1882.

Tolan, T.L. and Reidel, S.P. (comps.), 1989. Structure Map of a Portion of the Columbia River Flood-Basalt Province; enclosed in S.P. Reidel and P.R. Hooper (eds.), Volcanism and Tectonism in the Columbia River Flood-Basalt Province; Geological Society of America, Special Paper 239.

Tribe, S., 1993. Kunghit Island mylonite of the Louscoone Inlet fault system, Queen Charlotte Islands, British Columbia; in Current Research, Part E; Geological Survey of Canada, Paper 93-1E, p. 1-8.

VanDecar, J.C., 1991. Upper-Mantle Structure of the Cascadia Subduction Zone from Non-Linear Teleseismic Travel-Time Inversion; Ph.D. dissertation, Geophysics Program, University of Washington, Seattle, 165 p.

VanDecar, J.C., Crosson, R.S., and Creager, K.C., 1990. Teleseismic travel-time inversion for Cascadia subduction zone structure employing three-dimensional ray tracing; XXII General Assembly of the European Seismological Commission, Barcelona, Proceedings, unpaginated.

Van der Heyden, P., 1992. A Middle Jurassic to Early Tertiary Andean-Sierran arc model for the Coast Belt of British Columbia; Tectonics, v. 11, p. 82-97.

Vine, F.J. and Matthews, D.H., 1963. Magnetic anomalies southwest of Vancouver Island; Nature, v. 199, p. 947-949.

Vine, F.J. and Wilson, J.T., 1965. Magnetic anomalies over a young oceanic ridge off Vancouver Island; Science, v. 150, p. 485-489.

Von Huene, R., 1989. Continental margins around the Gulf of Alaska; in E.L. Winterer, D.M. Hussong, and R.W. Decker. (eds.), The Eastern Pacific Ocean and Hawaii; Geological Society of America, The Geology of North America, v. N, p. 383-401.

Von Huene, R., Shor, G.G., and Wageman, J., 1979. Continental margins of the eastern Gulf of Alaska and boundaries of tectonic plates; in J.S. Watkins, L. Montadert, and P.W. Dickerson (eds.), Geological and Geophysical Investigations of Continental Margins; American Association of Petroleum Geologists, Memoir 29, p. 273-290.

Wahlström, R. and Rogers, G.C., 1990. Relocation of Earthquakes Offshore Vancouver Island; Geological Survey of Canada, Open File 2268.

Wahlström, R. and Rogers, G.C., 1992. Relocation of earthquakes west of Vancouver Island, British Columbia, 1965-1983;

Canadian Journal of Earth Sciences, v. 29, p. 953-961.

Walcott, R.I., 1967. The Bouguer Anomaly Map of Southwestern British Columbia; University of British Columbia, Institute of Earth Sciences, Scientific Report 15.

Waldron, D.A., Clowes, R.M., and White, D.J., 1990. Seismic structure of a subducting oceanic plate off western Canada; in A.G. Green (ed.), Studies of Laterally Heterogeneous Structures Using Refraction and Reflection Data; Geological Survey of Canada, Paper 89-13, p. 105-113.

Watters, T.R., 1989. Periodically spaced anticlines of the Columbia Plateau; in S.P. Reidel and P.R. Hooper (eds.), Volcanism and Tectonism in the Columbia River Flood-Basalt Province; Geological Society of America, Special Paper 239, p. 283-292.

Weaver, C.C., Norris, R.D., and Jonientz-Trisler, C., 1990. Results of seismilogical monitoring in the Cascade Range, 1962-1989: earthquakes, eruptions, avalanches and other curiosities; Geoscience Canada, v. 17, p. 158-162.

Weber, M.H., 1990. Application of the Gaussian beam method to refraction data from the subduction zone beneath Vancouver Island and the North American mainland; in A.G. Green (ed.), Studies of Laterally Heterogeneous Structures Using Seismic Refraction and Reflection Data; Geological Survey of Canada, Paper 89-13, p. 43-52.

Wells, R.E. and Coe, R.S., 1985. Paleomagnetism and geology of Eocene volcanic rocks of southwest Washington, implications for mechanisms of tectonic rotation; Journal of Geophysical Research, v. 90, p. 1925-1947.

Werner, K.S., Graven, E.P., Berkman, T.A., and Parker, M.J., 1991. Direction of maximum horizontal compression in western Oregon determined by borehole breakouts; Tectonics, v. 10, p. 948-958.

West, D.O. and McCrumb, D.R., 1988. Coastline uplift in Oregon and Washington and the nature of Cascadia subduction-zone tectonics; Geology, v. 16, p. 169-172.

Whetten, J.T., Zartman, R.E., Blakely, R.J., and Jones, D.L., 1980. Allochthonous Jurassic ophiolite in northwest Washington; Bulletin of Geological Society of America, v. 91, p. 359-368.

White, J.M., 1990. Evidence of Paleogene sedimentation on Graham Island, Queen Charlotte Islands, west coast, Canada; Canadian Journal of Earth Sciences, v. 27, p. 533-538.

Wilson, D.S., 1986. A kinematic model for the Gorda Deformation Zone as a diffuse southern boundary of the Juan de Fuca plate; Journal of Geophysical Research, v. 91, p. 10,259-10,269.

Wilson, J.T., 1965. Transform faults, oceanic ridges and magnetic anomalies southwest of Vancouver Island; Science, v. 150, p. 482-485.

Wilt, M.J., Morrison, H.F., Lee, K.H., and Goldstein, N.E., 1989. Electromagnetic sounding in the Columbia Basin, Yakima, Washington; Geophysics, v. 54, p. 952-961.

Woodsworth, G.J., 1988. Karmutsen Formation and the east boundary of Wrangellia, Queen Charlotte Basin, British Columbia; in Current Research, Part E; Geological Survey of Canada, Paper 88-1E, p. 209-212.

Woodsworth, G.J. (ed.), 1991. Evolution and Hydrocarbon Potential of the Queen Charlotte Basin, British Columbia; Geological Survey of Canada, Paper 90-10, 569 p.

Woodsworth, G.J., Anderson, R.G., Brookfield, A., and Tercier, P., 1991. Distribution of Proterozoic to Miocene Plutonic Suites in the Canadian Cordillera; Geological Survey of Canada, Open File 1982 (set of 8 maps).

Woodsworth, G.J. and Orchard, M.J., 1985. Upper Paleozoic to lower Mesozoic strata and their conodonts, western Coast PLutonic Complex, British Columbia; Canadian Journal of Earth Sciences, v. 22, p. 1329-1344.

Yagishita, K., 1985. Evolution of a provenance as revealed by petrographic analyses of Cretaceous formations in the Queen Charlotte Islands, British Columbia, Canada; Sedimentology, v. 32, p. 671-684.

Yorath, C.J., 1980. The Apollo structure in Tofino Basin, Canadian Pacific continental shelf; Canadian Journal of Earth Sciences, v, 17, p. 758-775.

Yorath, C.J., 1987. Petroleum geology of the Canadian Pacific continental margin; in D.W. Scholl, A. Grantz, and J.G. Vedder (eds.), Geology and Resource Potential of the Continental Margin of Western North America and Adjacent Ocean Basins - Beaufort Sea to Baja California; Circum-Pacific Council for Energy and Mineral Resources, Earth Science Series, v. 6, p. 283-304.

Yorath, C.J. and Chase, R.L., 1981. Tectonic history of the Queen Charlotte Islands and adjacent areas - a model; Canadian Journal of Earth Sciences, v. 18, p. 1717-1739.

Yorath, C.J., Clowes, R.M., Green, A.G., Sutherland Brown, A., Brandon, M.T., Massey, N.W.D., Spencer, C., Kanasewich, E.R., and Hyndman, R.D., 1985a. Lithoprobe - Phase 1: southern Vancouver Island: preliminary analyses of reflection seismic profiles and surface geological studies; in Current Research, Part A; Geological Survey of Canada, Paper 85-1A, p. 543-554.

Yorath, C.J., Clowes, R.M., Macdonald, R.D., Spencer, C., Davis, E.E., Hyndman, R.D., Rohr, K., Sweeney, J.F., Currie, R.G., Halpenny, J.F., and Seemann, D.A., 1987. Marine Multichannel Seismic Reflection, Gravity, and Magnetic Profiles - Vancouver Island Continental Margin and Juan de Fuca Ridge; Geological Survey of Canada, Open-File Report 1661.

Yorath, C.J., Green, A.G., Clowes, R.M., Sutherland Brown, A., Brandon, M.T., Kanasewich, E.R., Hyndman, R.D., and Spencer, C., 1985b. Lithoprobe, southern Vancouver Island: seismic reflection sees through Wrangellia to the Juan de Fuca plate; Geology, v. 13, p. 759-762.

Yorath, C.J. and Hyndman, R.D., 1983. Subsidence and thermal history of Queen Charlotte Basin; Canadian Journal of Earth Sciences, v. 20, p. 135-159.

Yuan, T., Spence, G.D., and Hyndman, R.D., 1992. Structure beneath Queen Charlotte Sound from seismic-refraction and gravity interpretations; Canadian Journal of Earth Sciences, v. 29, p. 1509-1529.

Yuan, T., Spence, G.D., and Hyndman, R.D., 1994. Seismic velocities and inferred porosities in the accretionary wedge sediments at the Cascadia margin; Journal of Geophysical Research, v. 99, p. 4413-4427.

Zimmermann, G., Burkhardt, H., and Melchert, M., 1992. Estimation of porosity in crystalline rock by a multivariate statistical approach; Scientific Drilling, v. 3, p. 27-35.

Springer-Verlag
and the Environment

We at Springer-Verlag firmly believe that an international science publisher has a special obligation to the environment, and our corporate policies consistently reflect this conviction.

We also expect our business partners — paper mills, printers, packaging manufacturers, etc. — to commit themselves to using environmentally friendly materials and production processes.

The paper in this book is made from low- or no-chlorine pulp and is acid free, in conformance with international standards for paper permanency.

Lecture Notes in Earth Sciences